STUDY GUIDE
with Problems and Solutions for

# ORGANIC CHEMISTRY

STUDY GUIDE
with Problems and Solutions for

# ORGANIC CHEMISTRY

## *A Brief Course*

FOURTH EDITION by LINSTROMBERG/BAUMGARTEN

**Robert L. Zey**

*Central Missouri State University*

**Walter W. Linstromberg**

*University of Nebraska at Omaha*

D.C. HEATH AND COMPANY

Lexington, Massachusetts Toronto

Published simultaneously in Canada.

Printed in the United States of America.

International Standard Book Number: 0-669-00640-8

# To the Student

The great majority of students enrolled in a beginning organic chemistry course find it an interesting—even fascinating—subject. Few students find it a difficult subject to understand, but nearly all find it difficult to remember. In order to learn, and remember, organic chemistry you must be willing to work at it. To read your textbook, to attend the lectures, and to listen to your instructor however attentively is simply not enough. The learning of organic chemistry also involves a great amount of pencil-and-paper work on your part. This is where the problems and solutions in this study guide can be of assistance. Let us now see what this particular study guide is and how you might profit most from its use.

Specifically, this study guide is designed to assist you in understanding the facts and principles presented in *Organic Chemistry: A Brief Course,* fourth edition, by W. W. Linstromberg and H. E. Baumgarten. The study guide is divided into two parts. The first part contains nineteen chapters that parallel the nineteen chapters in the textbook by Linstromberg and Baumgarten. With few exceptions each chapter includes:

1. An introduction that explains the objective of the chapter and how the material it contains is pertinent to that in the following and/or preceding chapter.
2. Problems, including true-false, multiple choice, completion, matching, reaction sequence, and other miscellaneous types.
3. Answers to these problems. When considered necessary, explanations to the answers are also provided.

The second part of this study guide contains answers to all of the exercises in the textbook.

Begin your study of each chapter by first reading the Introduction. Here will be reviewed for you the general content of the corresponding chapter in the textbook. Next, read the true-false, completion, and matching exercises. Mark those questions for which you think you already know the answers. Now read the corresponding text chapter with the true-false, completion, and matching questions in mind. When you have finished reading the chapter, go over the Summary and the New Terms that you will find at the end of each chapter in the textbook. Now return to your study guide and and fill out all exercises. Check your answers with those given at the end of the chapter and make corrections. You may find it necessary to again read sections of the textbook to learn where you "went wrong" or "missed the point." Your experience with exercises in the first chapter or two will determine the best study procedure for you to use.

In addition to reading *Organic Chemistry: A Brief Course* carefully, you should also work out the answers to exercises that appear within the body of the text and at the ends of chapters.

Like any good learning tool, this study guide cannot do your work for you; it can only help you to do a good job when used to best advantage. Stick with it and good luck.

# Contents

# 1

# General Principles

## INTRODUCTION

This chapter begins the study of organic chemistry by pointing out how organic compounds are different from inorganic compounds. We learn what constitutes a chemical bond, how carbon is able to bond to itself as well as to other elements, and why, with rare exceptions, carbon always has four bonds. We learn to distinguish between empirical formulas and molecular formulas of organic compounds and how to determine such formulas from experimental data.

In addition to these fundamental concepts, which we already have encountered in our first course in chemistry, we read perhaps for the first time of the concept of **isomerism.** We soon discover in this chapter that one of the early facts to learn about organic chemistry is that a molecular formula has little meaning unless one also knows the structure of the compound it represents.

We see that organic compounds are classified into families and that there appears within the molecular architecture of each member of a family some characteristic structural feature (called a **functional group**) that is common to all.

We learn that organic reactions proceed through **transition states** and often form **intermediates** which may or may not be isolated, but which usually can be identified as integral parts of the **reaction mechanism.** A reaction mechanism gives a detailed description of a reaction in which all transition states and reaction intermediates are specified and it tells us *how* the reaction takes place. From a knowledge of reaction mechanisms we often are able to predict what the products of an organic reaction will be. An understanding of **acid-base theory** is important because many organic reactions are acid–base reactions.

A procedure is given to illustrate how **Lewis structures** can be drawn. A single Lewis structure often does not satisfactorily represent the physical and chemical properties of a molecule, ion, or radical. In this case all of the reasonable Lewis structures are drawn and a **resonance hybrid** of them all best describes the structure of the species.

The exercises and problems that follow should help to determine if you have accomplished what is so very important in learning any new subject–namely, getting off to a good start.

## PROBLEMS

### True–False

1. T F An orbital can be described as the space surrounding an atomic nucleus in which there is the greatest probability of finding an electron.
2. T F The *s* orbital is dumbbell-shaped.
3. T F Each shell has only one *s* orbital.
4. T F There can be no more than two electrons in any orbital and they must have opposite spin.
5. T F Compounds that are formed by ionic bonding are usually hard substances with high melting points.
6. T F Soluble organic compounds are completely ionized in solution.
7. T F The four carbon — hydrogen bonds in methane are equivalent.
8. T F There is equal sharing of the electrons in the covalent bond between hydrogen and chlorine.
9. T F The empirical formula and the molecular formula are always the same.
10. T F The structure of a molecule as well as the molecular weight are factors which influence melting and boiling points.
11. T F When two atomic orbitals from different atoms overlap, they form two molecular orbitals. The lower-energy molecular orbital is called the bonding orbital and the higher-energy orbital is called the antibonding orbital.
12. T F The equation for a reaction does not tell how a reaction takes place.

### Multiple Choice

13. Which of the following is not a characteristic of organic compounds?
    (a) Bonds which bind the atoms together are nearly always of the covalent type.
    (b) They usually have low melting points.
    (c) They usually are only slightly soluble or insoluble in water.
    (d) The unit particles are ions and not distinct molecules.
    (e) If water soluble they seldom conduct an electric current.

14. Which of the following is the most electronegative?
    (a) Bromine
    (b) Chlorine
    (c) Iodine
    (d) Fluorine
    (e) Oxygen

15. The element least likely to be found in an organic compound is

(a) Oxygen
(b) Silicon
(c) Nitrogen
(d) Sulfur
(e) Hydrogen

16. The empirical formula of a hydrocarbon was found to be $CH_2$. The molecular weight, determined by mass spectroscopy, is 84. The molecular formula is

(a) $CH_2$
(b) $C_2H_4$
(c) $C_4H_8$
(d) $C_5H_{10}$
(e) $C_6H_{12}$

17. Which of the following is not an isomer of

```
              CH3
              |
CH3—CH2—C—OH?
              |
              CH3
```

(a)
```
                 CH3
                 |
CH2=CH—C—OH
                 |
                 CH3
```

(b)
```
               CH3
               |
CH3—CH—CH—OH
       |
       CH3
```

(c)
```
       CH3
       |
CH3—C—CH2—OH
       |
       CH3
```

(d)
```
               CH3
               |
CH3—CH2—CH—O—CH3
```

(e) $CH_3—CH_2—CH_2—CH_2—CH_2—OH$

**Completion**

18. There can be no more than________ *p* orbitals in any shell.

19. Bonds formed by the complete transfer of electrons from one atom to another are called________________ bonds.

20. Bonds formed by a shared pair of electrons, to which each atom has donated one electron, are called______________ bonds.

21. A covalent bond in which the shared pair of electrons has been donated by one atom is called a __________ __________ bond.

22. The mixing of one 2*s* orbital with three 2*p* orbitals forms four equivalent hybridized orbitals. These orbitals are called____________orbitals and the angle between each orbital is______° ______'.

23. Compounds which have the same molecular formula, but different structures, are called ________.
24. A family of compounds in which one member differs from the next by a methylene group ($-CH_2-$) is called a ________ series.
25. The minimum energy necessary for a reaction to proceed from reactants to the transition state is called the ________ ________ ________ ________.

**Matching**

(Match the correct functional group with the class name.)

(a) $-OH$ (b) $-NH_2$ (c) $-\underset{\underset{O}{\|}}{C}-H$

(d) $-\overset{\overset{O}{\|}}{C}-$ (e) $-\overset{\overset{O}{\|}}{C}-O-H$ (f) $-\overset{|}{\underset{|}{C}}-O-\overset{|}{\underset{|}{C}}-$

26. ____ Carboxylic acids
27. ____ Ethers
28. ____ Amines
29. ____ Alcohols
30. ____ Aldehydes
31. ____ Ketones

(Match the reaction species with the proper identifying term.)

$$\underset{(g)}{CH_2{=}CH_2} + \underset{(h)}{H^+} \longrightarrow \underset{(i)}{CH_3{-}CH_2^+} \xrightarrow{Cl^-} \underset{(j)}{CH_3{-}CH_2{-}Cl}$$

32. ____ Lewis base
33. ____ Reaction product
34. ____ Substrate
35. ____ Electrophile
36. ____ Reagent
37. ____ Conjugate base of $CH_3{-}CH_2^+$
38. ____ Reaction intermediate

**Miscellaneous**

39. Draw the $2p_x$, $2p_y$, and $2p_z$ orbitals, using the axes below.

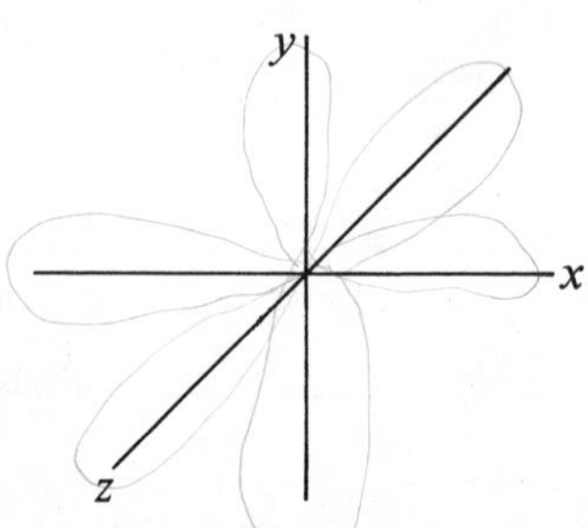

40. Using dots to represent the electrons of hydrogen and asterisks to represent the electrons of carbon, draw the Lewis structures for methane ($CH_4$), ethane ($C_2H_6$) and ethene ($C_2H_4$).

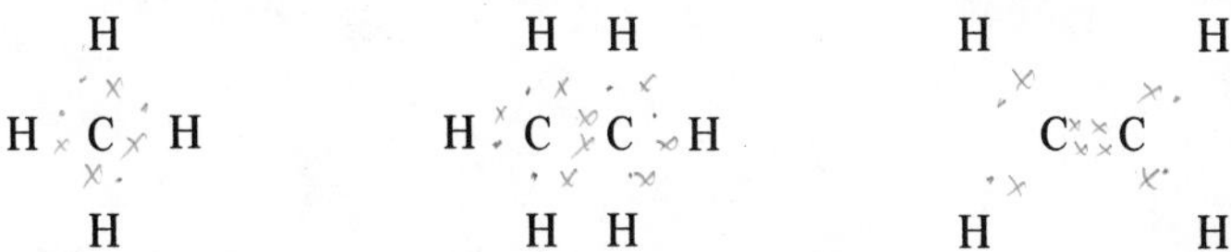

41. Carbon has six electrons. Using small arrows to show the spin of an electron, show the electronic configuration of carbon in its ground (normal) state and in the excited state.

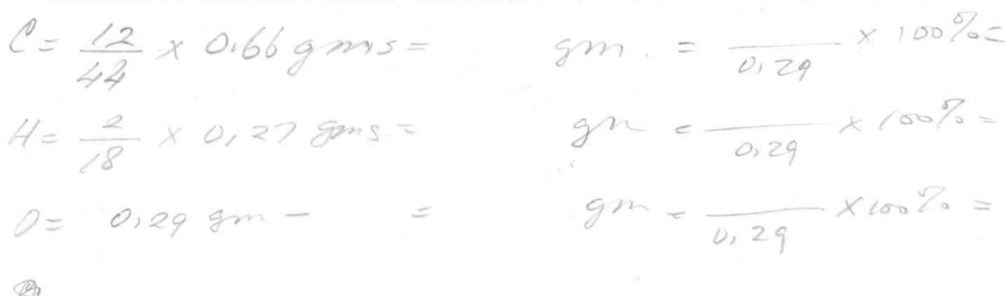


42. Upon complete combustion, 0.29 gram of an unknown organic compound containing carbon, oxygen, and hydrogen yielded 0.66 gram of carbon dioxide and 0.27 gram of water. What is the empirical formula? From this data can you determine the molecular formula?

43. A compound which is believed to have the molecular formula $C_4H_8O$ is submitted for elemental analysis. What percentage of carbon and hydrogen would you expect to be reported if $C_4H_8O$ is the correct formula?

44. Draw the four possible structures (isomers) for the molecular formula $C_4H_9Cl$.

45. Draw the structures (be sure to indicate the formal charge on the carbon atom, if any) and give the name of the methyl species which results from the indicated cleavage of the C—H bond of methane.

```
    H
    |
H—C—}H  ————————→  H⁺   +
    |
    H

    H
    |
H—C}—H  ————————→  :H⁻  +
    |
    H

    H
    |
H—C—}H  ————————→  H•   +
    |
    H
```

46. What is the conjugate acid of each of the following bases?

$H_2O$ ________ $CH_3O^-$ ________ $CH_3OH$ ________

$NH_3$ ________ $CH_2{=}CH_2$ ________ $CH_3NH_2$ ________

47. Should the Lewis structures below be connected by the symbolism [ ↔ ] or [ ⇌ ] ?

$$CH_2{=}CH{-}\overset{+}{C}H{-}CH_2\ddot{\underset{\cdot\cdot}{Br}}\!:\qquad \overset{+}{C}H_2{-}CH{=}CH{-}CH_2\ddot{\underset{\cdot\cdot}{Br}}\!:$$

## ANSWERS

### True–False

1. True
2. False. An *s* orbital is a sphere with its center at the nucleus of the atom. A *p* orbital is dumbbell-shaped.

3. True
4. True
5. True
6. False. Nearly all bonds which bind atoms together to form organic compounds are of the covalent type and the dissolved particle is a molecule. Compounds in which the bonding is ionic dissolve to give ions in solution.
7. True. All C—H bonds in methane are the same length (1.09 Å) and require the same amount of energy to break them.
8. False. Hydrogen and chlorine do not have equal attraction for electrons. Chlorine is more electronegative and the electron density of the bond is shifted toward the chlorine atom.

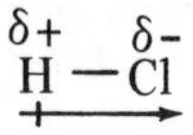

$$\overset{\delta+}{H}-\overset{\delta-}{Cl}$$
$$\longmapsto$$

9. False. The empirical formula is the simplest ratio that shows the relative numbers of atoms of the different elements present in the molecule. The molecular formula shows the actual number of atoms of the various elements present in the molecule. For example, benzene has the molecular formula $C_6H_6$. The empirical formula is CH.

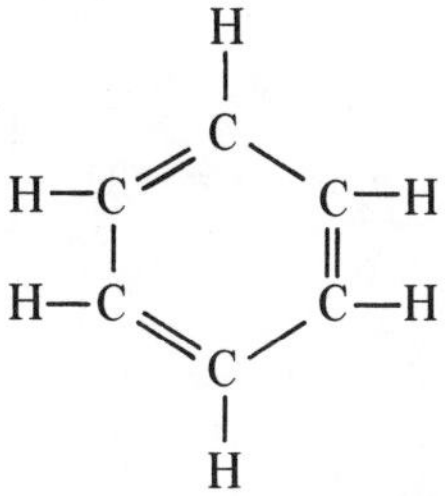

```
          H
          |
          C
      //     \
  H—C         C—H
     |         ||
  H—C         C—H
      \\     /
          C
          |
          H
```

10. True

| Structure | Boiling Point | Melting Point |
|---|---|---|
| $CH_3—CH_2—CH_2—CH_3$ | - 0.6° C | -135° C |
| $CH_3—\underset{\underset{CH_3}{\vert}}{CH}—CH_3$ | -10.2° C | -145° C |

11. True
12. True. The equation for a reaction tells what takes place, but not how it takes place. The reaction mechanism tells how it takes place.

**Multiple Choice**

13. (d)
14. (d) Fluorine is the most electronegative element. Electronegativities increase moving to the right and top of the periodic chart (the noble gases are excluded).
15. (b) Practically all organic compounds contain hydrogen. Oxygen and nitrogen are very common and sulfur is found to a lesser extent.

16. (e) The molecular formula must be some multiple of the empirical formula. How many $CH_2$ units are present in the molecular formula?

Weight of $CH_2$ unit = 12 + 2 = 14

How many $CH_2$ units will weigh 84?

$$\frac{\text{One } CH_2 \text{ unit}}{14} = \frac{x}{84}$$

$$x = \frac{1 \times 84}{14} = 6$$

Thus six $CH_2$ units must weigh 84 and the molecular formula is $6(CH_2)$ or $C_6H_{12}$.

17. (a) Isomers are compounds which have the same molecular formula, but different structures. The molecular formula of $CH_3{-}CH_2{-}C(CH_3)_2{-}OH$ is $C_5H_{12}O$.

The molecular formula of $CH_2{=}CH{-}C(CH_3)_2{-}OH$ is $C_5H_{10}O$.

## Completion

18. three. The K shell has no *p* orbitals. All other shells may have three *p* orbitals and six *p* electrons since each orbital may have two electrons.
19. ionic or electrovalent
20. covalent
21. coordinate covalent, or semipolar
22. $sp^3$, 109° 28′
23. isomers
24. homologous
25. free energy of activation

## Matching

| | | |
|---|---|---|
| 26. (e) | 28. (b) | 30. (c) |
| 27. (f) | 29. (a) | 31. (d) |
| 32. (g) | 33. (j) | 34. (g) |
| 35. (h) | 36. (h) | 37. (g) |
| 38. (i) | | |

**Miscellaneous**

39.

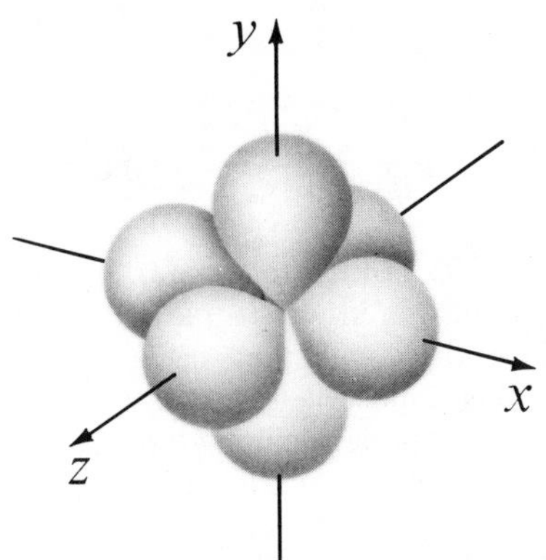


40.

```
    H              H H              H *       * H
H :C* H        H :C*:C* H           : C * * C :
    H              H H              H *       * H
```

41.

| | K shell | L shell | | | |
|---|---|---|---|---|---|
| | 1s | 2s | $2p_x$ | $2p_y$ | $2p_z$ |
| Normal | ↓↑ | ↓↑ | ↑ | ↑ | |
| Excited | ↓↑ | ↑ | ↑ | ↑ | ↑ |

42. $C_3H_6O$. Carbon comprises 12 g of the total 44 g, which is the molecular weight of carbon dioxide. If 12 g of carbon are needed to produce 44 g of carbon dioxide, how many grams of carbon are needed to produce 0.66 g of carbon dioxide?

$$\frac{12}{44} = \frac{x}{0.66}$$

$$x = \frac{0.66 \times 12}{44} = 0.18 \text{ g}$$

There is 0.18 g of carbon in 0.66 g of carbon dioxide. Hydrogen comprises 2 g of the total gram molecular weight of water, which is 18. If 2 g of hydrogen are needed to produce 18 g of water, how many grams of hydrogen are needed to produce 0.27 g of water?

$$\frac{2}{18} = \frac{x}{0.27}$$

$$x = \frac{2 \times 0.27}{18} = 0.03 \text{ g}$$

Thus there is 0.03 g of hydrogen in 0.27 g of water. The weight of carbon and hydrogen in the sample is 0.21 g (0.18 + 0.03). Since the total weight of the sample that was combusted was 0.29 g, 0.08 g of the sample weight must be due to oxygen.

The next step in determining the empirical formula is the determination of the number of gram-atoms (g–at) of each element present.

1 gram-atom of carbon = 12 g

How many gram-atoms are there in 0.18 g of carbon?

$$\frac{1 \text{ g-at}}{12 \text{ g}} = \frac{x}{0.18 \text{ g}}$$

$$x = \frac{0.18 \times 1}{12} = 0.015 \text{ g-at C}$$

1 gram-atom of hydrogen = 1 g

$$\frac{1 \text{ g-at}}{1 \text{ g}} = \frac{x}{0.03 \text{ g}}$$

$$x = \frac{0.03 \times 1}{1} = 0.03 \text{ g-at H}$$

1 gram-atom of oxygen = 16

$$\frac{1 \text{ g-at}}{16 \text{ g}} = \frac{x}{0.08 \text{ g}}$$

$$x = \frac{0.08 \times 1}{16} = 0.005 \text{ g-at 0}$$

The empirical formula is the smallest whole-number ratio of gram-atoms. This is found by dividing the number of gram-atoms of each element by the smallest number of gram-atoms present (this is oxygen–0.005).

$$\text{Carbon} \quad \frac{0.015}{0.005} = 3$$

$$\text{Hydrogen} \quad \frac{0.03}{0.005} = 6$$

$$\text{Oxygen} \quad \frac{0.005}{0.005} = 1$$

The empirical formula is $C_3H_6O$. Since we do not know the molecular weight of the compound we can not determine the molecular formula.

43. Carbon–66.7%; hydrogen–11.1%

The molecular weight of the compound, $C_4H_8O$, must be determined.

| | | | |
|---|---|---|---|
| Carbon | 4 atoms × 12 (atomic weight) | = | 48 |
| Hydrogen | 8 atoms × 1 (atomic weight) | = | 8 |
| Oxygen | 1 atom × 16 (atomic weight) | = | 16 |
| | Molecular Weight | | 72 |

Of the 72 g that one mole of the compound weighs, carbon is responsible for 48. The fraction of the total weight due to carbon when multiplied by 100, gives the the percentage of carbon in the compound.

$$\frac{48}{72} \times 100 = 66.7\%$$

Percentage of hydrogen

$$\frac{8}{72} \times 100 = 11.1\%$$

44. The simplest way to draw isomers is to begin with the longest continuous carbon chain possible and to decrease the chain length by one carbon after all the possible isomers for that particular chain length have been drawn.

$CH_3—CH_2—CH_2—CH_2—Cl$

$CH_3—CH(CH_3)—CH_2—Cl$

$CH_3—CH_2—CH(Cl)—CH_3$

$CH_3—C(Cl)(CH_3)—CH_3$

45. $H_3C—H \longrightarrow H^+ + H_3C:^-$ Methyl anion

$H_3C—H \longrightarrow :H^- + H_3C^+$ Methyl cation

$H_3C—H \longrightarrow H\cdot + H_3C\cdot$ Methyl radical

46.

$$\underset{\text{acid}}{H^+} + \underset{\text{base}}{B:} \longrightarrow \underset{\text{conjugate acid of B:}}{BH^+}$$

| | | |
|---|---|---|
| $H_3O^+$ | $CH_3OH$ | $CH_3OH_2^+$ |
| $NH_4^+$ | $CH_3{-}CH_2^+$ | $CH_3NH_3^+$ |

47. Lewis structures which can be equated by only shifting one or more pairs of electrons are resonance structures and are connected by the symbolism [ ↔ ].

$$CH_2{=}CH{-}\overset{+}{C}H{-}CH_2\ddot{B}r\colon \longleftrightarrow \overset{+}{C}H_2{-}CH{=}CH{-}CH_2\ddot{B}r\colon$$

# 2

# *The Alkanes, or Saturated Hydrocarbons*

## INTRODUCTION

In this chapter we meet our first "family" of compounds–the saturated hydrocarbons or alkanes. The study pattern established in this chapter–i.e., **structure, nomenclature, physical properties, preparation,** and **reactions** also will be followed in succeeding chapters to study other classes of compounds.

The nomenclature, or the naming, of alkanes is perhaps the most important part of this chapter because a number of organic compounds are named simply as substitution products of the alkanes. Thus, if we are to name correctly compounds that belong to other classes by their systematic names according to the IUPAC system, we first must build a foundation for such systematic names by learning to name the alkanes. The name of the parent alkane (with some modification) forms the stem from which names of other compounds are derived.

Examples:

If the parent hydrocarbon is

| | | Family |
|---|---|---|
| $CH_3—CH_2—CH_2—CH_3$ | Butane | Alkanes |
| ④ ③ ② ① | | |

the 4-carbon members of other families may be named as

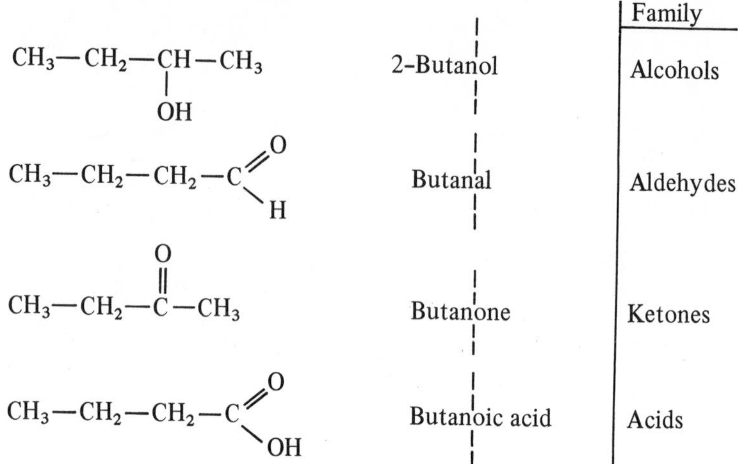

Family
$CH_3—CH_2—CH—CH_3$
OH
2-Butanol
Alcohols
$CH_3—CH_2—CH_2—C$
O
H
Butanal
Aldehydes
O
$CH_3—CH_2—C—CH_3$
Butanone
Ketones
$CH_3—CH_2—CH_2—C$
O
OH
Butanoic acid
Acids


## PROBLEMS

**True–False**

1. T F Propane has no isomers.
2. T F The names of the alkanes containing less than four carbon atoms are derived from the Greek prefix which is followed by the suffix -ane.
3. T F A tertiary carbon atom has three hydrogen atoms attached to it.
4. T F The single bond between carbon atoms in a molecule permits free rotation.
5. T F Alkanes are liquids at room temperature.
6. T F The odor of natural gas is due to impurities that have been added and not due to the alkanes.
7. T F Alkanes are insoluble in water.
8. T F It is seldom necessary to synthesize a hydrocarbon.
9. T F Grignard reagents are relatively inert.
10. T F A Grignard reagent is an example of an organometallic compound.
11. T F Incomplete combustion of alkanes results in the formation of carbon monoxide.
12. T F Chlorine does not react with alkanes at ordinary temperatures and in the dark.
13. T F When the reaction is catalyzed by high-frequency radiation, iodine reacts more rapidly with an alkane than does bromine.
14. T F Higher molecular weight hydrocarbons can be broken down to lower molecular weight hydrocarbons by a process known as pyrolysis.
15. T F Cyclopropane is more reactive than cyclohexane because of the ring strain (caused by the deviation of the bond angle from the normal tetrahedral angle) and eclipsed hydrogen strain.
16. T F The carbon atoms of all cycloalkanes lie in a plane.

17. T F The compounds represented by the following formulas are identical.

**Multiple Choice**

18. How many secondary carbon atoms does methylcyclopropane have?
    (a) none
    (b) one
    (c) two
    (d) three
    (e) four

19. What prefix is used to name compounds that have a methyl group attached to the next-to-last carbon atom in an otherwise continuous chain?

$$CH_3{-}\underset{\displaystyle CH_3}{\underset{|}{CH}}{-}CH_2{-}(CH_2)_n{-}CH_3$$

    (a) normal or *n*–
    (b) iso–
    (c) neo–
    (d) *sec*–
    (e) *tert*–

20. What type of an alkyl group is an isobutyl group?
    (a) primary
    (b) secondary
    (c) tertiary
    (d) none of these

21. The thermal decomposition of alkanes in the absence of air is known as
    (a) oxidation
    (b) hydrogenation
    (c) pyrolysis
    (d) combustion
    (e) substitution

22. The general designation for a Grignard reagent is
    (a) RMgX
    (b) $R_2Mg$
    (c) RXMg
    (d) $MgX_2$
    (e) RMgR

23. The bond angle between carbon atoms in cyclohexane is
    (a) 109° 28′
    (b) 60°
    (c) 90°
    (d) 120°
    (e) none of these

24. The general formula for a cycloalkane is
    (a) $C_nH_n$
    (b) $C_nH_{2n}$
    (c) $C_nH_{2n+2}$
    (d) $C_nH_{2n-2}$
    (e) $C_nH_{2n-4}$

25. The alkyl group known as the *sec*-butyl group is

(a) $CH_3{-}CH_2{-}CH_2{-}CH_2{-}$

(b) $$\begin{array}{c} CH_3{-}CH_2{-}\underset{|}{C}H{-}CH_3 \end{array}$$

(c) $$\begin{array}{l} CH_3{-}CH{-}CH_2{-} \\ \qquad\;\; | \\ \qquad CH_3 \end{array}$$

(d) $$\begin{array}{c} \quad\;\; CH_3 \\ \quad\;\; | \\ CH_3{-}C{-} \\ \quad\;\; | \\ \quad\;\; CH_3 \end{array}$$

26. Open chain hydrocarbons which have no double or triple bonds are sometimes called

(a) alkenes
(b) aromatic hydrocarbons
(c) alkynes
(d) paraffins
(e) olefins

27. What is the IUPAC name for

$$\begin{array}{ccccccc} & & CH_3 & & & & \\ & & | & & & & \\ & & CH_2 & & & CH_3 & \\ & & | & & & | & \\ CH_3{-} & CH_2{-} & CH & - & CH_2{-} & C & {-}CH_3 \\ & & & & & | & \\ & & & & & CH_2 & \\ & & & & & | & \\ & & & & & CH_3 & \end{array}$$

(a) 2-methyl-2,4-diethylhexane
(b) 2-ethyl-5,5-dimethylheptane
(c) 3,5-diethyl-5-methylhexane
(d) 5-ethyl-3,3-dimethylheptane
(e) 2,2-dimethyl-5-ethylheptane

**Completion**

28. Compounds containing only carbon and hydrogen are called ________________.

29. If the dihedral angle between the chlorine atoms in 1,2-dichloroethane is 0°, the compound would be in an ________________ conformation.

30. The general formula for alkanes is ________________.

31. Removal of one of the hydrogens of an alkane produces an ____________ group.

32. The general formula for an alkyl group is ________________.

33. A carbon bonded to three other carbons is called a ________________ carbon atom.

34. The only group that is an example of an iso group and a secondary group at the same time is the ________________ group.

35. The letters IUPAC is an abbreviation for ________ ________ ________ ________ ________ ________ ________.

36. The different arrangements possible by rotation about a single bond are called

______________.

37. A methyl group which has an odd, or unpaired, electron is called a

______________.

38. The general designation for an alkyl halide is________.

39. The two conformations of cyclohexane are called the ________ and ________ conformations or forms.

**Structural Formulas** (Write the correct structural formula for each of the following compounds or groups.)

40. *sec*-Butyl group

41. *n*-Propyl group

42. *n*-Pentane

43. Isopropyl group

44. Ethyl group

45. *tert*-Pentyl group

46. Isobutyl group

47. *tert*-Butyl group

48. Ethylmagnesium bromide

**Nomenclature** (Name each of the following according to IUPAC rules.)

49.

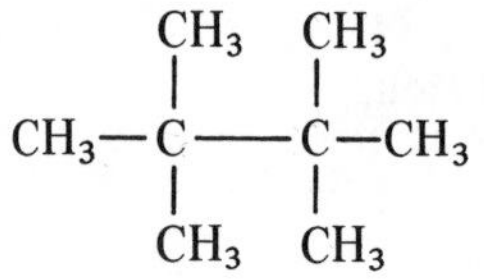

$$\begin{array}{ccccc} & & CH_3 & & CH_3 & & \\ & & | & & | & & \\ CH_3 & - & C & - & C & - & CH_3 \\ & & | & & | & & \\ & & CH_3 & & CH_3 & & \end{array}$$

________________________

50.

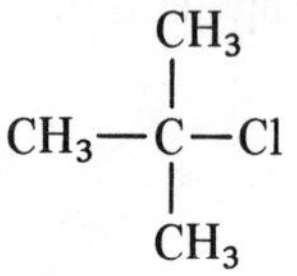

$$\begin{array}{ccccc} & & CH_3 & & \\ & & | & & \\ CH_3 & - & C & - & Cl \\ & & | & & \\ & & CH_3 & & \end{array}$$

________________________

51.

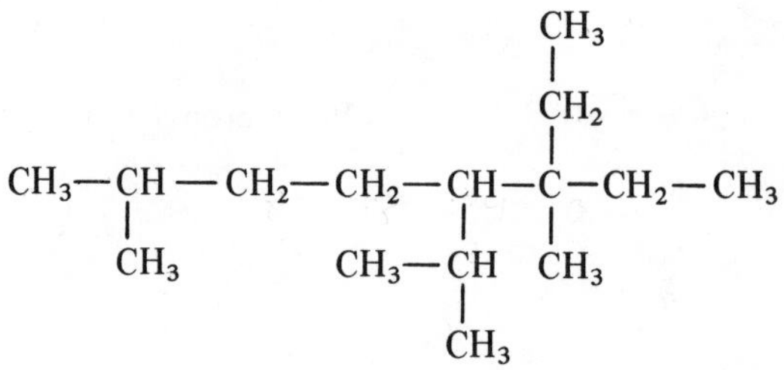

$$\begin{array}{ccccccccccccccc} & & & & & & & & & & CH_3 & & & & \\ & & & & & & & & & & | & & & & \\ & & & & & & & & & & CH_2 & & & & \\ & & & & & & & & & & | & & & & \\ CH_3 & - & CH & - & CH_2 & - & CH_2 & - & CH & - & C & - & CH_2 & - & CH_3 \\ & & | & & & & & & | & & | & & & & \\ & & CH_3 & & & & CH_3 & - & CH & & CH_3 & & & & \\ & & & & & & & & | & & & & & & \\ & & & & & & & & CH_3 & & & & & & \end{array}$$

________________________

52.

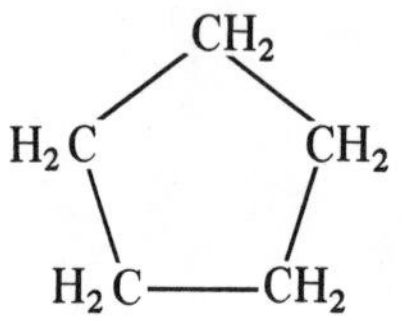

Cyclic structure: $CH_2$, $CH_2$, $CH_2$, $H_2C$, $H_2C$ (five-membered ring)

________________________

53.

Chair cyclohexane structure with substituents: H, H, H, H, H, H, H, H, H, $C_2H_5$, $C_2H_5$, $CH_3$

________________________

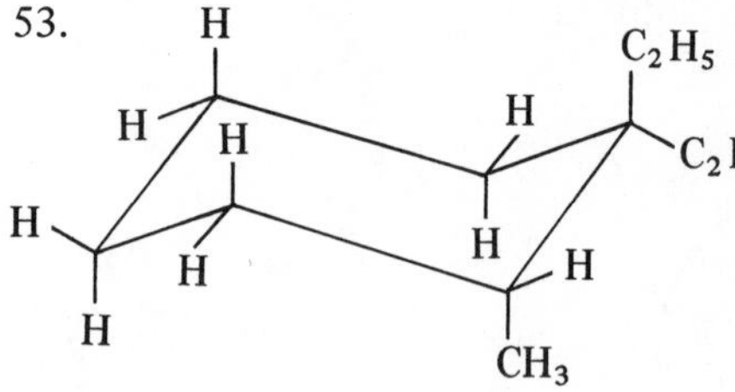

**Reactions** (Complete the following.)

54.

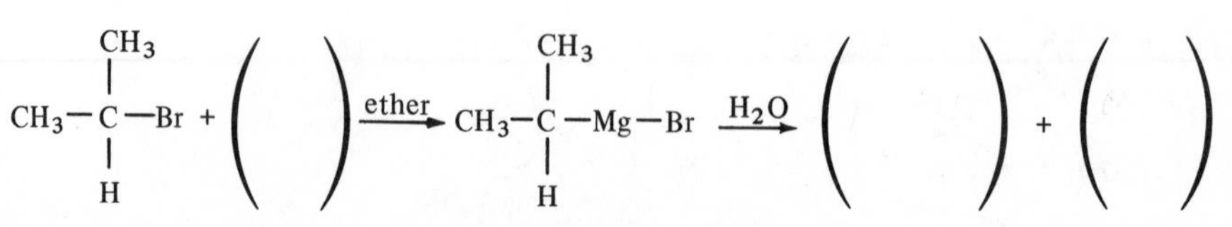

$$\begin{array}{c} CH_3 \\ | \\ CH_3-C-Br \\ | \\ H \end{array} + \left(\quad\right) \xrightarrow{\text{ether}} \begin{array}{c} CH_3 \\ | \\ CH_3-C-Mg-Br \\ | \\ H \end{array} \xrightarrow{H_2O} \left(\qquad\right) + \left(\quad\right)$$

55.

$$CH_3-\overset{\overset{CH_3}{|}}{CH}-CH_2-Cl + 2\ Li \xrightarrow{\text{ether}} \left(\qquad\qquad\right) + LiCl$$

56.

$$2\ CH_3-\overset{\overset{CH_3}{|}}{CH}-CH_2-Li + CuI \longrightarrow \left(\qquad\qquad\right) + LiI$$

57.

$$2\ CH_3-CH_2-CH_2-Br + Li\left(CH_3-\overset{\overset{CH_3}{|}}{CH}-CH_2\right)_2Cu \longrightarrow \left(\qquad\qquad\right) + LiBr + CuBr$$

58.

$$CH_4 + \left(\qquad\right) \xrightarrow{\text{sunlight}} CH_3Cl + HCl$$

59.

$$Br-CH_2-CH_2-CH_2-Br + Zn \longrightarrow \left(\qquad\qquad\right) + ZnBr_2$$

60.

$$\left(\qquad\qquad\right) + Br_2 \xrightarrow{CCl_4} Br-CH_2-CH_2-CH_2-Br$$

61.

$$2\left(\qquad\right) + Li\left(CH_3-\overset{\overset{CH_3}{|}}{CH}-CH_2\right)_2Cu \longrightarrow CH_3-\overset{\overset{CH_3}{|}}{CH}-CH_2-CH_2-CH_2-\overset{\overset{CH_3}{|}}{CH}-CH_3 + LiBr + CuBr$$

62.

$$2\ CH_3-\underset{\underset{CH_3}{|}}{CH}-CH_3 + 13\ O_2 \longrightarrow \left(\qquad\qquad\right) + \left(\qquad\qquad\right)$$

**Miscellaneous**

63. Draw a circle around each tertiary hydrogen atom.

```
    H        H        H        H        H
    |        |        |        |        |
H—C————C————C————C————C—H
    |        |        |        |        |
    H    H—C—H     H    H—C—H     H
             |                 |
             H                 H
```

64. Draw a circle around each primary carbon atom.

```
              H
              |
    H     H—C—H    H    H    H
    |         |        |    |    |
H—C————C————C—C—C—H
    |         |        |    |    |
    H     H—C—H    H    H    H
              |
              H
```

65. Write the mechanism for the free radical halogenation of an alkane (R—H).

66. How many monochlorinated isomers would result from the reaction of chlorine with *n*-butane in the presence of sunlight? Draw the structures for all of the monochlorinated isomers. The rate of abstraction of hydrogen atoms is always found to follow the sequence, tertiary > secondary > primary. Assuming the relative rates of substitution of hydrogen atoms are 5.0:3.8:1.0, respectively, predict the yields of the isomeric monochlorinated products.

67. Draw and name five substituted pentanes of formula $C_7H_{16}$.

C—C—C—C—C ______________________________

C—C—C—C—C ______________________________

C—C—C—C—C ______________________________

C—C—C—C—C ______________________________

C—C—C—C—C ______________________________

68. Draw all three possible secondary pentyl groups.

69.

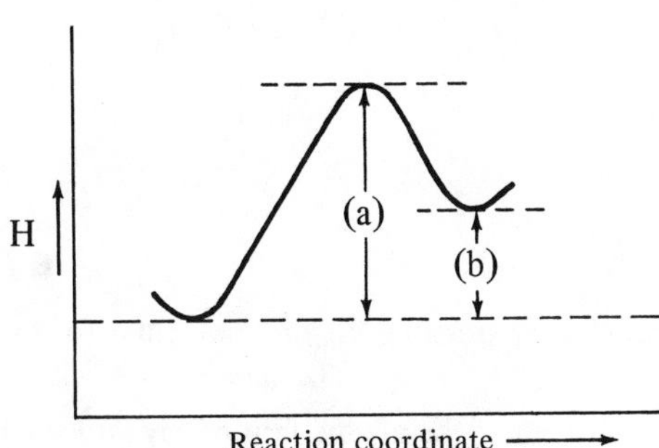


What is the energy term denoted by (a) called?
What is the energy term denoted by (b) called?
Is the reaction endothermic or exothermic, that is, was energy consumed or released by the reaction?

70. Draw the Newman projections of the anti and gauche conformations of 1,2-dichloroethane. Which form is more stable?

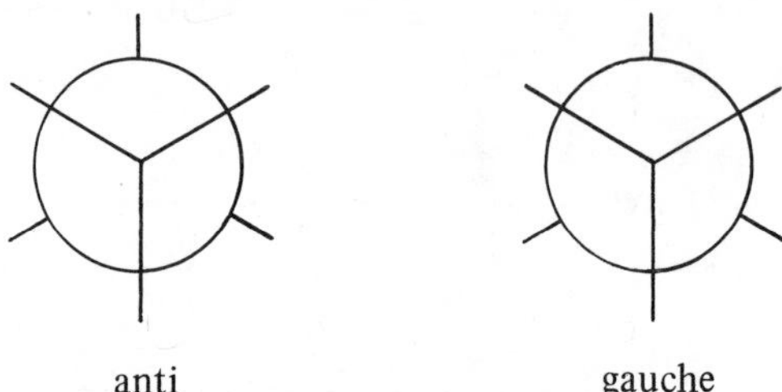


71. Draw the structures of the two chair conformations of *cis*-1,3-dimethylcyclohexane. Which form is more stable?

## ANSWERS

### True-False

1. True. There is no arrangement possible for the molecular formula $C_3H_8$ except $CH_3-CH_2-CH_3$.
2. False. The first four members of the alkane family have no single derivation, but in the higher members the number of carbon atoms are indicated by a Greek prefix.
3. False. A tertiary carbon atom has three carbon atoms attached to it.
4. True
5. False. The first four members of the alkane family are gases and the higher members are liquids or solids.

6. True. Alkanes are odorless when they are pure.
7. True. Like dissolve like. Water is polar and alkanes are nonpolar.
8. True
9. False. Grignard reagents are extremely reactive.
10. True
11. True
12. True. The reaction occurs by a free radical mechanism and high temperature or light is necessary to initiate the reaction.
13. False. Iodine reacts so slow as to be impractical. The trend among the halogens is $F_2 > Cl_2 > Br_2 > I_2$.
14. True
15. True
16. False. This is true for three-membered rings but not for the larger rings. Baeyer's theory was incorrect.
17. True

**Multiple Choice**

18. (c)

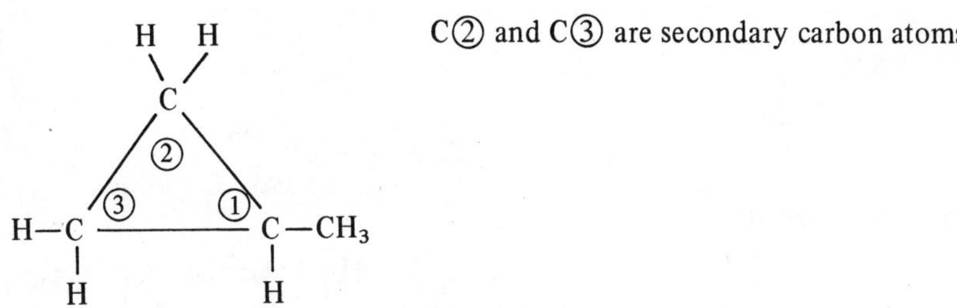


19. (b)
20. (a) $CH_3-\underset{\displaystyle CH_3}{\underset{|}{CH}}-CH_2-$ ← (only one carbon atom attached)
21. (c)
22. (a)
23. (a) Cycloalkanes of six or more carbon atoms can assume a "puckered" conformation to retain the tetrahedral bond angle.
24. (b)
25. (b)
26. (d)

27. (d)

$$\begin{array}{l} \qquad\qquad\quad CH_3 \\ \qquad\qquad\quad | \\ \qquad\qquad\quad CH_2 \qquad\quad CH_3 \\ \qquad\qquad\quad | \qquad\qquad\quad | \\ CH_3-CH_2-CH-CH_2-C-CH_3 \\ \;\;7 \qquad\;\; 6 \qquad\; 5 \qquad\; 4 \quad\; 3\,| \\ \qquad\qquad\qquad\qquad\qquad\quad 2\;CH_2 \\ \qquad\qquad\qquad\qquad\qquad\qquad | \\ \qquad\qquad\qquad\qquad\qquad\quad 1\;CH_3 \end{array}$$

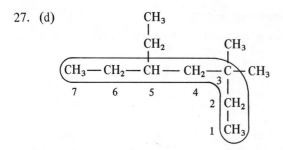

**Completion**

28. hydrocarbons
29. eclipsed
30. $C_nH_{2n+2}$
31. alkyl
32. $C_nH_{2n+1}$
33. tertiary
34. isopropyl $\begin{array}{c} \quad\;\; | \\ CH_3-CH \\ \quad\;\; | \\ \quad\;\; CH_3 \end{array}$
35. International Union of Pure and Applied Chemistry
36. conformations
37. methyl free radical $\begin{array}{c} H \\ \cdot * \\ H \overset{*}{\underset{\cdot}{}} C \cdot \\ * \cdot \\ H \end{array}$
38. RX
39. boat, chair

**Structural Formulas**

40. $\begin{array}{l} \qquad\qquad\; CH_3 \\ \qquad\qquad\;\; | \\ CH_3-CH_2-CH- \end{array}$
41. $CH_3-CH_2-CH_2-$
42. $CH_3-CH_2-CH_2-CH_2-CH_3$
43. $\begin{array}{l} \qquad\;\;\; H \\ \qquad\;\;\; | \\ CH_3-C- \\ \qquad\;\;\; | \\ \qquad\;\; CH_3 \end{array}$
44. $CH_3-CH_2-$
45. $\begin{array}{l} \qquad\qquad\quad\; CH_3 \\ \qquad\qquad\qquad | \\ CH_3-CH_2-C- \\ \qquad\qquad\qquad | \\ \qquad\qquad\quad\; CH_3 \end{array}$
46. $\begin{array}{l} \qquad\;\; CH_3 \\ \qquad\;\;\; | \\ CH_3-CH-CH_2- \end{array}$
47. $\begin{array}{l} \qquad\;\; CH_3 \\ \qquad\;\;\; | \\ CH_3-C- \\ \qquad\;\;\; | \\ \qquad\;\; CH_3 \end{array}$
48. $CH_3-CH_2-MgBr$

## Nomenclature

49. 2,2,3,3-tetramethylbutane
50. 2-chloro-2-methylpropane (*tert*-butyl chloride is the common name)
51. 6-ethyl-5-isopropyl-2,6-dimethyloctane. According to the older nomenclature rules, the parent chain would have been numbered in the opposite direction and this compound would have been named 3-ethyl-4-isopropyl-3,7-dimethyloctane. Also note that the prefix di- is not used in establishing alphabetical order, but the prefix iso- is used.
52. cyclopentane
53. 1,1-diethyl-2-methylcyclohexane

## Reactions

54.

$$CH_3-\underset{H}{\overset{CH_3}{\underset{|}{\overset{|}{C}}}}-Br + (Mg) \xrightarrow{\text{ether}} CH_3-\underset{H}{\overset{CH_3}{\underset{|}{\overset{|}{C}}}}-Mg-Br \xrightarrow{H_2O} \left(CH_3-\underset{H}{\overset{CH_3}{\underset{|}{\overset{|}{C}}}}-H\right) + (HOMgBr)$$

A Grignard reaction.

55.

$$CH_3-\overset{CH_3}{\overset{|}{CH}}-CH_2-Cl + 2\ Li \xrightarrow{\text{ether}} \left(CH_3-\overset{CH_3}{\overset{|}{CH}}-CH_2-Li\right) + LiCl$$

56.

$$2\ CH_3-\overset{CH_3}{\overset{|}{CH}}-CH_2-Li + CuI \longrightarrow \left(Li\left(CH_3-\overset{CH_3}{\overset{|}{CH}}-CH_2\right)_2Cu\right) + LiI$$

57.

$$2\ CH_3-CH_2-CH_2-Br + Li\left(CH_3-\overset{CH_3}{\overset{|}{CH}}-CH_2\right)_2Cu \longrightarrow \left(CH_3-\overset{CH_3}{\overset{|}{CH}}-CH_2-CH_2-CH_2-CH_3\right)$$
$$+ LiBr + CuBr$$

58.

$$CH_4 + (Cl_2) \xrightarrow{\text{sunlight}} CH_3Cl + HCl$$

59.

$$Br-CH_2-CH_2-CH_2-Br + Zn \longrightarrow \left(\text{cyclopropane: } CH_2,\ H_2C-CH_2\right) + ZnBr_2$$

60.

$$\left(\text{cyclopropane: } CH_2,\ H_2C-CH_2\right) + Br_2 \xrightarrow{CCl_4} Br-CH_2-CH_2-CH_2-Br$$

61.

$$2\left(\begin{matrix} & CH_3 & & \\ & | & & \\ CH_3- & CH & -CH_2-CH_2-Br \end{matrix}\right) + Li\left(\begin{matrix} & CH_3 & \\ & | & \\ CH_3- & CH & -CH_2 \end{matrix}\right)_2Cu \rightarrow \begin{matrix} & CH_3 & & CH_3 & \\ & | & & | & \\ CH_3- & CH & -CH_2-CH_2-CH_2- & CH & -CH_3 \end{matrix}$$

$$+\ LiBr + CuBr$$

62.

$$2\ \begin{matrix} CH_3- & CH & -CH_3 \\ & | & \\ & CH_3 & \end{matrix} + 13\ O_2 \longrightarrow \left(8\ CO_2\right) + \left(10\ H_2O\right)$$

**Miscellaneous**

63.

```
      H     (H)     H     (H)     H
      |      |      |      |      |
   H—C——————C——————C——————C——————C—H
      |      |      |      |      |
      H   H—C—H     H   H—C—H     H
             |             |
             H             H
```

64.

```
               H
               |
      H     H—(C)—H    H   H   H
      |        |       |   |   |
   H—(C)———————C———————C———C——(C)—H
      |        |       |   |   |
      H     H—(C)—H    H   H   H
               |
               H
```

65.

(a) $Cl_2 \xrightarrow{\text{sunlight}} 2\ Cl\cdot$

(b) $Cl\cdot + R{-}H \longrightarrow HCl + R\cdot$

(c) $R\cdot + Cl_2 \longrightarrow RCl + Cl\cdot$

then b, c, b, c, etc.

66. There are two sets of nonequivalent hydrogens and therefore two monochlorinated isomers are possible.

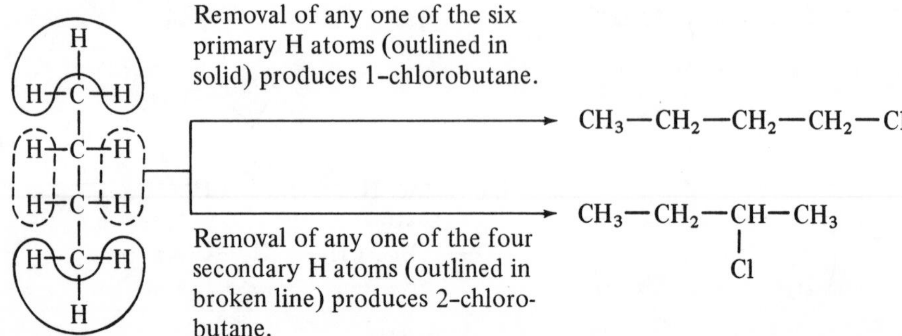

Removal of any one of the six primary H atoms (outlined in solid) produces 1-chlorobutane.
CH3—CH2—CH2—CH2—Cl
CH3—CH2—CH—CH3
Cl
Removal of any one of the four secondary H atoms (outlined in broken line) produces 2-chloro-butane.

$$\frac{\begin{pmatrix}\text{no. of specific}\\ \text{type of H}\end{pmatrix}\begin{pmatrix}\text{reactivity of}\\ \text{that hydrogen}\end{pmatrix}}{\begin{pmatrix}\text{no. of}\\ H_A\end{pmatrix}\begin{pmatrix}\text{Reactivity}\\ \text{of } H_A\end{pmatrix}+\begin{pmatrix}\text{no. of}\\ H_B\end{pmatrix}\begin{pmatrix}\text{Reactivity}\\ \text{of } H_B\end{pmatrix}+\cdots}\times 100 = \%\ \text{yield}$$

Removal of a primary hydrogen gives 1-chlorobutane.

$$\frac{(\text{no. of } 1^\circ\text{ H})\ (\text{reactivity of } 1^\circ\text{ H})}{\begin{pmatrix}\text{no. of}\\ 1^\circ\text{ H}\end{pmatrix}\begin{pmatrix}\text{Reactivity}\\ \text{of } 1^\circ\text{ H}\end{pmatrix}+\begin{pmatrix}\text{no. of}\\ 2^\circ\text{ H}\end{pmatrix}\begin{pmatrix}\text{Reactivity}\\ \text{of } 2^\circ\text{ H}\end{pmatrix}}\times 100 = \%\ \text{1-chlorobutane}$$

$$\frac{6\times 1.0}{(6\times 1.0)+(4\times 3.8)}\times 100 = 28.3\%$$

Similarly for 2-chlorobutane

$$\frac{4\times 3.8}{(6\times 1.0)+(4\times 3.8)}\times 100 = 71.7\%$$

67.

```
                 CH3
                 |
CH3—CH2—CH—CH—CH3          2,3-dimethylpentane
                     |
                     CH3

CH3—CH—CH2—CH—CH3          2,4-dimethylpentane
        |               |
        CH3          CH3

                          CH3
                          |
CH3—CH2—CH2—C—CH3          2,2-dimethylpentane
                          |
                          CH3

                 CH3
                 |
CH3—CH2—C—CH2—CH3          3,3-dimethylpentane
                 |
                 CH3

CH3—CH2—CH—CH2—CH3         3-ethylpentane
                 |
                 CH2
                 |
                 CH3
```

Note the systematic method in which the isomers are drawn. First, all the isomers are drawn which have the two methyl groups on different carbon atoms. Next, all the isomers are drawn which have both methyl groups on the same carbon atom. Finally, an ethyl group is substituted on the pentane chain.

68.

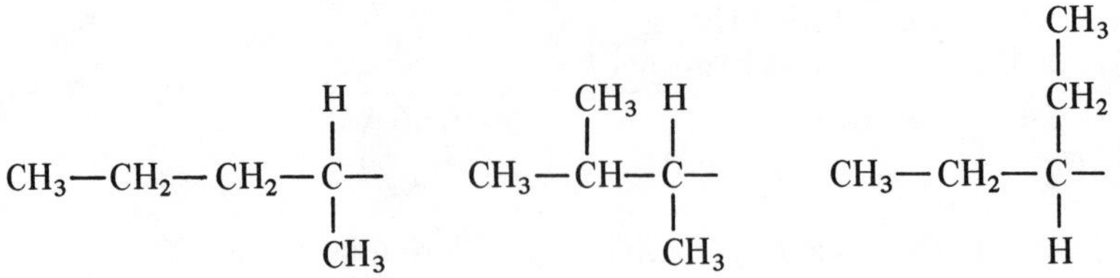


69. (a) heat of activation ($\Delta H^{\ddagger}$)
    (b) heat of reaction ($\Delta H$)

The reaction is endothermic; energy is consumed by the reaction.

70.

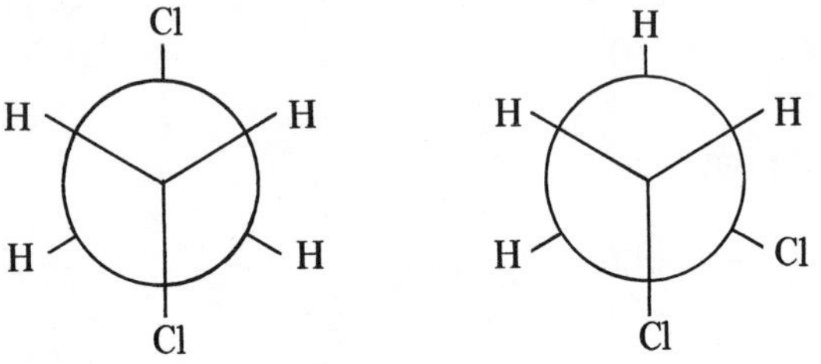


anti gauche

Because of their size and high electron density, the chlorine atoms try to get as far apart as possible. This occurs in the anti conformation and it is more stable.

71.

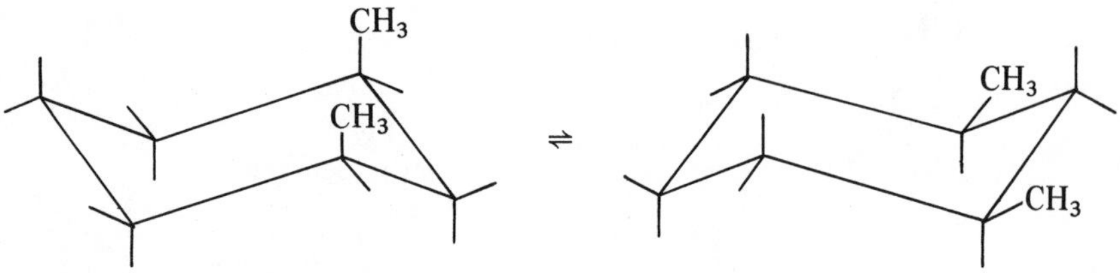


diaxial diequatorial

The diequatorial conformation is more stable because there is less interaction between the methyl groups.

# 3

# *The Unsaturated Hydrocarbons — Alkenes and Alkynes*

## INTRODUCTION

The alkenes (olefins) and the alkynes (acetylenes) comprise two classes of organic compounds also composed of only carbon and hydrogen atoms. Unlike the alkanes, members of these two classes of hydrocarbons do not contain the maximum number of hydrogen atoms possible for each carbon in the molecule and for this reason are said to be "unsaturated". We shall find it necessary to expand our rules of nomenclature in order to name members of these two classes.

It is important for us to understand that carbon—carbon multiple bonds, —C=C— or —C≡C—, are not simply two or three identical bonds, but that such bonds are made up of one **sigma** ($\sigma$) carbon—carbon bond and one or two **pi** ($\pi$) carbon-carbon bonds. It is also important that we recognize differences in the geometry of the molecule at the site of these multiple bonds. For example a double bond between two carbon atoms imposes a planar geometry at this site and makes possible a variation in the space arrangements of groups attached to the doubly bonded carbon atoms. This results in a kind of isomerism known as *cis,trans*-isomerism. Acetylene can not show this type of isomerism because it is a linear molecule.

We will note that the electron pair comprising a $\pi$ bond is not held as firmly between two carbon atoms as the electron pair is in the $\sigma$ bond, and because of this the unsaturated hydrocarbons are more reactive than the alkanes. The most important reactions of the unsaturated hydrocarbons are those that result in addition of reagents at the double or triple bond. Such additions are not made in a haphazard manner but are governed by factors which will allow us to predict exactly what the products will be.

## PROBLEMS

### True–False

1. T F Alkenes and alkynes are similar to alkanes in reactivity.
2. T F More than one pair of electrons is shared between two carbon atoms in unsaturated compounds.
3. T F The general formula for an alkene containing only one double bond is the same as that for a cycloalkane containing the same number of carbon atoms.
4. T F A double bond consists of two equivalent bonds called pi ($\pi$) bonds.
5. T F Stereoisomers possible because of the restricted rotation about a carbon-carbon double bond are called *cis,trans*-isomers.
6. T F All of the atoms in ethylene lie in the same plane.
7. T F The pi bond in ethylene is more easily broken than is the sigma bond.
8. T F The distance between doubly bonded carbon atoms is greater than the distance between singly bonded carbon atoms.
9. T F Fewer isomers are possible for $C_5H_{10}$ than for $C_5H_{12}$.
10. T F The principal reactions of the alkenes are those in which reagents add to the carbons having the double bond.
11. T F Alkyl groups are electron-withdrawing.
12. T F Chlorine is the positive part of the reactant when hypochlorous acid is added to an alkene.
13. T F "Isoprene", the fundamental building block of many natural products, is the common name for 2-methyl-1,3-butadiene.
14. T F The acetylene molecule is linear.
15. T F A hydrogen atom bonded to an acetylenic carbon exhibits some acidic character.
16. T F Differences in chemical behavior between alkynes and alkenes is due, in part, to the acidity of the acetylenic hydrogen atom.
17. T F Neoprene is a synthetic rubber made by the polymerization of chloroprene.

### Multiple Choice

18. The reaction least likely to occur is

(a) $CH_3-CH=CH_2 + HBr \longrightarrow CH_3-\underset{\displaystyle Br}{\underset{|}{CH}}-CH_3$

(b) $CH_3-CH=CH_2 + H_2SO_4 \longrightarrow CH_3-\underset{\displaystyle OSO_3H}{\underset{|}{CH}}-CH_3$

(c) $CH_3-CH=CH_2 + HOBr \longrightarrow CH_3-\underset{\displaystyle Br}{\underset{|}{CH}}-CH_2OH$

*(cont'd)*

(d) $CH_3{-}CH{=}CH_2 + HCl \longrightarrow CH_3{-}CH(Cl){-}CH_3$

19. An example of hydrogenation is

(a) $CH_2{=}CH_2 + H_2O \xrightarrow{\text{catalyst}} CH_3{-}CH_2{-}OH$

(b) $2\,CH_3OH \xrightarrow{\text{catalyst}} CH_3{-}O{-}CH_3 + H_2O$

(c) $CH_2{=}CH_2 + H_2 \xrightarrow{\text{catalyst}} CH_3{-}CH_3$

(d) $CH_3{-}CH_2{-}CH_2{-}CH_3 + \text{heat} \xrightarrow{\text{catalyst}} CH_3{-}CH_2{-}CH{=}CH_2 + H_2$

(e) $CH_3{-}C({=}O){-}O{-}CH_3 + H_2O \xrightarrow{\text{catalyst}} CH_3{-}C({=}O){-}OH + CH_3OH$

20. The reaction (at right) can be brought about by heating the starting material with

(a) Mg
(b) Zn(Hg) and HCl
(c) $SOCl_2$
(d) $H_3PO_4$
(e) $H_2$ and Pd on charcoal

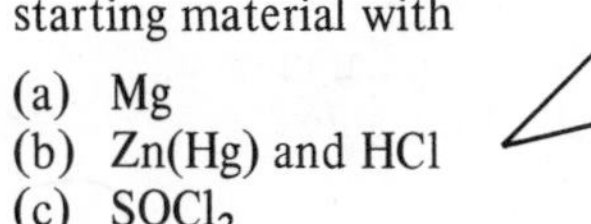

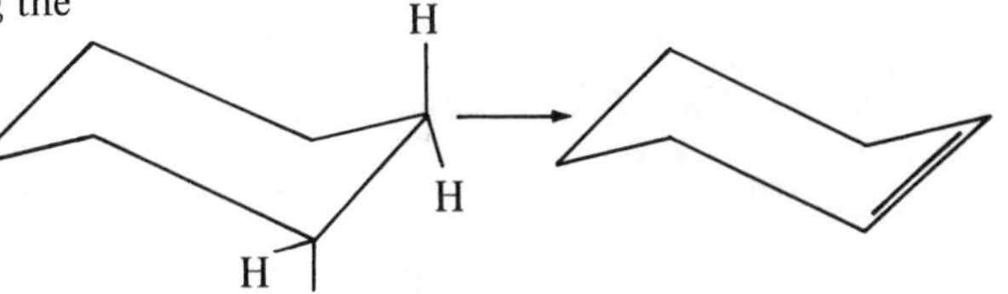


21. Which compound may exhibit *cis,trans*-isomerism?

(a) $CH_3{-}C(H)(CO_2H){-}NH_2$

(b) $CH_3{-}CH_2{-}C(H)(CH_3){-}CH_2{-}CH_3$

(c) $CH_3{-}CH{=}CH_2$

(d) $CH_3{-}CH_2{-}CH_2{-}CH_2{-}OH$

(e) $CH_3{-}CH{=}CH{-}CH_3$

22. An unknown alkyl halide reacts with alcoholic potassium hydroxide to produce a hydrocarbon $C_4H_8$. Oxidation of the hydrocarbon gives propionic acid ($CH_3CH_2COOH$) and $CO_2$. The structure of the alkyl halide is

(a) $CH_3{-}CH_2{-}CH_2{-}CH_2{-}Br$

(b) $CH_3{-}CH(Br){-}CH(Br){-}CH_3$

(c) $CH_3{-}CH_2{-}CH(Br){-}CH_3$

(d) $CH_3{-}CH(Br){-}CH_2{-}CH_2{-}Br$

(e) $Br{-}CH_2{-}CH_2{-}CH_2{-}CH_2{-}Br$

23. According to rules of systematic nomenclature, the triple bond is indicated by the suffix
    (a) ane
    (b) ol
    (c) ine
    (d) ene
    (e) yne

24. The triply bonded carbon atom is hybridized to give
    (a) 1 *sp* orbital
    (b) 2 $sp^2$ orbitals
    (c) 2 *sp* orbitals
    (d) 4 $sp^3$ orbitals
    (e) none of these

25. A triple bond consists of
    (a) 2 sigma bonds and 1 pi bond
    (b) 3 sigma bonds
    (c) 1 sigma bond and 2 pi bonds
    (d) 3 pi bonds
    (e) none of these

26. 1-Butyne will react with which of the following reagents?
    (a) $HgSO_4$ and $H_2SO_4$ (aq)
    (b) $Cu(NH_3)_2Cl$
    (c) HBr
    (d) Na, liquid ammonia
    (e) all of the above

27. The disappearance of the characteristic purple color of $KMnO_4$ in its reaction with an alkene is known as the______________for unsaturation.
    (a) Markovnikov test
    (b) Baeyer test
    (c) Wurtz test
    (d) Grignard test
    (e) pyrolytic test

28. Which of the following free radicals is the most stable?
    (a) $\cdot CH_3$
    (b) $\cdot CH_2{-}CH_3$
    (c) 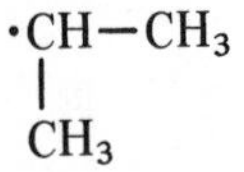 $\cdot CH{-}CH_3$ with $CH_3$ on the CH
    (d) 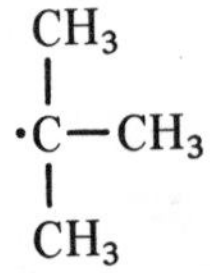 $\cdot C{-}CH_3$ with $CH_3$ above and $CH_3$ below the C

29. The most stable alkene presented is:
    (a) 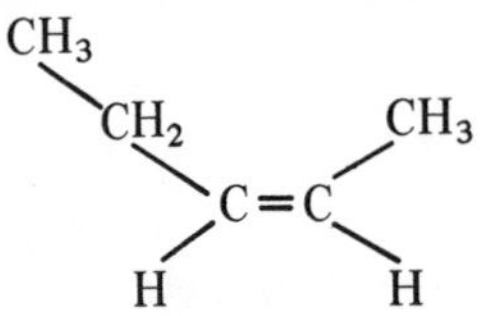 $CH_3{-}CH_2(H)C{=}C(H){-}CH_3$ (H below each C, H trans to ... as drawn: $CH_3CH_2$ upper left, H lower left; $CH_3$ upper right, H lower right)
    (b) $CH_3{-}CH_2{-}CH_2(H)C{=}CH_2$ ($CH_3CH_2CH_2$ upper left, H lower left; H upper right, H lower right)
    (c) $(CH_3)(H)C{=}C(CH_3)_2$ ($CH_3$ upper left, H lower left; $CH_3$ upper right, $CH_3$ lower right)
    (d) 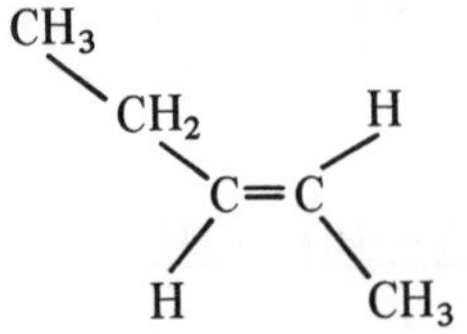 $CH_3{-}CH_2(H)C{=}C(H)CH_3$ ($CH_3CH_2$ upper left, H lower left; H upper right, $CH_3$ lower right)
    (e) 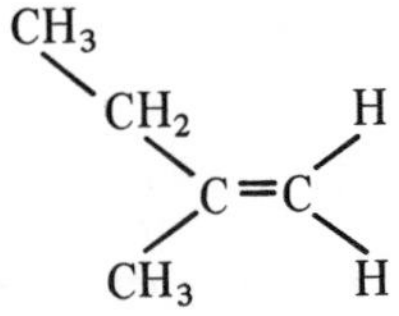 $(CH_3CH_2)(CH_3)C{=}CH_2$ ($CH_3CH_2$ upper left, $CH_3$ lower left; H upper right, H lower right)

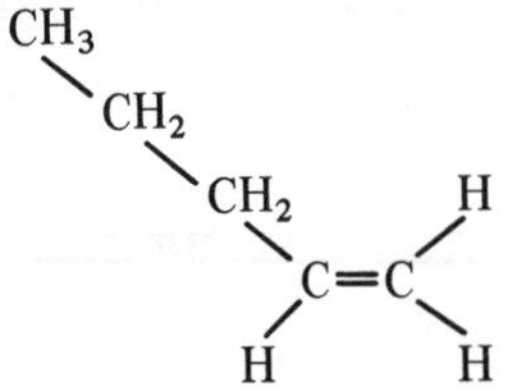

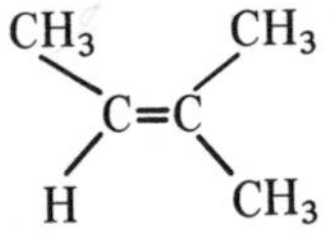

## Completion

30. Two classes of compounds included in the unsaturated hydrocarbons are ______________ and ______________.

31. Olefins is another name used for the ______________.

32. Acetylenes is another name used for the ______________.

33. Alkenes are named systematically by replacing the suffix *-ane* of the corresponding alkane with ______________ and numerically indicating the position of the double bond.

34. The *cis,trans*-isomer that has two identical groups on the same side of a planar or doubly bonded molecule is called the ______________ isomer.

35. The indirect addition of the elements of water to an alkene is known as ______________.

36. A compound that has two —OH groups on adjacent carbon atoms is known as a ______________.

37. Huge molecules, when formed through the repeated addition or condensation of many simple molecules, are called ______________.

38. A carbon chain containing alternate double and single bonds is said to be ______________.

39. The general formula for an alkyne is ______________ if only one triple bond is present in the structure.

**Matching** (Match each of the following terms with the statement below which best defines or explains it.)

(a) Dehydrohalogenation
(b) Halohydrin
(c) *cis,trans*-Isomerism
(d) Polymerization
(e) $sp^2$ Hybridization
(f) Vicinal dihalide
(g) Pyrolysis
(h) Markovnikov's rule
(i) Catalytic hydrogenation

40. ______ A variation in the spacial arrangement of groups attached to the double bond made possible by restricted rotation about the bond.

41. ______ The mixing of one 2*s* and two 2*p* orbitals to form three $sp^2$ orbitals.

42. ______ The high-temperature decomposition of organic compounds.

43. ______ The elimination of the elements of hydrogen halide (HX) from adjacent carbons.

44. ______ The addition of hydrogen under pressure and in the presence of a catalyst.

45. ______ An empirical rule regarding the mode of addition of Lewis acids to unsymmetrical olefins.

46. _____ An organic compound having a halogen and hydroxy group on adjacent carbon atoms.

47. _____ The repeated addition or condensation of small molecules resulting in the formation of huge molecules.

48. _____ An organic compound that has two halogen atoms on adjacent carbon atoms.

**Structural Formulas** (Write the correct structural formula for each of the following compounds or groups.)

49. *cis*-2-Butene

50. Dichlorocarbene

51. A tertiary carbonium ion

52. A bromonium ion

53. Simmons-Smith reagent

54. 1,2-Propanediol (Propylene glycol)

55. Isoprene

56. Acetylene

57. Vinyl chloride

58. Chloroprene

59. 1,2-Dibromopropane (Propylene bromide)

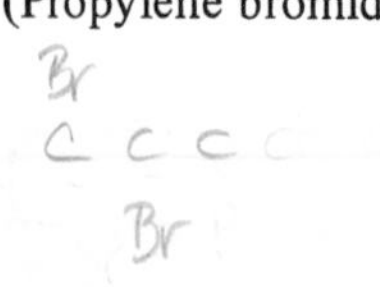

60. *tert*-Butyl hydrogen sulfate

61. The ozonide of 2,3-dimethylbutane

**Nomenclature** (Name each of the following according to IUPAC rules.)

62.

```
        H   H
        |   |
        C = C
       /     \
   H—C         C—H
     ‖         ‖
   H—C         C—H
       \     /
        C = C
        |   |
        H   H
```

1-3-5-7 cyclooctatetraene

63.

```
         H  CH3
         |  |
  H2C = C — C — CH3
            |
            CH3
```

3,3-dimethyl-1-butene

64. $CH_3—CH—C{\equiv}C—CH_2—CH_3$ (with $CH_3$ on the second carbon)

2-methyl-3-hexyne

65.

```
CH3
   \
    CH2       H
       \     /
        C = C
       /     \
      H       CH2
                 \
                  CH3
```

66. $CH_2{=}CH—CH_2—CH{=}CH_2$

**Reactions** (Complete the following.)

67. ( $H_3C-C(H)(H)-C(H)(OH)-H$ ) $\xrightarrow[\text{heat}]{H_2SO_4,}$ $CH_3—CH{=}CH_2 + H_2O$

68.

```
        CH3
        |
  CH3—C—CH2—CH3
        |
        Cl
```

$\xrightarrow[\text{KOH } \Delta]{\text{alcoholic}}$ ( $H_3C-C(CH_3){=}C(H)-CH_3$ ) $+ KCl + H_2O$

(main product)

69.

```
        H2
        C
      /   \
  H2C       C—H
   |        ||
  H2C       C—H
      \   /
        C
        H2
```

$\xrightarrow{H_2,\ \text{catalyst, heat, pressure}}$ (   )

70.

$$CH_3—\underset{\underset{CH_3}{|}}{C}=CH_2 \xrightarrow{HOCl} (\quad)$$

71.

$$CH_3—\underset{\underset{H_3C}{|}}{C}=\underset{\underset{CH_3}{|}}{C}—CH_3 + KMnO_4$$

→ cold, dilute → (   )

→ hot, conc. → (   )

72.

$$CaC_2 + 2H_2O \longrightarrow (\quad) + Ca(OH)_2$$

73.

$$H—C{\equiv}C{:}^- Na^+ + (\quad) \longrightarrow H—C{\equiv}C—CH_2—CH_3 + NaBr$$

74.

$$CH_2{=}CH—CH{=}CH_2 + HI \longrightarrow (\quad) + (\quad)$$

1,2-addition product    1,4-addition product

75. $I—CH_2—I + Zn(Cu) \longrightarrow [\quad]$

$$[\quad] + (CH_3)_2C{=}C(CH_3)_2 \longrightarrow (\quad) + ZnI_2$$

**Syntheses** (Outline syntheses leading to the preparation of the following compounds from the suggested starting materials. Use any other reagents that are necessary.)

76. Prepare 1,2-dibromopropane from isopropyl alcohol $(CH_3{-}\underset{\underset{\textstyle OH}{|}}{CH}{-}CH_3)$.

77. Prepare 2-bromopropane from 1-bromopropane.

78. Prepare 2-propanol $(CH_3{-}\underset{\underset{\textstyle OH}{|}}{CH}{-}CH_3)$ from 1-propanol $(CH_3{-}CH_2{-}CH_2{-}OH)$.

79. Prepare 2-chloroethanol $(Cl{-}CH_2{-}\underset{\underset{\textstyle OH}{|}}{CH_2})$ from ethyl alcohol.

80. Prepare acetone ($\mathrm{CH_3{-}\underset{\underset{\textstyle O}{\|}}{C}{-}CH_3}$) from *tert*-butyl bromide.

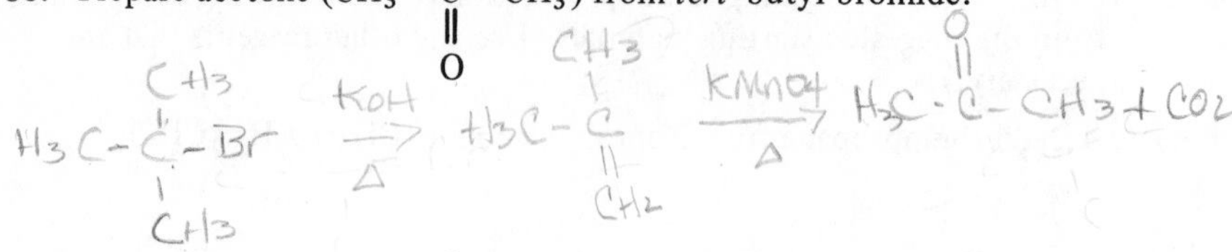

81. Prepare propyne from propene.

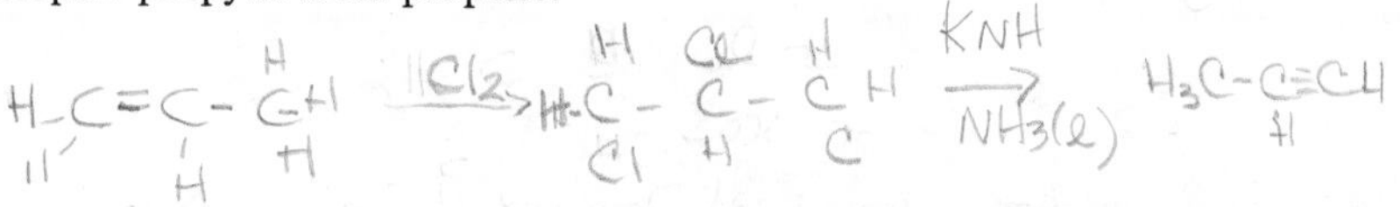

82. Prepare 6-methyl-3-heptyne from 1-butyne and any alkyl halide that is necessary.

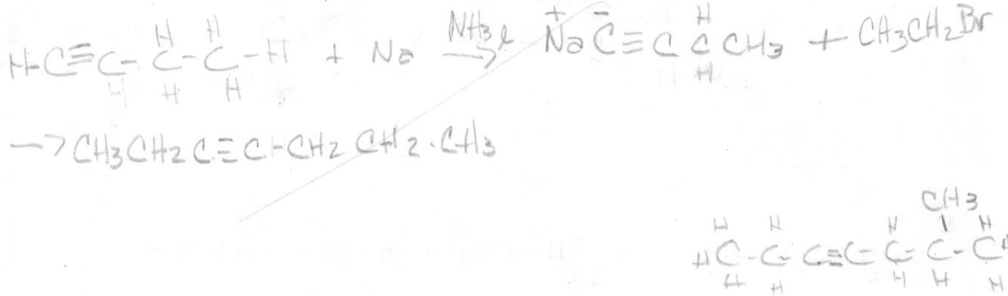

83. Prepare 2,2-dibromopropane from 1,2-dibromopropane.

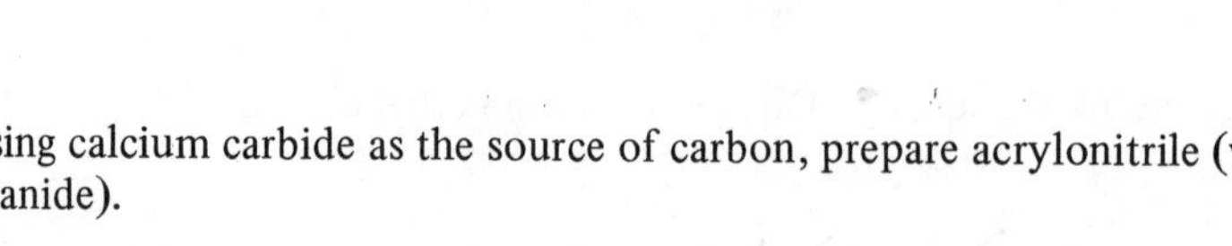

84. Using calcium carbide as the source of carbon, prepare acrylonitrile (vinyl cyanide).

85. Starting with acetylene, prepare 2-butanone ($CH_3-\underset{\underset{O}{\|}}{C}-CH_2-CH_3$).

2 H-C≡C-H

## Miscellaneous

86. Why is a different product obtained in each of the following reactions? Discuss from a mechanistic standpoint.

$$CH_3-CH=CH_2 \xrightarrow{HBr}$$

$$CH_3-CH=CH_2 \xrightarrow[\text{peroxide}]{HBr,}$$

87. The labels from bottles of hexane, cyclohexene, and 1-hexyne have come off and fallen to the floor. Describe simple tests that would enable you to correctly relabel the bottles.

88. Deuterium ($D_2$) in the presence of a catalyst behaves very much like hydrogen. Which product would be formed in the following reaction?

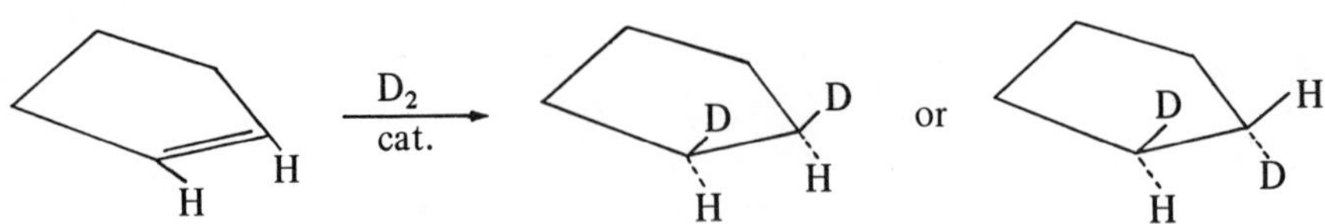


Does bromine add in the same manner? Which product is formed in the bromination of cyclopentene?

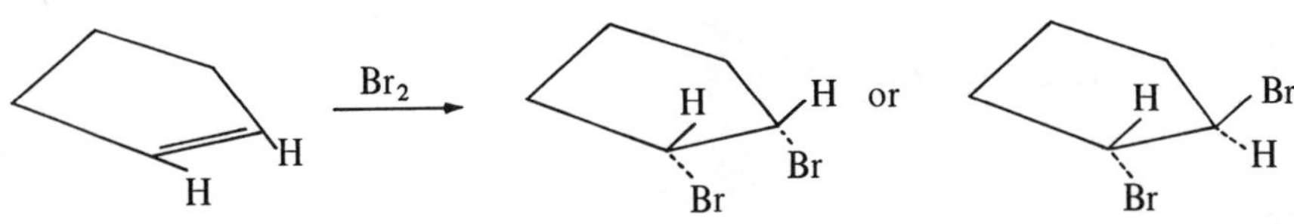


89. An unknown hydrocarbon had a molecular formula of $C_{12}H_{20}$. It absorbed three moles of hydrogen upon catalytic hydrogenation. Upon ozonolysis, it gave two moles each of the following compounds.

$$CH_3-\overset{\overset{\displaystyle CH_3}{|}}{C}=O \qquad H-\overset{\overset{\displaystyle O}{\|}}{C}-CH_2-\overset{\overset{\displaystyle O}{\|}}{C}-H$$

What is the structure of the unknown?

What products would be formed if the unknown was reacted with hot, concentrated potassium permanganate?

90. Write the mechanism for the free radical polymerization of propylene using R—O—O—R as the catalyst in the chain initiation step.

91. Draw the structure of the alkene you would use in a reaction with dichlorocarbene to prepare 1,1-dichloro-*cis*-2,3-dimethylcyclopropane.

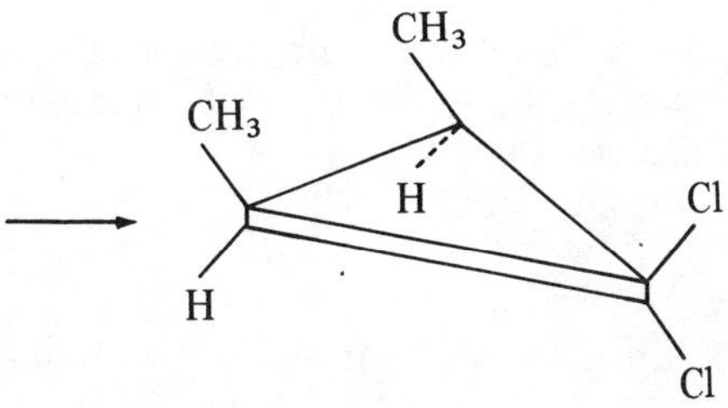

92. Give the structure of the diene and the dienophile which could be used to synthesize each of the following compounds using the Diels-Alder reaction.

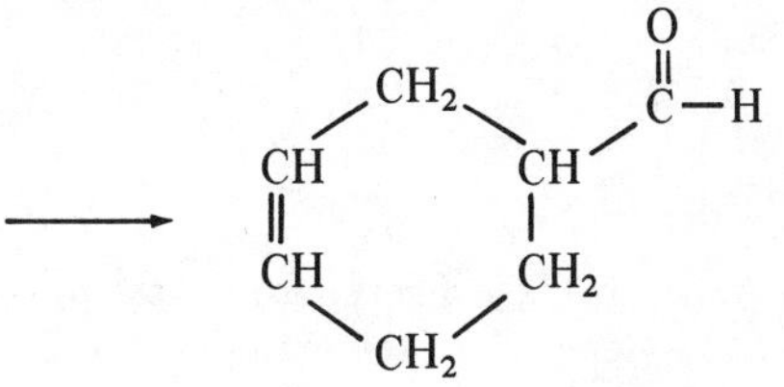

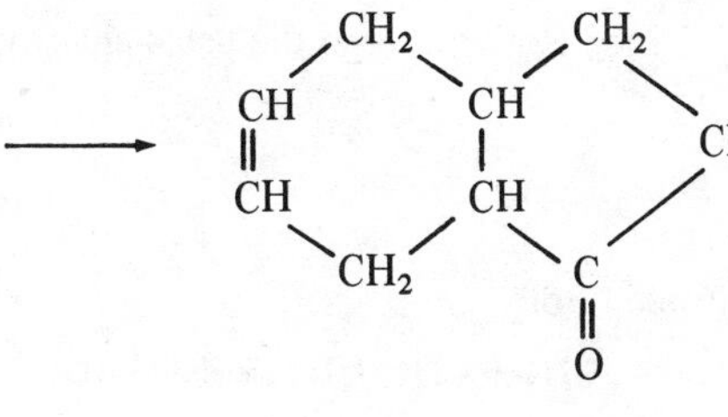

## ANSWERS

### True-False

1. False. Alkanes are relatively inert. Alkenes and alkynes, because of the unsaturation, are reactive compounds.
2. True. Alkenes share two pair of electrons and alkynes share three pair.

3. True. It is $C_nH_{2n}$ for both.
4. False. A double bond is made of a sigma bond and a pi bond.
5. True
6. True
7. True. The electron cloud of a $\pi$ orbital is more accessible.
8. False. C—C > C=C > C≡C (bond length)
9. False. A greater number of isomers is possible for an alkene than for an alkane of the same number of carbon atoms because the position of the double bond may vary and there is also the possibility of *cis,trans*-isomerism.
10. True
11. False. Alkyl groups are electron-releasing by inductive effects. Laws of electrostatics tell us the stability of a charged system is increased by dispersal of the charge. Thus,

$$\begin{array}{c}R\\\downarrow\\R\rightarrow C+\\\uparrow\\R\end{array} > \begin{array}{c}H\\|\\R\rightarrow C+\\\uparrow\\R\end{array} > \begin{array}{c}H\\|\\R\rightarrow C+\\|\\H\end{array}$$

(stability of carbonium ions: 3° > 2° > 1°)

12. True
13. True
14. True
15. True. It takes a very strong base to remove the acetylenic hydrogen however. Acetylene is less acidic than $H_2O$.

$$R-C\equiv C^-Na^+ + H_2O \longrightarrow R-C\equiv C-H + NaOH$$

16. True. Both undergo the usual electrophilic addition reactions.
17. True

**Multiple Choice**

18. (c) $CH_3-CH=CH_2 + \overset{-}{HO}\,|\,\overset{+}{Br} \longrightarrow CH_3-\underset{\displaystyle OH}{\underset{|}{CH}}-CH_2-Br$

19. (c)
20. (d) The dehydration reaction also can be brought about with $Al_2O_3$ and $H_2SO_4$.
21. (e) *Cis,trans*-isomerism is made possible by restricted rotation, such as in the case of a double bond. However, when two groups (or atoms) attached to a doubly bonded carbon are identical *cis,trans*-isomerism is not possible.

22. (a) $C_4H_8$ is the general formula for an alkene. The unknown alkyl halide must therefore have been a monohalogenated hydrocarbon.

$$CH_3-CH_2-CH=CH_2 \xrightarrow{\text{oxidation}} CH_3-CH_2-\overset{\overset{\displaystyle O}{\|}}{C}-OH + CO_2$$

Thus, *n*-butyl bromide must have been the unknown alkyl halide.

23. (e) 24. (c) 25. (c) 26. (e) 27. (b)

28. (d) The relative stability of free radicals is: tertiary > secondary > primary.

29. (c) The more highly substituted alkene is generally more stable.

**Completion**

30. alkenes (olefins) and alkynes (acetylenes)
31. alkenes
32. alkynes
33. *-ene*
34. *cis*
35. hydration
36. glycol
37. polymers
38. conjugated
39. $C_nH_{2n-2}$

**Matching**

40. (c)
41. (e)
42. (g)
43. (a)
44. (i)
45. (h)
46. (b)
47. (d)
48. (f)

**Structural Formulas**

49. 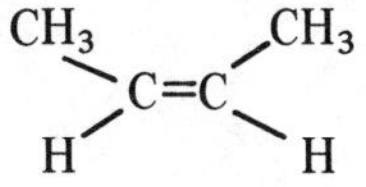 $CH_3(H)C=C(H)CH_3$ (trans: $CH_3$ upper left, H lower left; $CH_3$ upper right, H lower right)

50. 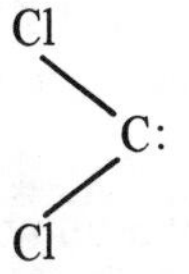 $Cl_2C{:}$

51. 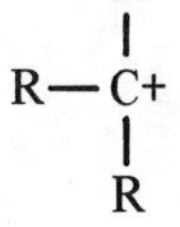 $R_3C^+$

52. 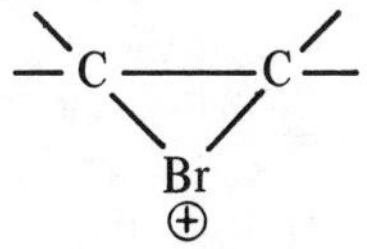 Bromonium ion: $>C—C<$ bridged by $Br^{\oplus}$

53. $[I-CH_2-Zn-I]$

54. 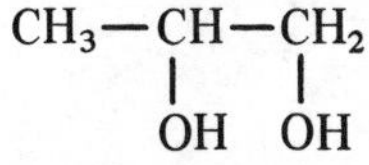 $CH_3-\underset{\displaystyle OH}{\underset{|}{CH}}-\underset{\displaystyle OH}{\underset{|}{CH_2}}$

55. 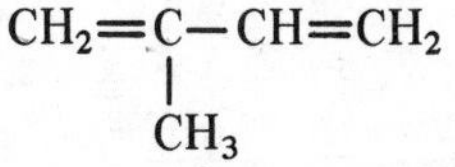 $CH_2=\underset{\displaystyle CH_3}{\underset{|}{C}}-CH=CH_2$

56. $H-C\equiv C-H$

57. 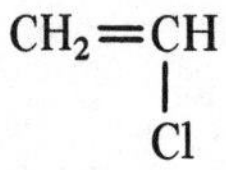 $CH_2{=}CH(Cl)$ (vinyl chloride: $CH_2{=}CH$ with Cl on the CH)

58. 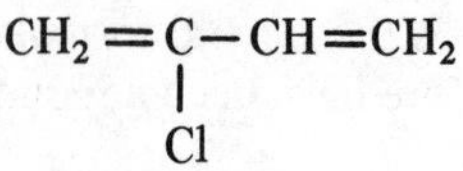 $CH_2{=}C(Cl){-}CH{=}CH_2$

59. 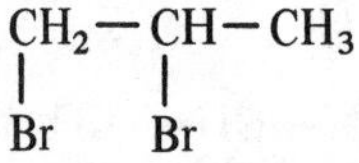 $CH_2(Br){-}CH(Br){-}CH_3$

60. 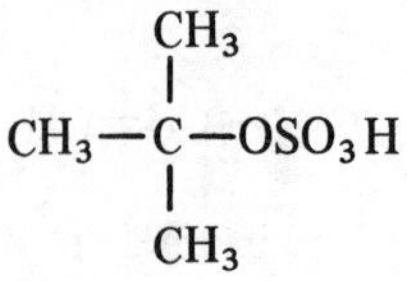 $(CH_3)_3C{-}OSO_3H$

$$\begin{array}{c} CH_3 \\ | \\ CH_3{-}C{-}OSO_3H \\ | \\ CH_3 \end{array}$$

61. 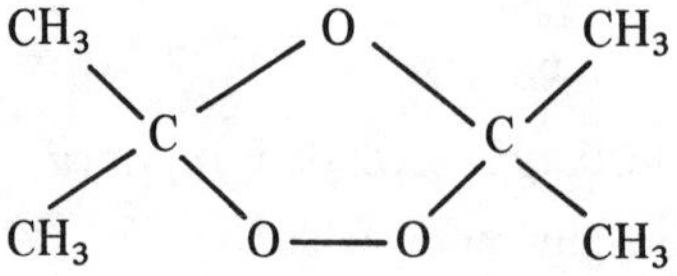 Cyclic ozonide: two $C(CH_3)_2$ carbons joined through a ring O and through an O—O bridge

$$(CH_3)_2C\overset{O}{\frown}C(CH_3)_2,\ \text{with } O{-}O \text{ bridge}$$

## Nomenclature

62. 1,3,5,7-cyclooctatetraene
63. 3,3-dimethyl-1-butene
64. 2-methyl-3-hexyne
65. *trans*-3-hexene
66. 1,4-pentadiene

## Reactions

67.

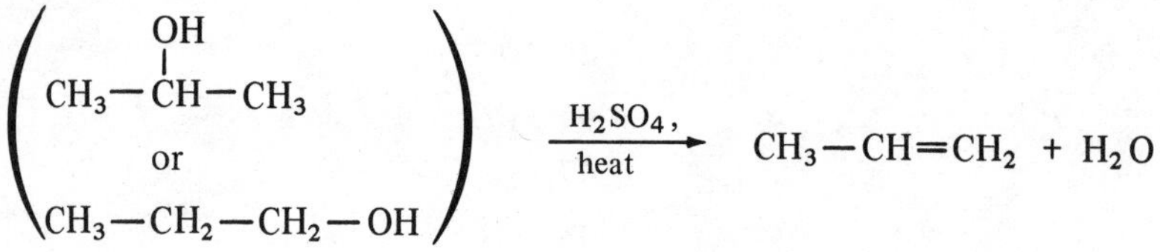

$$\left(\begin{array}{c} CH_3{-}CH(OH){-}CH_3 \\ \text{or} \\ CH_3{-}CH_2{-}CH_2{-}OH \end{array}\right) \xrightarrow[\text{heat}]{H_2SO_4,} CH_3{-}CH{=}CH_2 + H_2O$$

68.

$$CH_3{-}C(CH_3)(Cl){-}CH_2{-}CH_3 \xrightarrow[\text{KOH}]{\text{alcoholic}} \left(CH_3{-}C(CH_3){=}CH{-}CH_3\right) + KCl + H_2O$$

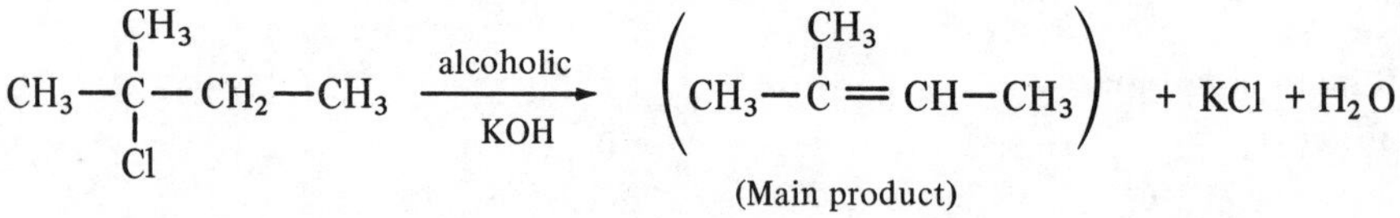

(Main product)

The alkene with the greater number of alkyl groups attached to the doubly bonded carbons is the more stable one and thus the main product.

69.

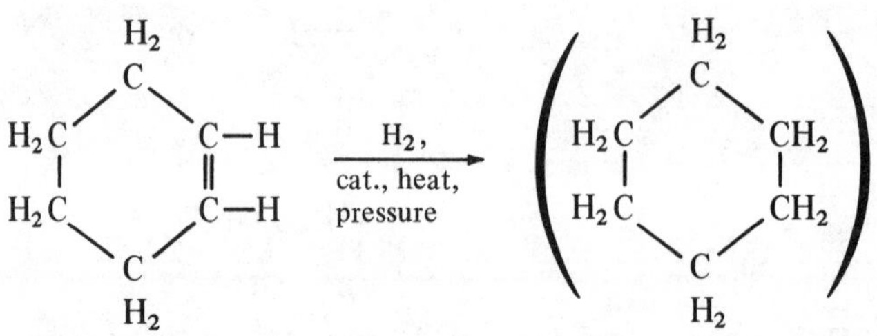

$$\text{cyclohexene } (H_2C, H_2C, CH_2, CH_2, C{-}H{=}C{-}H) \xrightarrow[\text{cat., heat, pressure}]{H_2,} \left(\text{cyclohexane } (CH_2)_6\right)$$

70.

$$CH_3-\underset{\underset{CH_3}{|}}{C}=CH_2 \xrightarrow{HOCl} \left( CH_3-\overset{\overset{OH}{|}}{\underset{\underset{CH_3}{|}}{C}}-CH_2-Cl \right)$$

71.

$$CH_3-\underset{\underset{H_3C}{|}}{C}=\underset{\underset{CH_3}{|}}{C}-CH_3 + KMnO_4 \xrightarrow{\text{cold, dilute}} \left( CH_3-\overset{\overset{OH}{|}}{\underset{\underset{H_3C}{|}}{C}}-\overset{\overset{OH}{|}}{\underset{\underset{CH_3}{|}}{C}}-CH_3 \right)$$

$$\xrightarrow{\text{hot, conc.}} \left( 2\ CH_3-\underset{\underset{CH_3}{|}}{C}=O \right)$$

72. $CaC_2 + 2H_2O \longrightarrow (H-C\equiv C-H) + Ca(OH)_2$

73. $H-C\equiv C\!:^- Na^+ + (CH_3-CH_2-Br) \longrightarrow H-C\equiv C-CH_2-CH_3 + NaBr$

74. $CH_2=CH-CH=CH_2 + HI \longrightarrow \left( CH_3-\underset{\underset{I}{|}}{CH}-CH=CH_2 \right) + \left( CH_3-CH=CH-CH_2-I \right)$

1,2-addition product 1,4-addition product

75. $I-CH_2-I + Zn(Cu) \longrightarrow [I-CH_2-Zn-I]$

$$[I-CH_2-Zn-I] + (CH_3)_2C=C(CH_3)_2 \longrightarrow \left( CH_2\!\!<\!\!\begin{array}{c} C(CH_3)_2 \\ | \\ C(CH_3)_2 \end{array} \right) + ZnI_2$$

## Syntheses

76. A synthesis involving more than one step requires a thorough knowledge of the methods of preparation and the reactions of each family of compounds. Usually you should work backwards, thinking of the methods that can be used to prepare the desired product. The next step is to see if the intermediate can be prepared from the specified starting material. There can be more than one correct synthetic pathway.

Vicinal dihalides, such as $CH_3-\overset{\overset{Br}{|}}{CH}-\overset{\overset{Br}{|}}{CH_2}$, are easily prepared by the addition of bromine to the alkene. In this example, the intermediate would be $CH_3-CH=CH_2$. Can this intermediate be prepared from the specified starting material? The answer is yes, it can be prepared by dehydration. Thus, we have a method of synthesis.

$$CH_3-\underset{\underset{OH}{|}}{CH}-CH_3 \xrightarrow[\text{heat}]{H_2SO_4,} CH_3-CH=CH_2 \xrightarrow{Br_2} CH_3-\underset{\underset{Br}{|}}{CH}-\underset{\underset{Br}{|}}{CH_2}$$

77. 2-Bromopropane can be prepared by:
   (a) free radical substitution of propane by bromine.
   (b) electrophilic addition of HBr to propene.

   Which of these two intermediates can be easily prepared from 1-bromopropane? Reduction of *n*-propyl magnesium bromide yields propane. Dehydrohalogenation of 1-bromopropane yields propene. Thus, at least two methods are available.

$$CH_3-CH_2-CH_2-Br \xrightarrow[\text{ether}]{Mg,} CH_3-CH_2-CH_2-Mg-Br \xrightarrow{H_2O} CH_3-CH_2-CH_3$$

$$CH_3-CH_2-CH_3 \xrightarrow{\text{light, } Br_2} CH_3-\underset{\underset{Br}{|}}{CH}-CH_3$$

$$CH_3-CH_2-CH_2-Br \xrightarrow[\text{KOH}]{\text{alcoholic}} CH_3-CH=CH_2 \xrightarrow{HBr} CH_3-\underset{\underset{Br}{|}}{CH}-CH_3$$

78. How many ways do you know to prepare alcohols? So far, the only method is the hydration of alkenes. Can the necessary intermediate ($CH_3-CH=CH_2$) be prepared from *n*-propyl alcohol? Yes, by dehydration.

$$CH_3-CH_2-CH_2-OH \xrightarrow[\text{heat}]{H_2SO_4,} CH_3-CH=CH_2 \xrightarrow{H_2SO_4} CH_3-\overset{\overset{OSO_3H}{|}}{CH}-CH_3$$

$$CH_3-\overset{\overset{OSO_3H}{|}}{CH}-CH_3 \xrightarrow{H_2O} CH_3-\underset{\underset{OH}{|}}{CH}-CH_3$$

79. $$CH_3-CH_2-OH \xrightarrow[\text{heat}]{H_2SO_4,} CH_2=CH_2 \xrightarrow{HOCl} \underset{\underset{OH}{|}}{CH_2}-\underset{\underset{Cl}{|}}{CH_2}$$

80. $$CH_3-\underset{\underset{CH_3}{|}}{\overset{\overset{CH_3}{|}}{C}}-Br \xrightarrow[\text{KOH}]{\text{alc.}} CH_3-\overset{\overset{CH_3}{|}}{C}=CH_2 \xrightarrow[\text{heat}]{KMnO_4,} CH_3-\overset{\overset{O}{\|}}{C}-CH_3 + K_2CO_3$$

81. $$CH_3-CH=CH_2 \xrightarrow{Br_2} CH_3-\underset{\underset{Br}{|}}{CH}-\underset{\underset{Br}{|}}{CH_2} \xrightarrow[\text{KOH}]{\text{alc.}} CH_3-CH=\underset{\underset{Br}{|}}{CH} \xrightarrow{NaNH_2} CH_3-C\equiv C-H$$

82.

$$CH_3-CH_2-C\equiv C-H \xrightarrow[\text{liq. } NH_3]{\text{Na,}} CH_3-CH_2-C\equiv C^- Na^+$$

$$\xrightarrow{CH_3-CH(CH_3)-CH_2-Cl} CH_3-CH_2-C\equiv C-CH_2-CH(CH_3)-CH_3$$

83.

$$CH_3-CH(Br)-CH_2-Br \xrightarrow[\text{KOH}]{\text{alcoholic}} CH_3-CH{=}CH(Br) \xrightarrow{NaNH_2} CH_3-C\equiv CH \xrightarrow{2\ HBr} CH_3-CBr_2-CH_3$$

84.

$$CaC_2 \xrightarrow{H_2O} HC\equiv CH \xrightarrow[\text{CuCl, } NH_4Cl\text{-}HCl]{\text{HCN,}} H_2C{=}C(H)-CN$$

85.

$$HC\equiv CH \xrightarrow[\text{liq. } NH_3]{\text{Na,}} HC\equiv C^- Na^+ \xrightarrow{CH_3-CH_2-Cl} CH_3-CH_2-C\equiv CH$$

$$\xrightarrow{H_2O,\ H_2SO_4,\ HgSO_4} \left[CH_3-CH_2-C(OH){=}CH_2\right] \rightleftharpoons CH_3-CH_2-C(=O)-CH_3$$

**Miscellaneous**

86.

$$CH_3-CH{=}CH_2 \xrightarrow{HBr} CH_3-CH(Br)-CH_3$$

Electrophilic addition reaction. The $H^+$ adds first to form the more stable carbonium ion ($CH_3-\overset{+}{C}H-CH_3$)

$$CH_3-CH{=}CH_2 \xrightarrow{\text{HBr, peroxide}} CH_3-CH_2-CH_2Br$$

Free radical addition reaction. The Br• adds first to form the more stable free radical ($CH_3-\dot{C}H-CH_2-Br$).

87.

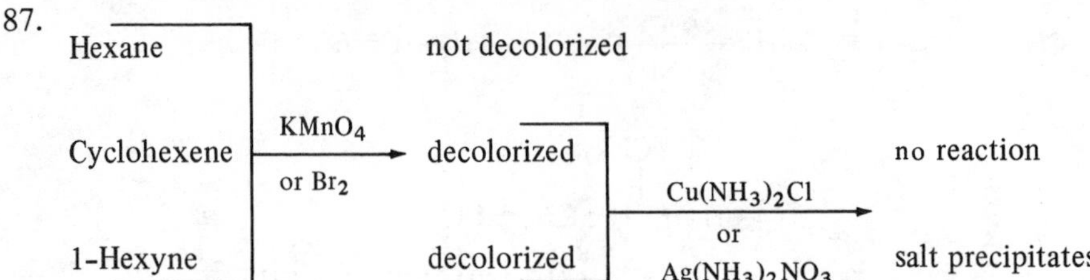


88. The *cis* product is formed. The $H_2$ or $D_2$ atoms add simultaneously and from the same side of the double bond.

Bromine does not add in the same manner as does $H_2$. Bromination results in formation of the *trans* isomer. The reaction is stepwise and takes place through a bromonium ion intermediate.

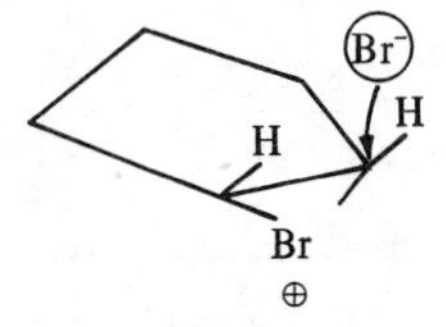


The bromine anion then attacks from the opposite side.

89. $CH_3{-}C(CH_3){=}CH{-}CH_2{-}CH{=}CH{-}CH_2{-}CH{=}C(CH_3){-}CH_3$ (unknown)

hot conc. $KMnO_4$ ↓

$$2\ (CH_3)_2C{=}O + 2\ {}^{+}K^{-}O{-}\overset{O}{\overset{\|}{C}}{-}CH_2{-}\overset{O}{\overset{\|}{C}}{-}O^{-}K^{+}$$

90. $R{-}O{-}O{-}R \xrightarrow{\text{heat}} 2\ RO\cdot$

$$RO\cdot + CH_3{-}CH{=}CH_2 \longrightarrow CH_3{-}\dot{C}H{-}CH_2{-}OR$$

$$CH_3{-}\dot{C}H{-}CH_2{-}OR + CH_3{-}CH{=}CH_2 \longrightarrow CH_3{-}\dot{C}H{-}CH_2{-}CH(CH_3){-}CH_2{-}OR$$

$$CH_3{-}\dot{C}H{-}CH_2{-}CH(CH_3){-}CH_2{-}OR + CH_3{-}CH{=}CH_2 \longrightarrow CH_3{-}\dot{C}H{-}CH_2{-}CH(CH_3){-}CH_2{-}CH(CH_3){-}CH_2{-}OR$$

and so on until two free radicals finally couple or disproportionate. Note that the more stable free radical is formed in each addition to the double bond.

91. 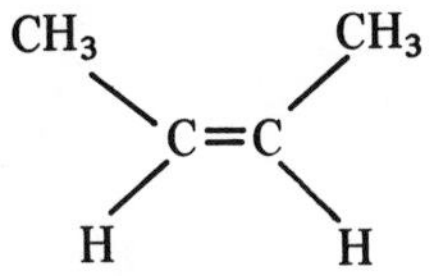 $CH_3$ $CH_3$ / C=C / H H

Dichlorocarbene adds stereospecifically *cis* to alkenes. Had you used *trans*-2-butene as a starting material, the product would have been 1,1-dichloro-*trans*-2,3-dimethylcyclopropane.

92.

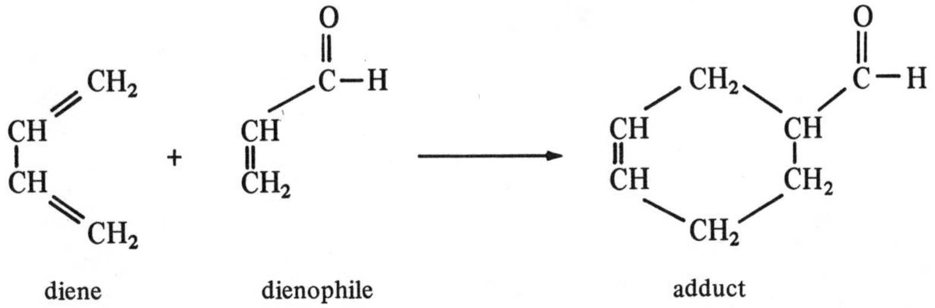

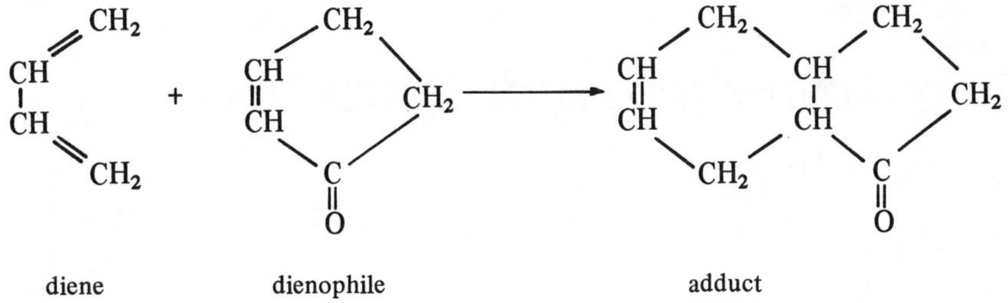

$CH_2$=CH–CH=$CH_2$ + $CH_2$=CH–C(=O)–H ⟶ adduct (ring: CH=CH–$CH_2$–CH(C(=O)–H)–$CH_2$–$CH_2$)

diene dienophile adduct

$CH_2$=CH–CH=$CH_2$ + cyclopentenone (CH=CH–C(=O)–$CH_2$–$CH_2$) ⟶ adduct (CH=CH–$CH_2$–CH–CH–$CH_2$ fused with $CH_2$–$CH_2$–C=O)

diene dienophile adduct

# 4

# The Aromatic Hydrocarbons, or Arenes

## INTRODUCTION

In this chapter we study the structure and the physical and chemical properties of the aromatic hydrocarbons. These are cyclic compounds that have a planar structure and a total of $4n + 2$ $\pi$ electrons where $n$ is an integer. Benzene is the most important aromatic hydrocarbon and most of this chapter is devoted to benzene and its derivatives. The structure of benzene is considered to be that of a resonance hybrid whose true structure is intermediate to that of the two Kekulé resonance forms shown below. However, either of the Kekulé structures usually is used to represent benzene,

```
          H                          H
          |                          |
          C                          C
        //  \                      /  \\
   H-C        C-H            H-C        C-H
     |        ||   <------>     ||        |
   H-C        C-H            H-C        C-H
        \\  /                      \  //
          C                          C
          |                          |
          H                          H
```

shown below as a regular hexagon with alternate double and single bonds.

We shall learn that benzene and other aromatic hydrocarbons do not undergo addition reactions that we have come to expect for unsaturated hydrocarbons. Instead, aromatic hydrocarbons undergo mainly **electrophilic substitution** reactions in which one or more of the ring hydrogen atoms is replaced.

Certain groups, when attached to the benzene ring, have an activating effect upon the ring and enable the ring to undergo electrophilic substitution reactions more readily than the unsubstituted benzene ring. These activating groups direct the attacking species to either the *ortho* or *para* position. Groups on the benzene ring which deactivate it to further substitution are *meta* directors. An important exception to this general rule is found in the halogens which are *ortho-para* directors, but deactivators.

Y

*o* *o*

*m* *m*

*p*

*o* = positions *ortho* to Y

*m* = positions *meta* to Y

*p* = position *para* to Y

## PROBLEMS

### True–False

1. T F The chemical behavior of benzene is very similar to that of alkenes.
2. T F An organic compound which is classified as "aromatic" usually is very stable.
3. T F The bond distances between adjacent carbon atoms in benzene are all the same.
4. T F Except for halogens, most *ortho-para* directors activate the ring.
5. T F The group that attacks the benzene ring determines its orientation in the final product.
6. T F In general, if the atom bonded to the benzene nucleus is bonded to another atom by a double or triple bond, the group will be a *meta* director.
7. T F Oxidation of a monoalkylbenzene with potassium dichromate and sulfuric acid will yield benzoic acid.
8. T F If a substituent bonded to a benzene ring is deactivating, the compound will be more reactive than benzene toward electrophilic aromatic substitution reactions.
9. T F Approximately the same amount of heat is evolved in the catalytic hydrogenation of one mole of benzene as is evolved in the catalytic hydrogenation of three moles of cyclohexene.

### Multiple Choice

10. Which of the following reactions will produce cumene (isopropylbenzene)?

(a) $C_6H_6 + CH_3{-}\underset{\displaystyle Cl}{\underset{|}{CH}}{-}CH_3 + AlCl_3 \longrightarrow$

(b) benzene $+ CH_3—CH{=}CH_2 + HCl + AlCl_3 \longrightarrow$

(c) benzene $+ CH_3—CH(OH)—CH_3 + H_2SO_4 \longrightarrow$

(d) benzene $+ CH_3—CH_2—CH_2—Cl + AlCl_3 \longrightarrow$

(e) All of the above produce cumene.

11. 3-Chlorophenol is structure ________.

(a) benzene ring with OH and Cl on adjacent carbons (OH, Cl)

(c) benzene ring with OH and Cl on opposite carbons (Cl, OH)

(b) benzene ring with OH and Cl in meta positions (OH, Cl)

12. The reaction: benzene $+ {}^{\oplus}NO_2 \longrightarrow$ C₆H₅–$NO_2$ $+ H^{\oplus}$ is a(n)

(a) addition reaction
(b) elimination reaction
(c) electrophilic substitution reaction
(d) nucleophilic substitution reaction
(e) reduction reaction

13. The reaction least likely to occur is

(a) benzene $+ HNO_3 \xrightarrow{H_2SO_4}$ C₆H₅–$NO_2$

(b) benzene $+ Cl_2 \xrightarrow[\text{light}]{\text{ultraviolet}}$ C₆H₅–Cl

(c) benzene $+ H_2SO_4 \longrightarrow$ C₆H₅–$SO_3H$

(d) benzene $+ Br_2 \xrightarrow{Fe}$ C₆H₅–Br

(cont'd)

(e) 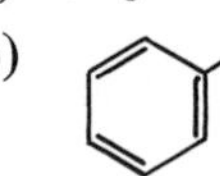 + $CH_3-CH_2-Br \xrightarrow{AlCl_3}$ $C_6H_5-CH_2-CH_3$

14. An aryl group is
    (a) $CH_3-$
    (b) $C_6H_5-$
    (c) $\cdot CH_3$
    (d) $CH_3-C(=O)-$
    (e) $CH_2=C(H)-$

15. $C_6H_6$ + $CH_3-CH_2-Br \longrightarrow C_6H_5-CH_2-CH_3$ + HBr

    The catalyst used to complete the above reaction is
    (a) Na
    (b) Mg
    (c) $AlCl_3$
    (d) KOH
    (e) $H_2$

16. The electrophilic species which is considered to be the active agent in the nitration of the aromatic ring is
    (a) $NO_2^-$
    (b) $NO^+$
    (c) $HNO_2^+$
    (d) $NO_2^+$
    (e) $ONO_2^+$

17. The carbon-carbon bond distance in benzene is
    (a) longer than a C−C single bond
    (b) longer than a C=C double bond
    (c) shorter than a C=C double bond
    (d) shorter than a C≡C triple bond

18. The name of this compound is

    (a) naphthalene
    (b) benzene
    (c) anthracene
    (d) triphenyl
    (e) phenanthrene

## Completion

19. The empirical formula for benzene is __________.

20. A resonance hybrid is more stable than any of the contributing resonance forms. This extra stability is called __________ __________.

21. The principal reactions of aromatic hydrocarbons are __________ __________ reactions.

22. A reagent which seeks an unshared pair of electrons is called an ______________ reagent.

**Matching** (Match the two words that name the same compound.)

(a) methylbenzene
(b) *m*-xylene
(c) hydroxybenzene
(d) phenylethane
(e) *o*-bromohydroxybenzene
(f) *p*-nitrotoluene
(g) 1,4-dimethylbenzene
(h) 4-nitrobenzoic acid
(i) 1-bromo-2-nitrobenzene
(j) isopropylbenzene

23. ______ *p*-xylene
24. ______ 1,3-dimethylbenzene
25. ______ 4-nitrotoluene
26. ______ phenol
27. ______ *o*-bromonitrobenzene
28. ______ ethylbenzene
29. ______ cumene
30. ______ toluene
31. ______ 2-bromophenol
32. ______ *p*-nitrobenzoic acid

**Structural Formulas** (Write the correct structural formula for each of the following compounds or groups.)

33. Phenyl group

34. Benzyl bromide

35. Naphthalene

36. 2-Chlorotoluene

37. 1,4-Dinitrobenzene

38. 1,2,4-Tribromobenzene

**Reactions** (Complete the following.)

39.

$C_6H_5{-}O{-}CH_3 + Br_2 \xrightarrow{FeBr_3} (\quad) + (\quad) + HBr$

40.

$C_6H_5{-}\overset{O}{\overset{\|}{C}}{-}OH + HNO_3 \xrightarrow{H_2SO_4} (\quad) + H_2O$

41.

$C_6H_5{-}Br + Mg \xrightarrow[\text{ether}]{\text{anhydrous}} (\quad) \xrightarrow{H_2O} (\quad)$

$+ Mg(OH)Br$

42.

$CH_3{-}C_6H_4{-}CH_2{-}CH_3 \xrightarrow{K_2Cr_2O_7,\ H_2SO_4,\ \text{heat}} (\quad) + CO_2 + H_2O$

43.

$C_6H_6 + CH_3{-}CH_2{-}CH_2{-}CH_2{-}Cl \xrightarrow{AlCl_3} (\quad) + HCl$

44.

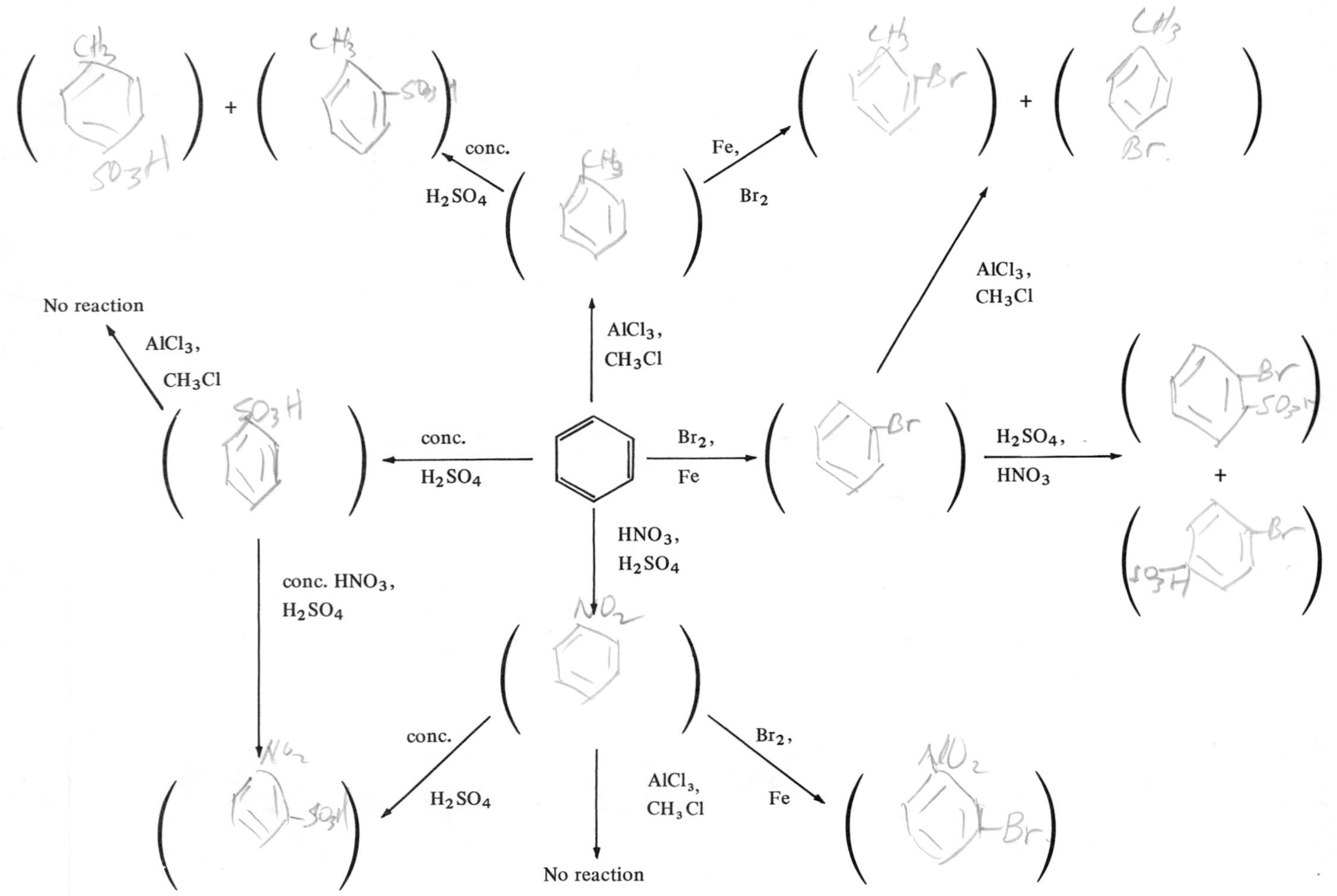
conc. H$_2$SO$_4$
Fe, Br$_2$
+
+
AlCl$_3$, CH$_3$Cl
AlCl$_3$, CH$_3$Cl
No reaction
AlCl$_3$, CH$_3$Cl
conc. H$_2$SO$_4$
Br$_2$, Fe
H$_2$SO$_4$, HNO$_3$
+
HNO$_3$, H$_2$SO$_4$
conc. HNO$_3$, H$_2$SO$_4$
conc. H$_2$SO$_4$
AlCl$_3$, CH$_3$Cl
Br$_2$, Fe
No reaction

**Miscellaneous**

45. The actual structure of benzene is a hybrid of what two Kekulé structures? (Draw the structures below.)

46. Label each of the positions, relative to the methyl group, as *ortho, para,* or *meta.*

$CH_3$

47. A dibromobenzene yields only one mono-substituted product when it is nitrated. Is the compound *ortho*–, *meta*–, or *para*–dibromobenzene?

48. Starting with benzene, show by equations how you would prepare (a) 1-bromo-4-nitrobenzene and (b) 1-bromo-3-nitrobenzene.

49. Place a check in the box which appropriately describes the influence of each substituent toward electrophilic substitution.

| | Orientation | | Reactivity Relative to Benzene | |
|---|---|---|---|---|
| Substituent | Ortho-Para | Meta | Activating | Deactivating |
| $-Cl$ | | | | |
| $-CH_2-CH_3$ | | | | |
| $-C(=O)-CH_3$ | | | | |
| $-NH_2$ | | | | |
| $-NH_3^+$ | | | | |
| $-OH$ | | | | |
| $-O-CH_2-CH_3$ | | | | |
| $-C(=O)-O-H$ | | | | |
| $-SO_3H$ | | | | |
| $-NO_2$ | | | | |

50. Draw the cationic intermediates (sigma complex) formed by the addition of a nitronium ion ($NO_2^+$) to the *ortho, meta,* and *para* positions of chlorobenzene. Which intermediate do you predict would require the highest energy of activation for its formation?

## ANSWERS

### True–False

1. False. The principal reactions of the alkenes are *additions* to the carbons having the double bond. The principal reactions of benzene and other aromatic compounds are electrophilic aromatic *substitutions.*
2. True. The $4n + 2$ $\pi$ electrons are delocalized.
3. True
4. True
5. False. The substituent already present on the ring determines the orientation of the incoming group.
6. True

7. True

$$C_6H_5\text{—R} \xrightarrow[H_2SO_4]{K_2Cr_2O_7,} C_6H_5\text{—}CO_2H$$

Regardless of its length, the alkyl side chain is degraded to benzoic acid.

8. False. A deactivating substituent makes the compound less reactive than benzene.
9. False. The heat evolved in the reduction of one mole of benzene is less than that evolved in the reduction of three moles of cyclohexene. The energy difference is called the stabilization energy or resonance energy of benzene.

**Multiple Choice**

10. (e) The cation ($\oplus CH_2{-}CH_2{-}CH_3$) formed in (d) undergoes rearrangement to the more stable secondary carbonium ion ($CH_3{-}\overset{}{\underset{\oplus}{CH}}{-}CH_3$).
11. (b) Also called *m*-chlorophenol.
12. (c) It is a substitution reaction because the nitro group, $NO_2$, is substituted for a hydrogen atom. It is electrophilic substitution because the attacking species is a positive group seeking a center of high electron density.
13. (b)
14. (b) The removal of a hydrogen from an arene produces an aryl group.
15. (c) This is an example of a Friedel-Crafts reaction.
16. (d) This is called a nitronium ion.
17. (b) The structure of benzene is a hybrid of the resonance forms.

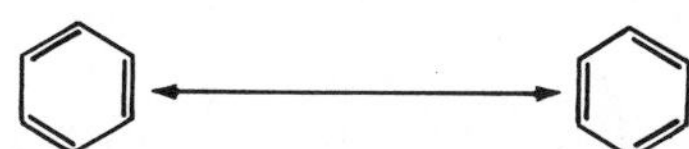

and the carbon-carbon bond distance is shorter than a C—C single bond, but longer than a C=C double bond. Each bond has some single and some double bond character.

18. (c)

**Completion**

19. CH
20. stabilization or resonance energy
21. electrophilic substitution
22. electrophilic (also a Lewis acid)

**Matching**

23. (g)
24. (b)
25. (f)
26. (c)
27. (i)
28. (d)
29. (j)
30. (a)
31. (e)
32. (h)

## Structural Formulas

33. H, H, H, H, H

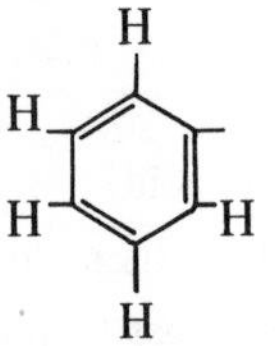

34. $CH_2-Br$

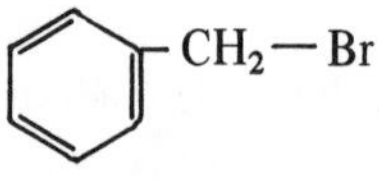

35.

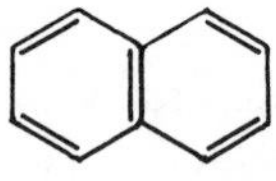

36. $CH_3$, Cl

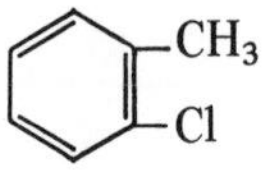

37. $O_2N$, $NO_2$

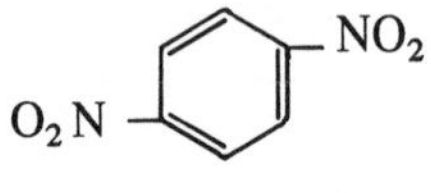

38. Br, Br, Br

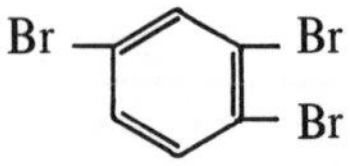

## Reactions

39. $O-CH_3$ + $Br_2$ $\xrightarrow{FeBr_3}$ ( $OCH_3$, Br ) + ( Br, $OCH_3$ ) + HBr

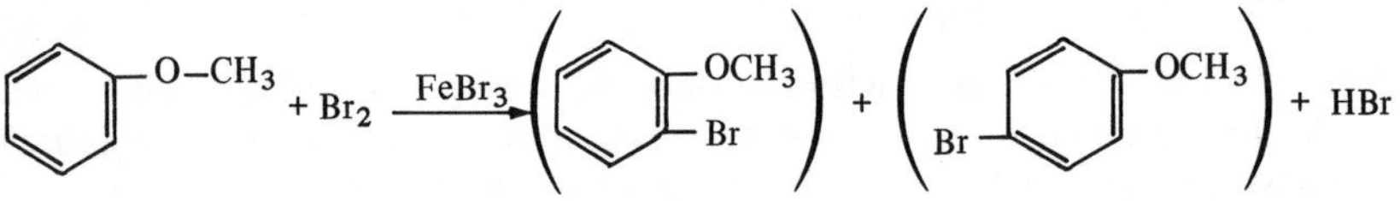

40. O, $\|$, C—OH + $HNO_3$ $\xrightarrow{H_2SO_4}$ ( $CO_2H$, $NO_2$ ) + $H_2O$

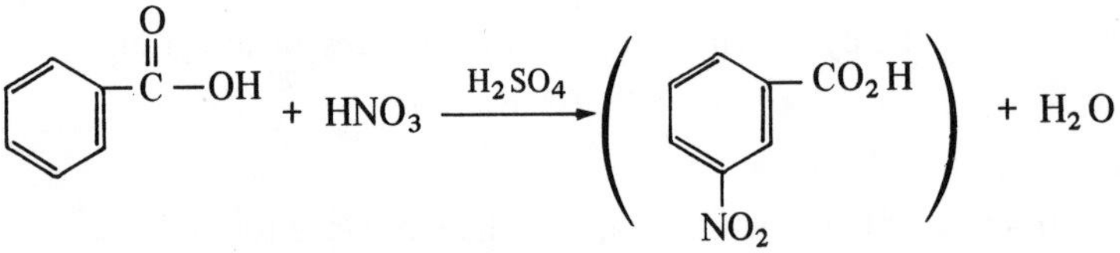

41. Br + Mg $\xrightarrow[\text{ether}]{\text{anhydrous}}$ ( Mg—Br ) $\xrightarrow{H_2O}$ ( ) + Mg(OH)Br

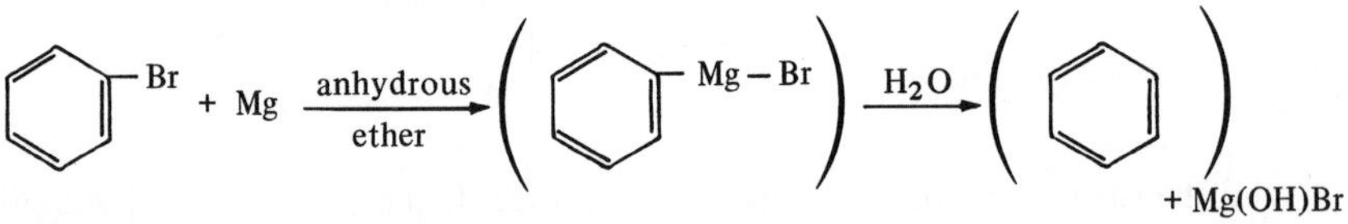

42. $CH_3$, $CH_2$—$CH_3$ $\xrightarrow[\text{H}_2\text{SO}_4\text{, heat}]{\text{K}_2\text{Cr}_2\text{O}_7\text{,}}$ ( $CO_2H$, $CO_2H$ ) + $CO_2$ + $H_2O$

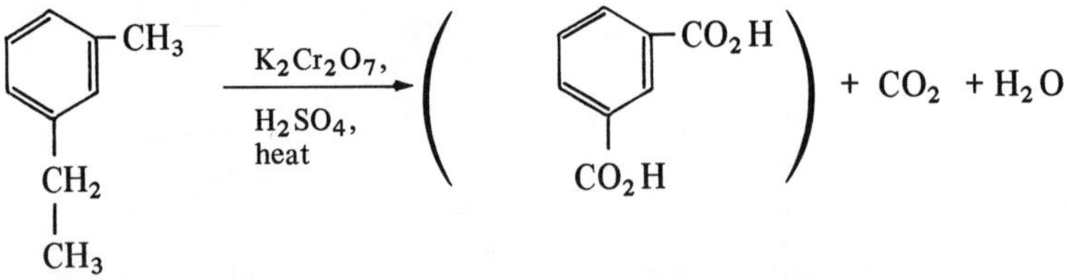

43. + $CH_3-CH_2-CH_2-CH_2-Cl$ $\xrightarrow{AlCl_3}$ ( $CH_3$, $CH-CH_2-CH_3$ ) + HCl

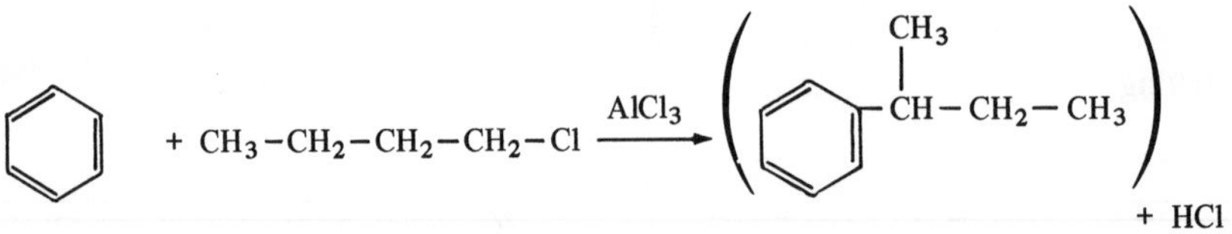

The primary carbonium ion, formed originally, rearranges to the more stable secondary carbonium ion before attacking the benzene ring.

44.

Benzene $\xrightarrow{AlCl_3,\ CH_3Cl}$ toluene ($C_6H_5CH_3$)

Toluene $\xrightarrow{Fe,\ Br_2}$ (o-bromotoluene) + (p-bromotoluene)

Toluene $\xrightarrow{conc.\ H_2SO_4}$ (p-$HO_3S$–$C_6H_4$–$CH_3$) + (o-$CH_3$–$C_6H_4$–$SO_3H$)

Benzene $\xrightarrow{Br_2,\ Fe}$ bromobenzene ($C_6H_5Br$)

Bromobenzene $\xrightarrow{AlCl_3,\ CH_3Cl}$ (o-bromotoluene) + (p-bromotoluene)

Bromobenzene $\xrightarrow{H_2SO_4,\ HNO_3}$ (o-Br–$C_6H_4$–$NO_2$) + (p-$O_2N$–$C_6H_4$–Br)

Benzene $\xrightarrow{HNO_3,\ H_2SO_4}$ nitrobenzene ($C_6H_5NO_2$)

Nitrobenzene $\xrightarrow{Br_2,\ Fe}$ (m-Br–$C_6H_4$–$NO_2$)

Nitrobenzene $\xrightarrow{AlCl_3,\ CH_3Cl}$ No reaction

Nitrobenzene $\xrightarrow{conc.\ H_2SO_4}$ (m-$HO_3S$–$C_6H_4$–$NO_2$)

Benzene $\xrightarrow{conc.\ H_2SO_4}$ benzenesulfonic acid ($C_6H_5SO_3H$)

Benzenesulfonic acid $\xrightarrow{conc.\ HNO_3,\ H_2SO_4}$ (m-$HO_3S$–$C_6H_4$–$NO_2$)

Benzenesulfonic acid $\xrightarrow{AlCl_3,\ CH_3Cl}$ No reaction

**Miscellaneous**

45.

46.

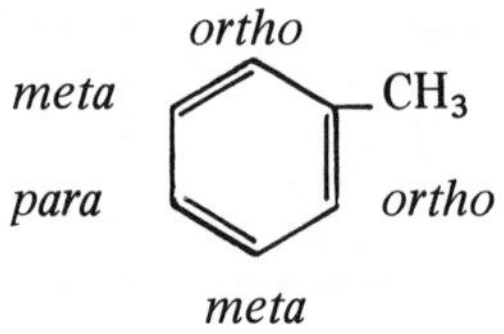

*ortho*
*meta* $CH_3$
*para* *ortho*
*meta*

47. The compound is *p*-dibromobenzene, because the only ring positions open are all *ortho* to one or the other bromine atoms.

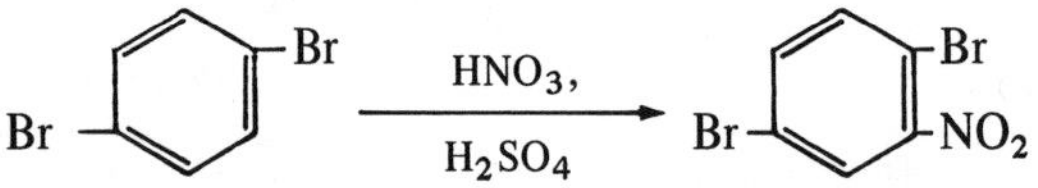

Br, Br — $HNO_3$, $H_2SO_4$ → Br, Br, $NO_2$

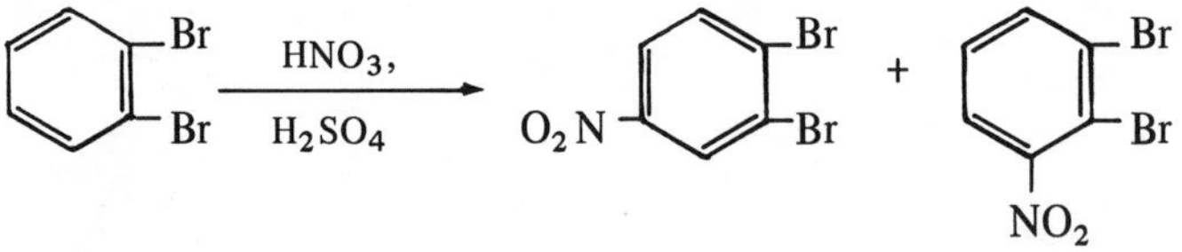

Br, Br — $HNO_3$, $H_2SO_4$ → $O_2N$, Br, Br + Br, Br, $NO_2$

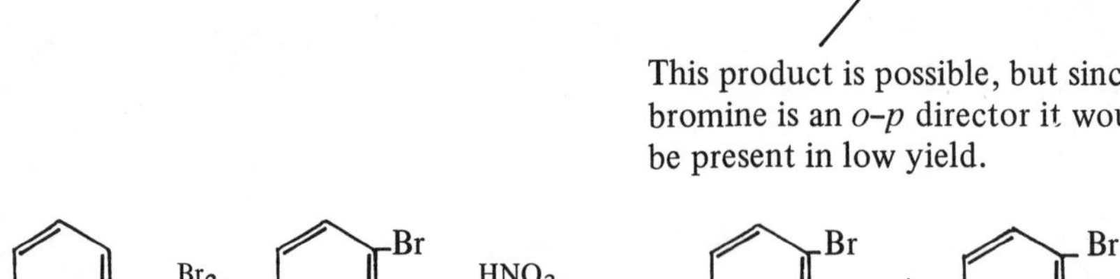

Br, Br — $HNO_3$, $H_2SO_4$ → Br, $NO_2$, Br + $NO_2$, Br, Br + $O_2N$, Br, Br

This product is possible, but since bromine is an *o–p* director it would be present in low yield.

48. $Br_2$, Fe → Br — $HNO_3$, $H_2SO_4$ → $O_2N$, Br + Br, $NO_2$

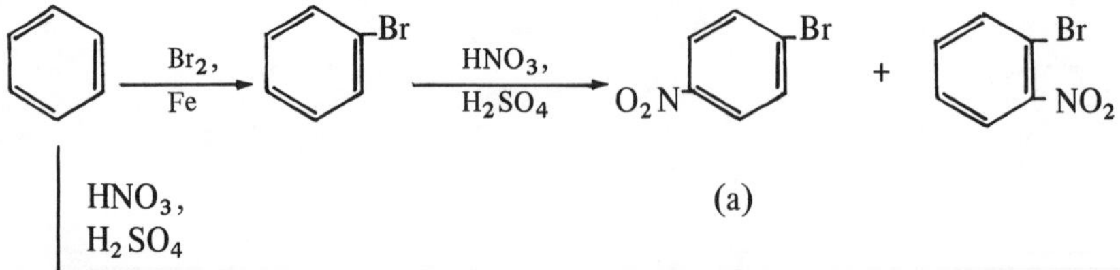

(a)

$HNO_3$, $H_2SO_4$

(*cont'd next page*)

The *para* isomer is usually less soluble and can be separated by repeated recrystallizations.

$NO_2$-benzene $\xrightarrow{Br_2, Fe}$ m-bromonitrobenzene (Br meta to $NO_2$)

(Note that in the preparation of disubstituted benzenes, the sequence of reaction is important in order to obtain the desired isomer.)

(b) *m*-Bromonitrobenzene

49. Place a check in the box which appropriately describes the influence of each substituent toward electrophilic substitution.

| | **Orientation** | | **Reactivity Relative to Benzene** | |
|---|---|---|---|---|
| **Substituent** | ***Ortho-Para*** | ***Meta*** | **Activating** | **Deactivating** |
| $-Cl$ | ✓ | | | ✓ |
| $-CH_2-CH_3$ | ✓ | | ✓ | |
| $-\overset{O}{\overset{\|\|}{C}}-CH_3$ | | ✓ | | ✓ |
| $-NH_2$ | ✓ | | ✓ | |
| $-NH_3^+$ | | ✓ | | ✓ |
| $-OH$ | ✓ | | ✓ | |
| $-O-CH_2-CH_3$ | ✓ | | ✓ | |
| $-\overset{O}{\overset{\|\|}{C}}-O-H$ | | ✓ | | ✓ |
| $-SO_3H$ | | ✓ | | ✓ |
| $-NO_2$ | | ✓ | | ✓ |

50.

*ortho* ↔ ↔ ↔ extra form

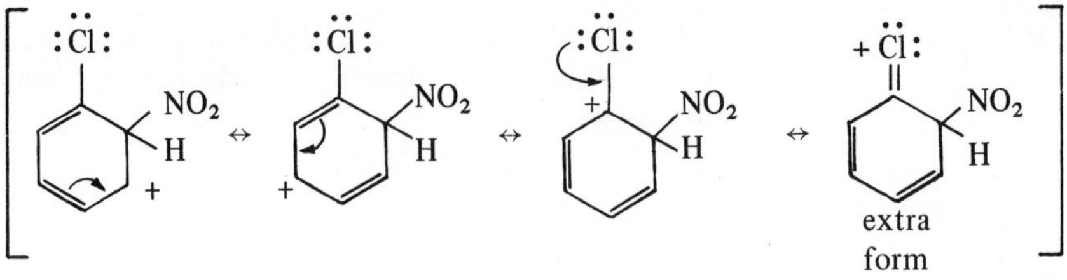

*meta* ↔ ↔

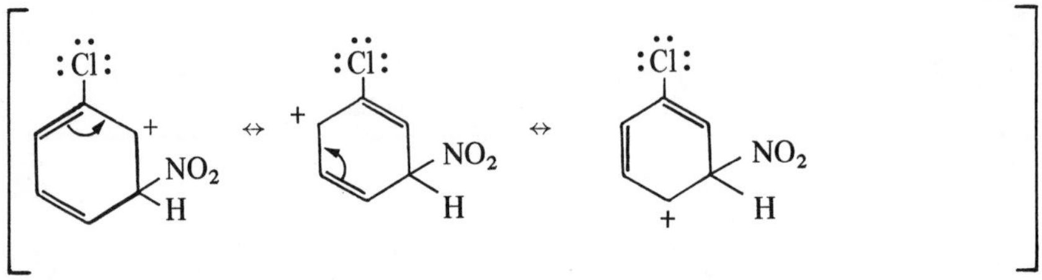

*para* ↔ ↔ ↔ extra form

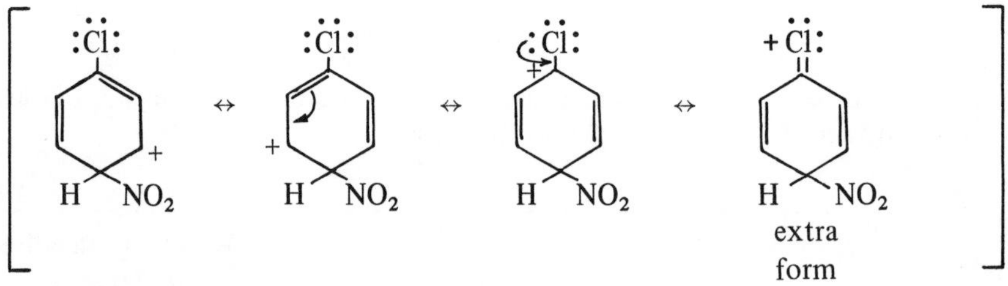

The dispersal of charge is greater in the *ortho* and *para* intermediates and therefore they are more stable than the *meta* intermediate. The energy of activation is greater for the formation of the *meta* intermediate. Since the chloro group is deactivating, all energies of activation are greater than that for the nitration of benzene.

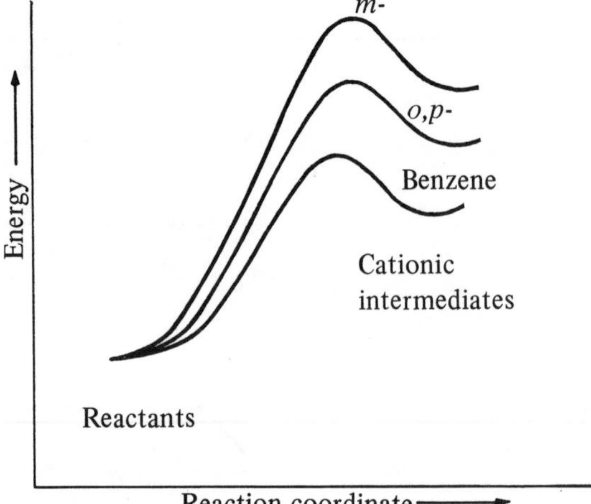

# 5

# *Stereoisomerism*

## INTRODUCTION

Stereoisomers are isomers made possible by different spatial arrangements of groups within a molecular structure. This chapter is devoted to stereoisomers which differ in **configuration** as opposed to **conformation**, that is, they can not be interconverted without breaking bonds and rearranging groups. Previously, we studied a type of stereoisomerism called *cis,trans*-isomerism. We will learn a different system of designation for this type of isomer: the (*E*), (*Z*)-system. Most of the discussion in this chapter is concerned with another type of stereoisomerism called **optical isomerism.** Optical isomers differ from each other as the left hand differs from the right. This difference in "handedness" is called **chirality**. Thus, chiral molecules exist in two spatial arrangements which are **mirror images** of each other. These mirror image isomers are called **enantiomers** or **enantiomorphs** and have the property of rotating the plane of **plane-polarized light** by the same number of degrees, but in opposite directions. We shall see how these mirror image isomers may be prepared, how they differ from each other, and how mixtures of equal parts of such isomers **(racemates)** can be resolved or separated from each other **(resolution)**. In addition to these new terms we also will want to become acquainted with other terms: *levo, dextro, meso,* **diastereoisomers**, D and L configurations, and (*R*) and (*S*) configurations.

## PROBLEMS

### True–False

1. T F Molecules which differ in conformation are interconverted by the expenditure of relatively small amounts of energy.

2. T F A sodium lamp generates plane-polarized light.
3. T F Nicol prisms are made of calcite crystals, $CaCO_3$, and are used as both the polarizer and analyzer in instruments called polarimeters.
4. T F The specific rotation of a compound depends upon the number of molecules of the optically active material that are in the path of the plane-polarized light.
5. T F Optical isomerism is possible whenever an asymmetric carbon atom is part of a molecular structure.
6. T F The D form of any compound will rotate plane-polarized light to the right (clockwise).
7. T F Racemates are optically inactive by external compensation.
8. T F Most laboratory syntheses that lead to the production of an asymmetric carbon atom produce a racemic mixture.
9. T F A great number of natural products are optically active and stereospecific in their reactions.
10. T F *Meso* forms are optically inactive because any effect that half of the molecule has on plane-polarized light is exactly compensated for by the other half of the molecule.
11. T F A common method of resolution involves reacting the racemate with an optically active reagent to produce diastereoisomers. After separation of the diastereoisomers, they are reconverted to the original optically active material.
12. T F D-Glyceraldehyde has the same configuration as (*R*)-glyceraldehyde.
13. T F Enantiomorphs differ only in their arrangement in space, direction of rotation of plane-polarized light, and speed of reaction with an optically active reagent.
14. T F A molecule which contains two asymmetric carbon atoms may not be chiral.
15. T F All molecules that contain only one asymmetric carbon atom are chiral, but all chiral molecules do not contain an asymmetric carbon atom.

**Matching** (Match the proper term with the statement which defines or explains it.)

(a) stereoisomers
(b) *cis,trans*-isomers
(c) monochromatic
(d) plane-polarized
(e) polarimeter
(f) asymmetric
(g) enantiomorphs
(h) racemic mixture
(i) diastereoisomers
(j) glyceraldehyde
(k) *meso*
(l) resolution
(m) chirality

16. _____ Structural feature of optically active compounds that makes possible their optical activity.
17. _____ Light of only one wavelength.
18. _____ Stereoisomers which are mirror images.

19. _____ Stereoisomers made possible by the restricted rotation about a double bond.

20. _____ Optically inactive mixture of equal amounts of enantiomorphs.

21. _____ Instrument used to measure the optical activity of a compound.

22. _____ Separation of a racemic mixture into its optically active forms.

23. _____ Compound with two similar asymmetric carbon atoms which is optically inactive by reason of internal compensation.

24. _____ Isomers that differ in the manner in which their atoms are arranged in space.

25. _____ Light of one wavelength vibrating in a single plane.

26. _____ Optically active stereoisomers that are not enantiomorphs.

27. _____ Standard to which optically active compounds are related.

28. _____ Carbon atom with four unlike groups attached.

**Miscellaneous**

**Questions 29–32** (Indicate in the space provided which pairs of isomers differ in configuration and which in conformation.)

29. ______________

Cl Cl    Cl Cl

30. ______________

H H Cl H Cl H    H H H Cl Cl H

C—C    C—C

31. ______________

$CH_3$ C Br Cl H    $CH_3$ C Cl Br H

32. ______________

$CH_3$ $CH_3$ C=C H H    $CH_3$ H C=C H $CH_3$

**Questions 33-42** (In the first column, place a check in front of the statements which indicate how diastereoisomers differ. Check the second column if the statements apply to differences in enantiomorphs, and the third column if the statements apply to differences in *cis,trans*-isomers.)

| Diastereoisomers | Enantiomorphs | *cis,trans*-Isomers | |
|---|---|---|---|
| | | | 33. Configuration |
| | | | 34. Conformation |
| | | | 35. Orientation of molecules in space |
| | | | 36. Melting points and boiling points |
| | | | 37. Speed of reaction with an optically inactive material |
| | | | 38. Speed of reaction with an enantiomer |
| | | | 39. Dipole moment |
| | | | 40. Molecular formula |
| | | | 41. Solubility |
| | | | 42. Direction, but not angle, of rotation of plane-polarized light |

**Questions 43–48** (Classify as *cis,trans*-isomers, enantiomorphs, *meso* forms, and diastereoisomers. Write your answer above each pair of formulas.)

43. ______________________ 44. ______________________

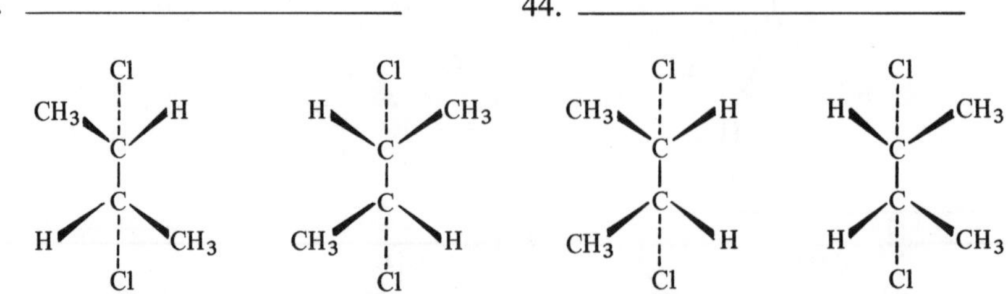

45. ______________________

Cl
$CH_3$ C H
C
H C $CH_3$
Br

Cl
H C $CH_3$
C
$CH_3$ C H
Br

46. ______________________

Cl
$CH_3$ C H
C
H C $CH_3$
Br

Cl
H C $CH_3$
C
H C $CH_3$
Br

47. ______________________

$CH_3$ C=C $CH_3$
H H

$CH_3$ C=C H
H $CH_3$

48. ______________________

$CH_3$
H—C—Cl
Br

$CH_3$
Cl—C—H
Br

49. What do the designations *dextro, d*, (+), and D mean?

**Questions 50-53** (Draw structures of all of the stereoisomers possible. Classify each pair of isomers as *cis,trans*-isomers; enantiomorphs; and diastereoisomers. Identify any *meso* forms present.)

50. $CH_3-CH=\underset{\displaystyle Cl}{\underset{|}{C}}-CH=CH_2$

51. $CH_3-\underset{\displaystyle OH}{\underset{|}{CH}}-CO_2H$

52. $CH_3-\underset{\displaystyle Br}{\underset{|}{CH}}-\underset{\displaystyle Br}{\underset{|}{CH}}-CO_2H$

53. $HO_2C-\underset{\displaystyle OH}{\underset{|}{CH}}-\underset{\displaystyle OH}{\underset{|}{CH}}-CO_2H$

54. Identify the configurations of each of the following as (*R*) or (*S*).

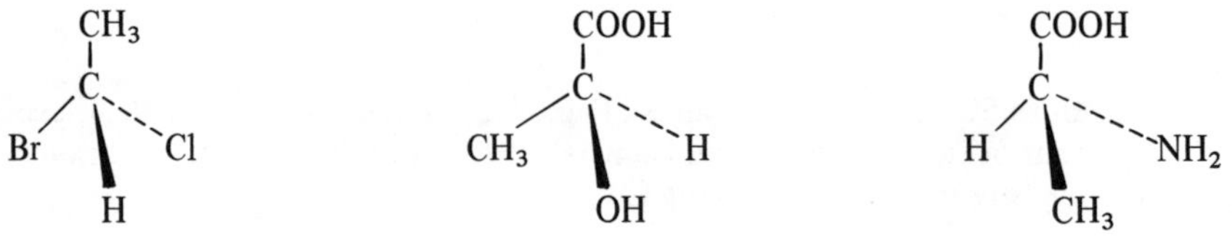

$CH_3$ C Br Cl H

COOH C $CH_3$ H OH

COOH C H $NH_2$ $CH_3$

55. Identify each of the following as (*E*) or (*Z*).

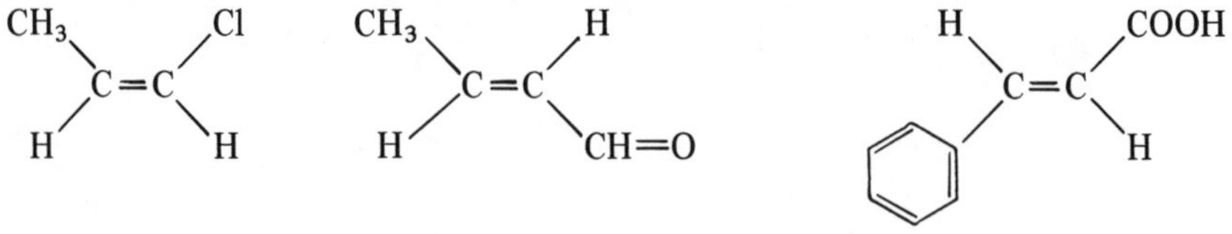

$CH_3$ Cl C=C H H

$CH_3$ H C=C H CH=O

H COOH C=C H

## ANSWERS

### True–False

1. True
2. False. A sodium lamp generates monochromatic light, but the light must be passed through a polarizer to make it plane-polarized.
3. True

4. True.

$$\text{Specific rotation } [\alpha] = \frac{\text{Observed rotation in degrees}}{\text{Length of sample tube in dm.} \times \text{Concentration (g/ml)}}$$

5. True. An asymmetric carbon atom is a sufficient condition for optical isomerism, but not a necessary one.
6. False. The designation "D" refers to configuration and relation to D-glyceraldehyde. It does not refer to the direction of rotation of plane-polarized light.
7. True. The effect of one enantiomorph on plane-polarized light is nullified by the presence of the other enantiomorph.
8. True
9. True
10. True
11. True. Enantiomorphs have the same physical properties except for rotation of plane-polarized light. Diastereoisomers can be separated because they have different physical properties, a difference in solubility being one of the most useful.
12. True
13. True
14. True. A compound having two asymmetric carbon atoms which hold identical groups may have a plane of symmetry. *meso*-2,3-Dibromobutane would be an example of such a compound.
15. True

**Matching**

| | | | | |
|---|---|---|---|---|
| 16. (m) | 17. (c) | 18. (g) | 19. (b) | 20. (h) |
| 21. (e) | 22. (l) | 23. (k) | 24. (a) | 25. (d) |
| 26. (i) | 27. (j) | 28. (f) | | |

**Miscellaneous**

29. configuration. The isomers cannot be interconverted without breaking bonds and rearranging groups.
30. conformation. The isomers can be interconverted by rotation around the carbon-carbon bond. This requires a relatively small expenditure of energy.
31. configuration. These are enantiomers.
32. configuration. These are *cis,trans*-isomers.

| | Diastereoisomers | Enantiomorphs | *cis,trans*-Isomers |
|---|---|---|---|
| 33. | ✓ | ✓ | ✓ |
| 34. | | | |
| 35. | ✓ | ✓ | ✓ |
| 36. | ✓ | | ✓ |
| 37. | ✓ | | ✓ |
| 38. | ✓ | ✓ | ✓ |
| 39. | ✓ | | ✓ |
| 40. | | | |
| 41. | ✓ | | ✓ |
| 42. | | ✓ | |

Diastereoisomers will differ in angle and perhaps direction of rotation of plane-polarized light. *cis,trans*-Isomers are not optical isomers.

43. enantiomorphs. The exchange of any two groups on each and every asymmetric carbon atom converts the molecule into its enantiomorph.
44. *meso* form. There is a plane of symmetry. The two molecules are identical.
45. enantiomorphs. See the answer to question 43.
46. diastereoisomers. In a molecule with two or more asymmetric carbon atoms, the exchange of any two groups on any one (and only one) asymmetric carbon atom converts the molecule into one of its diastereoisomers.
47. *cis,trans*-isomers
48. enantiomorphs. See the answer to question 43.

49. *dextro, d,* and (+); all refer to the direction of rotation of plane-polarized light. All mean rotation to the right (clockwise). The designation "D" refers to the configuration.

50.

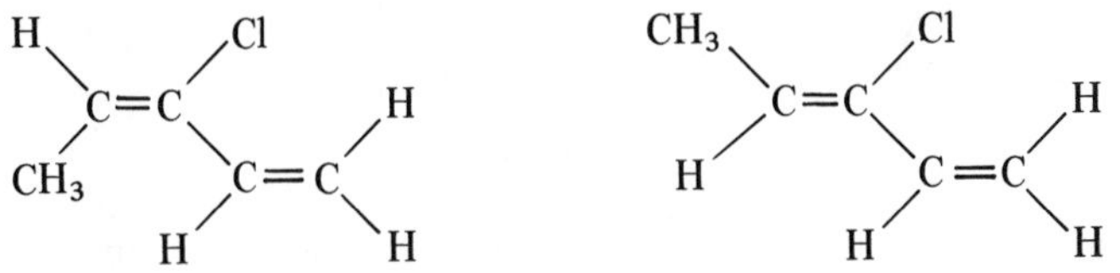

*cis,trans*-Isomers

51.

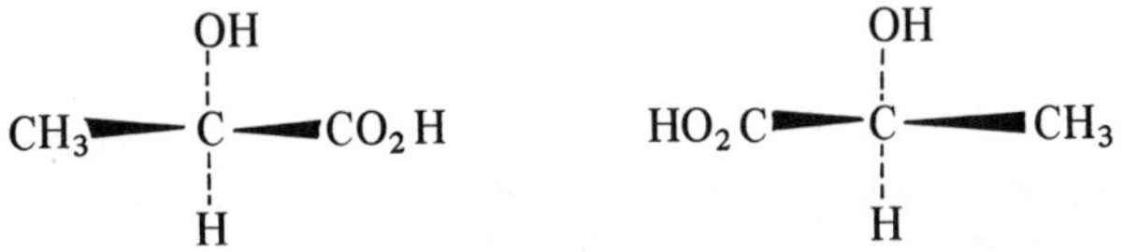

Enantiomorphs

52.

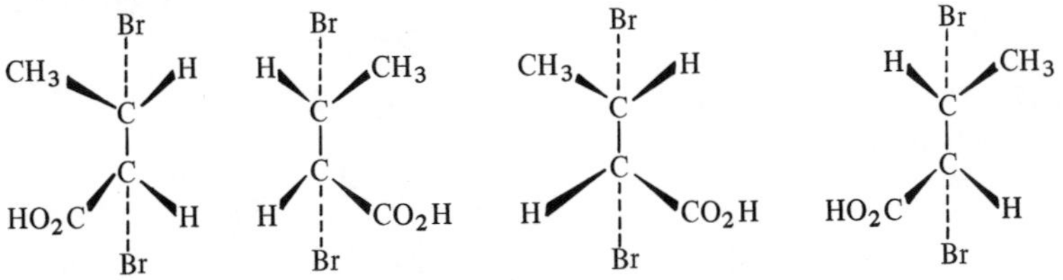

(a) (b) (c) (d)

Enantiomorphs Enantiomorphs

(a) and (c), (a) and (d), (b) and (c), and (b) and (d) are diastereoisomers

53.

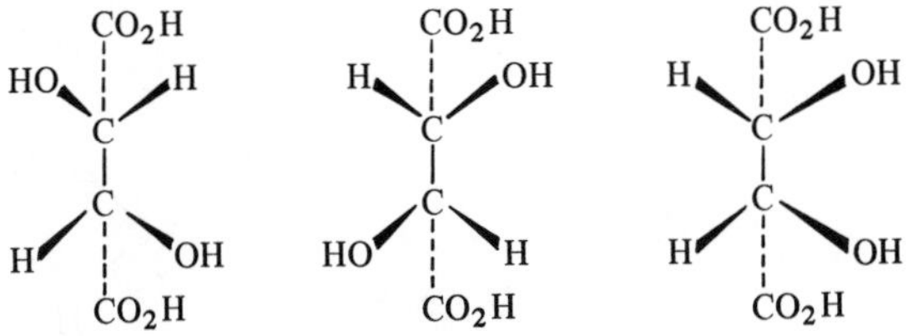

(a) (b) (c)

(a) and (c), and (b) and (c), are diastereoisomers

Enantiomorphs *meso* form

54. a. (*R*) configuration. The priority of the groups in diminishing order is: $Br > Cl > CH_3 > H$.

The group of lowest priority must occupy a position which is the most remote from us as we view the asymmetric carbon atom. Thus, we must turn the molecule without changing its configuration.

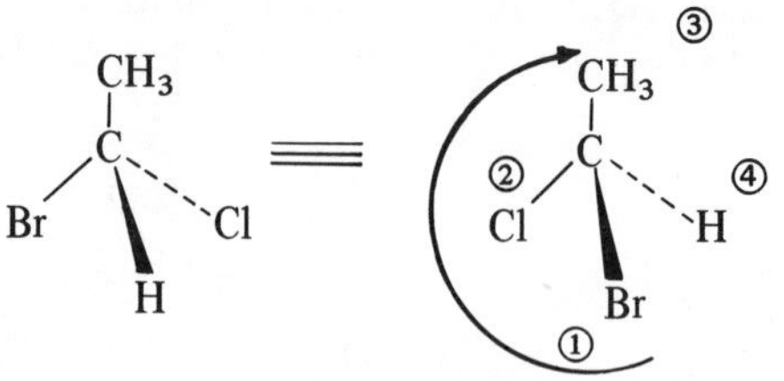

The order of the groups in diminishing priority is clockwise and thus the configuration is (*R*).

We could, of course, arrive at the same answer using Fischer projection formulas.

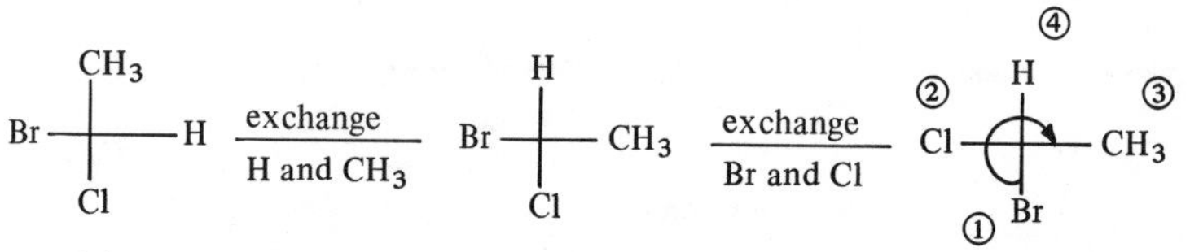

(a.) — exchange H and $CH_3$ — enantiomer of (a.) — exchange Br and Cl — same configuration as (a.)

b. OH > COOH > $CH_3$ > H, (*S*) configuration

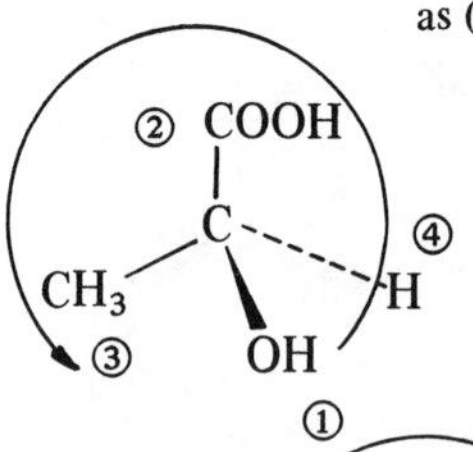

c. $NH_2$ > COOH > $CH_3$ > H, (*S*) configuration

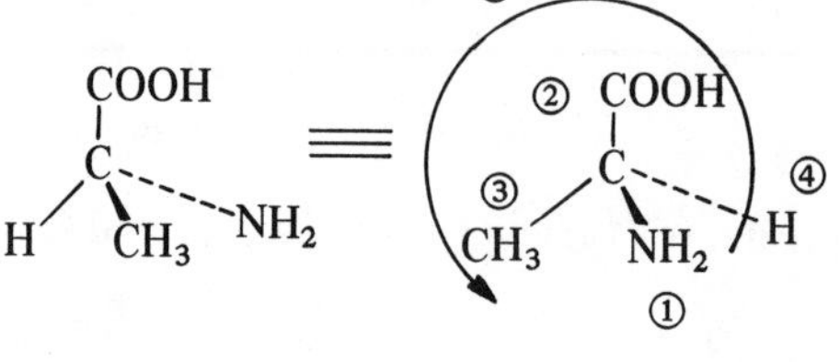

55. (*E*)–across, (*Z*)–together

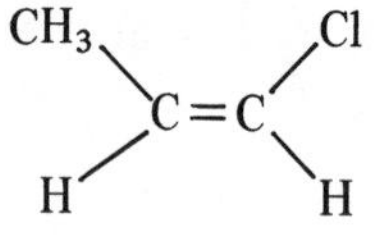

(*Z*) 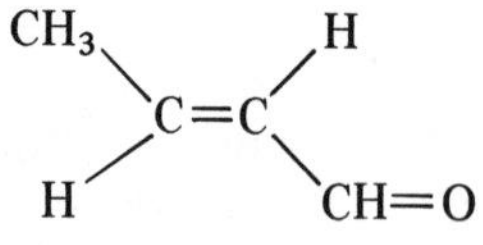 (*E*) 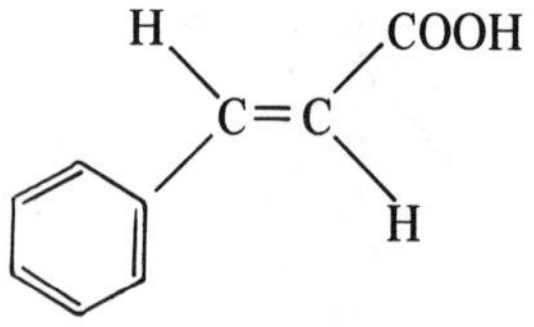 (*E*)

# 6

# *Determination of Molecular Structure; Spectroscopy*

## INTRODUCTION

The determination of molecular structure has always been a very important part of organic chemistry. Within the past thirty years, spectroscopy has provided the chemist with a powerful tool to help simplify this task. This chapter is concerned primarily with **absorption spectroscopy**, techniques based on the measurement of the amount and the specific frequencies of electromagnetic radiation selectively absorbed by a molecule. The instruments used to make these measurements are called **spectrometers** and a plot of the results is called a **spectrum**. Energy absorbed by a molecule is quantized and is related to the frequency ($\nu$) of the electromagnetic wave by the equation $E = h\nu$, where $h$ is Planck's constant.

**Nuclear magnetic resonance spectroscopy, nmr,** measures the energy, in the high radiofrequency portion of the electromagnetic spectrum, necessary to change the alignment of magnetic nuclei in a magnetic field. Inasmuch as nearly all organic molecules contain hydrogen, the hydrogen nucleus is of particular interest to the organic chemist and this type of spectroscopy is specifically called **proton nmr spectroscopy (pmr)**. The pmr spectrum provides us with four kinds of information about a molecular structure: a) the *number of signals* tells us how many different kinds (nonequivalent) of protons are present in the molecule, b) the *chemical shift* tells us something about the electronic environment of the protons, c) the *intensity of the signal,* as measured by an electronic integrator, tells us the ratio of the number of protons in each set of nonequivalent protons, and d) the *splitting of a signal* tells us the number of protons on adjacent carbon atoms.

**Infrared spectroscopy (ir)** is most useful in determining the functional groups present in an organic molecule. We learn that bonds in organic molecules are *stretching* and *bending* in a quantized fashion and that these *vibrations* are related to absorption of infrared radiation, that portion of the electromagnetic spectrum lying beyond (lower frequency, less energy) the red end of the visible region.

**Ultraviolet-visible spectroscopy (uv-vis)** is related to the different electronic energy states associated with the movement of an electron from a bonding ($\sigma$ or $\pi$) or nonbonding ($n$) molecular orbital to an antibonding ($\sigma^*$ or $\pi^*$) orbital. This type of spectroscopy is most useful in determining the nature and extent of conjugation in a molecule.

Although not discussed extensively in this chapter, we learn that **carbon-13 nmr (cmr)**, **mass spectroscopy**, and **x-ray crystallography** also are very useful techniques for the determination of molecular structure.

## PROBLEMS

### True–False

1. T F A molecule can exist only in discrete energy states and thus its energy transitions are not continuous, but instead are quantized.
2. T F The ${}^{12}_{6}C$ isotope gives an nmr spectrum, but it is not used in determining structure as often as proton (${}^{1}_{1}H$) nmr.
3. T F In nmr spectroscopy, we may either change the radiofrequency until we have the energy necessary to cause some of the nuclei aligned with the magnetic field to reorient themselves against the field or we may maintain a constant radiofrequency and change the strength of the magnetic field to bring about the resonance condition.
4. T F The gyromagnetic ratio is a constant characteristic of the particular nucleus.
5. T F If the nuclei of protons were not shielded they would all be in resonance at 60 MHz in a magnetic field of 14,092 gauss.
6. T F Tetramethylsilane, TMS, is commonly used as a standard in nmr measurements because of the intense single peak given by its twelve equivalent protons and because the protons, being highly shielded, are in resonance at a higher field than most protons in organic compounds.
7. T F Protons of a specific compound will have the same chemical shift in $\delta$ units regardless of whether the spectrum is obtained on a 60 MHz, 100 MHz, or 220 MHz spectrometer.
8. T F The area under each peak in the nmr spectrum is approximately proportional to the number of protons responsible for the peak. Electronic integrators can be used to determine the ratio of the number of protons in each set of nonequivalent protons.
9. T F The chemical shift of the proton on a hydroxyl group (—OH) is variable ($\delta$ 1-6) because the proton is rapidly moving from one molecule to another and its chemical shift is sensitive to experimental conditions. One way to establish which nmr peak is due to this proton is to treat the compound with $D_2O$.
10. T F The coupling constant $J$ is dependent on the spectrometer frequency.

11. T F In the nmr spectrum, first-order splitting is not generally observed unless the difference in resonance frequencies of the coupled protons ($\Delta\nu$) is approximately ten times larger than the coupling constant ($J$), i.e. $\Delta\nu > 10 \times J$.

12. T F In most solvents the OH proton of an alcohol ($R—CH_2—OH$) and the protons on the adjacent carbon atom do not split each other.

13. T F The region below 1500 $cm^{-1}$ is sometimes called the "fingerprint" region.

14. T F The most common infrared sample cells are constructed with sodium chloride plates.

15. T F Absorption peaks in the uv-vis region are usually broader than those in the ir region.

16. T F In the uv-vis region the intensity of absorption is independent of the number of absorbing molecules in the light path.

17. T F The most useful transitions in the uv-vis region are the $n \rightarrow \pi^*$ and $\pi \rightarrow \pi^*$ transitions.

18. T F Because the intensity of absorption in the uv-vis region is proportional to the number of molecules in the light path, uv-vis spectroscopy has many analytical applications.

19. T F Determination of the mass of a molecular ion, as done is mass spectroscopy, to four decimal places can give the molecular formula of a compound.

**Multiple Choice**

20. How many sets of nonequivalent protons are there in

$$Cl—CH_2—\underset{\displaystyle CH_3}{\underset{|}{CH}}—CH_2—CH_3?$$

(a) 1 (d) 4
(b) 2 (e) 5
(c) 3

21. How many sets of nonequivalent protons are there in

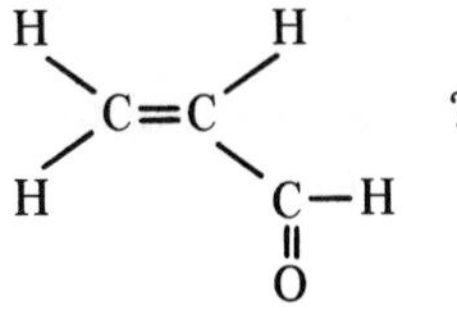


?

(a) 1 (d) 4
(b) 2 (e) 5
(c) 3

22. Which of the following is the least desirable solvent for use in nmr spectroscopy?

(a) $CCl_4$ (d) $CDCl_3$
(b) $D_2O$ (e) $D_3C—\underset{\displaystyle O}{\underset{\|}{C}}—CD_3$
(c) $CH_3CH_2OH$

23. The farthest downfield shift will be exhibited by the proton or protons in which compound?
    (a) $R-CH_3$
    (b) $Ar-H$
    (c) $R-C(=O)-H$
    (d) $R-CH_2-Cl$
    (e) $R-C(=O)-OH$

24. The multiplicity of the nmr signal for $H_b$ will be

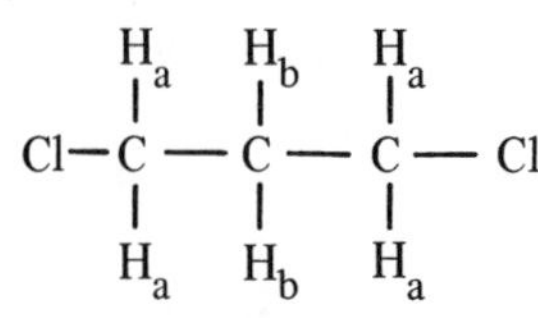


    (a) 1
    (b) 2
    (c) 3
    (d) 4
    (e) 5

25. Which stretching vibration occurs at the highest frequency?
    (a) C≡C  (b) C=C  (c) C–C

26. Electromagnetic radiation in the ultraviolet region is of (higher, lower) energy than electromagnetic radiation in the infrared region.

27. The longer the wavelength (λ) of light the (greater, less) the energy.

28. The greater the frequency ($\nu$) of light the (greater, less) the energy.

29. As the radiofrequency of the nmr spectrometer is increased, (for example 60 MHz to 100 MHz) the magnetic strength of the magnet must be (increased, decreased).

30. An increase in shielding leads to an (upfield, downfield) shift in the nmr spectrum.

31. The $\sigma \rightarrow \sigma^*$ electronic transitions are at (longer, shorter) wavelengths than $n \rightarrow \sigma^*$ transitions.

32. A compound was changed from $R-C(=O)-CH_2-CH_3$ to $R-C(=O)-CH{=}CH_2$.
    Absorption in the uv region would change to a (shorter, longer) wavelength.

33. A bathochromic shift is a shift toward the (red, violet) end of the spectrum.

## Completion

34. A plot or graph that records the amount of light absorbed by a sample as a function of the frequency or wavelength of electromagnetic radiation is called a

    ____________________.

35. The mathematical relationship between the energy absorbed and the
    (a) frequency ($\nu$) of electromagnetic radiation is ____________________
    (b) wavenumber ($\tilde{\nu}$) of electromagnetic radiation is ____________________

(c) wavelength ($\lambda$) of electromagnetic radiation is ______________________

36. The promotion of an electron, in an organic molecule, from its lower energy state, bonding ($\sigma$ or $\pi$) or nonbonding ($n$) molecular orbital, to a higher energy state, antibonding ($\sigma^*$ or $\pi^*$) orbital, is associated with the ______________ or ______________ regions of the electromagnetic spectrum.

37. Absorption of ______________ radiation can be correlated with the transitions of atoms, in a molecule, between two different vibration or rotational energy levels.

38. The position of a peak in nmr relative to a standard such as TMS is called the ______________ ______________ .

39. $\delta$ 4.4 = $\tau$ ____

40. The phenomenon in which the magnetic field of a proton interacts with a neighboring proton to cause the nmr peaks of each proton to be split is called ______________________.

41. If the multiplicity of an nmr signal is four, the area ratio of peaks will be approximately____ - ____ - ____ - ____.

42. The three basic units of a typical ir or uv–vis spectrometer are ______________, ______________, and ______________.

43. In ir spectroscopy, solid samples are often ground with __________ __________ and pressed into a transparent pellet.

44. The unit commonly used to define the nature of the electromagnetic radiation in the ir region is __________ and in the uv–vis region it is __________.

45. In mass spectroscopy, the radical cation produced by the loss of an electron from a molecule is called the __________ ______________.

46. A technique used for the determination of the precise geometry of molecules in the crystalline state is __________ ______________.

**Matching** (Match each infrared region with the group or groups that would absorb radiation in that region.)

(a) 3700–3100 $cm^{-1}$

(b) 3100–3000 $cm^{-1}$

(c) 3000–2800 $cm^{-1}$

(d) 2400–2000 $cm^{-1}$

(e) 1870–1630 $cm^{-1}$

(f) 1680–1620 $cm^{-1}$

(g) 1615–1515 $cm^{-1}$

(h) 1300–1000 $cm^{-1}$

(i) 1250–1150 $cm^{-1}$

47. ____ C—C

48. ____ C=C

49. ____ C≡C, C≡N

50. ____ aromatic ring

51. _____ —C—H

52. _____ =C—H, ≡C—H, aromatic CH

53. _____ C—O

54. _____ C=O

55. _____ OH, NH

**Miscellaneous**

56. Draw arrows to indicate in which direction the specific items increase.

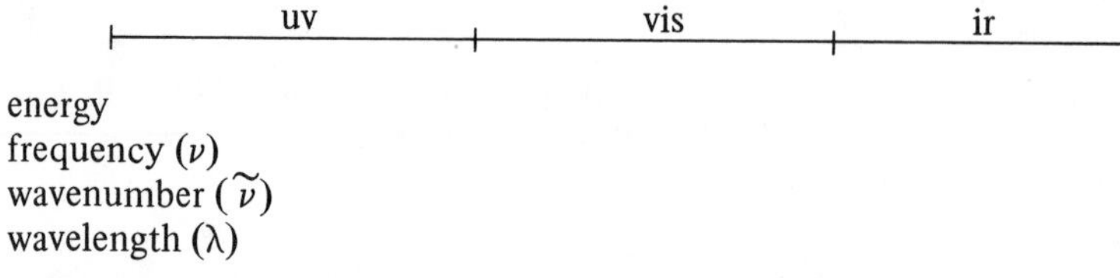


energy
frequency ($\nu$)
wavenumber ($\tilde{\nu}$)
wavelength ($\lambda$)

57. Assume the stretching vibration of a C=C occurs at 1650 $cm^{-1}$ (ir region) and the $\pi \rightarrow \pi^*$ electronic transition occurs at 175 nm (uv region). Calculate the amount of energy (kcal/mole) necessary for each transition. (Let Planck's constant X velocity of light = $2.8590 \times 10^{-3} \frac{\text{kcal cm}}{\text{mole}}$.)

58. Draw arrows to indicate the proper direction for the specified terms.

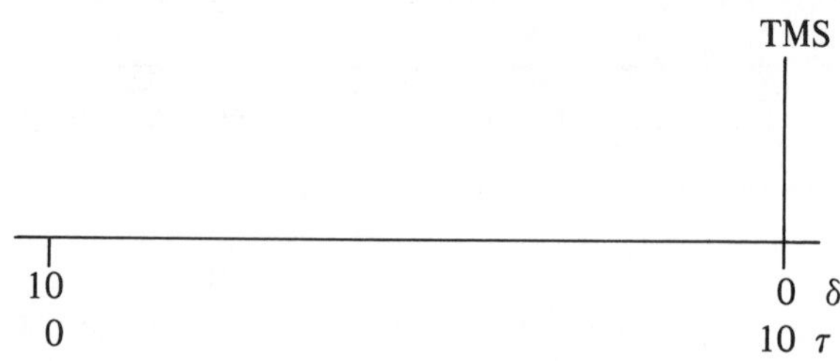


upfield
shielding
greater applied field

59. Proton chemical shifts cover a range of about 1000 Hz at 60 MHz. What would the range be in $\delta$ units?

60. Using lines of appropriate height and spacing, sketch the nmr spectrum expected for

$$X-\underset{X}{\overset{H_a}{C}}-\underset{X}{\overset{H_b}{C}}-H_b.$$

$H_a$ is deshielded relative to $H_b$.
$\Delta\nu_{ab}$ = 20 Hz (The difference in resonance frequencies between $H_a$ and $H_b$)
$J_{ab}$ = 2 Hz

downfield ←

61. Draw a splitting diagram for $H_b$ in each of the following, showing approximate height and spacing of the lines. ($J_{ab}$ = 1 Hz, $J_{bc}$ = 3 Hz)

(a)

$$-\underset{H_a}{\overset{|}{C}}-\underset{H_b}{\overset{|}{C}}-$$

(b)

$$-\underset{H_a}{\overset{|}{C}}-\underset{H_b}{\overset{|}{C}}-\underset{H_a}{\overset{|}{C}}-$$

(c)

$$-\underset{H_a}{\overset{|}{C}}-\underset{H_b}{\overset{|}{C}}-\underset{H_c}{\overset{|}{C}}-$$

(d)

$$-\underset{H_a}{\overset{|}{C}}-\underset{H_b}{\overset{|}{C}}-\underset{H_c}{\overset{|}{C}}-H_c$$

62. Determine ⊖ (the number of rings and/or double bonds) for a molecule of the formula, $C_7H_{12}NOCl$.

## ANSWERS

### True–False

1. True
2. False. $^{12}_{6}C$ has an even number of protons and neutrons. Its spin quantum number is zero and it has no magnetic properties. $^{13}_{6}C$ has an odd number of neutrons and a spin quantum number of ½. It has magnetic properties and the use of $^{13}C$ nmr is rapidly increasing, but it is not as common as proton nmr.
3. True
4. True
5. True. Because the nuclei are shielded the magnetic field felt by the nuclei (effective field) is less than the applied magnetic field and either the applied magnetic field must be increased or the frequency decreased to bring the nuclei into resonance.
6. True
7. True. Chemical shifts reported in frequency units, hertz (Hz) (formerly cycles per second or cps), are dependent on the spectrometer frequency. Thus a proton

$$\delta = \frac{\text{chemical shift in Hz}}{\text{spectrometer frequency in MHz}}$$

   with a chemical shift of 5 δ corresponds to a chemical shift of 300 Hz on a 60 MHz spectrometer and 500 Hz on a 100 MHz spectrometer. (Note that the unit of δ is parts per million (ppm), but it is not usually specified.)
8. True. Note that this does not give us the exact number of protons in each set. If we know the formula of the compound, however, we can determine the exact number of protons in each set.
9. True. A proton attached to oxygen, nitrogen, or sulfur can often be exchanged for deuterium by treating the compound with $D_2O$. Deuterium resonance peaks do not occur in the pmr region. Also, the chemical shift of the proton on a hydroxyl group can often be shifted by changing the concentration of the sample.
10. False
11. True. Spectra that are complex on a 60 MHz spectrometer are often simplified to first-order spectra on a 100 MHz instrument because $\Delta\nu$ is dependent on the spectrometer frequency while $J$ is not.

    If $\Delta\nu$ is 6 Hz on a 60 MHz spectrometer, it is 10 Hz on a 100 MHz instrument.

$$\frac{\Delta\nu \text{ in HZ}}{\text{spectrometer frequency in MHz}} = \frac{\Delta\nu \text{ in Hz}}{\text{spectrometer frequency in MHz}}$$

$$\frac{6}{60} = \frac{x}{100}$$

12. True. The rapid exchange of protons between OH groups eliminates the effects of the coupling. Purified deuterated dimethyl sulfoxide or acetone slow the rate

of exchange and splitting usually occurs in these solvents.

13. True
14. True. Thus, the solvents used in these cells must not dissolve or react with sodium chloride.
15. True. Transitions between two vibrational energy levels can involve several rotational levels, but transitions between two electronic energy levels can involve several vibrational and rotational levels.
16. False. The intensity of absorption is proportional to the number of absorbing molecules in the light path.
17. True
18. True
19. True

**Multiple Choice**

20. (e)
21. (d)

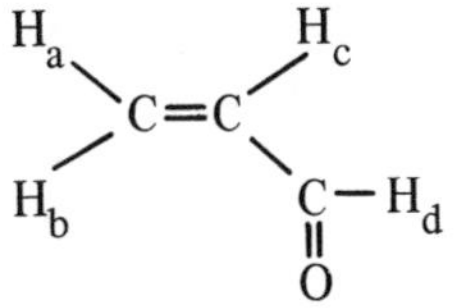

$$\begin{array}{ccc} H_a & & H_c \\ & \diagdown C{=}C \diagup & \\ & \diagup \quad\; \diagdown & \\ H_b & & \overset{\displaystyle C - H_d}{\underset{O}{\|}} \end{array}$$

Even though $H_a$ and $H_b$ are on the same carbon atom they are not chemically equivalent. There is restricted rotation around the C=C and $H_a$ does not have the same electronic environment as does $H_b$.

22. (c) Nonprotonic solvents are usually used to avoid blocking out a particular region of the nmr spectrum where the protonic solvent would absorb.
23. (e)

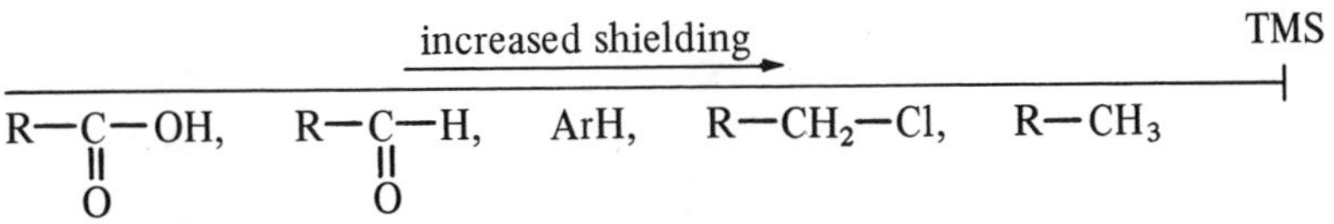

increased shielding → TMS

$$\underset{\overset{\|}{O}}{R-C-OH}, \quad \underset{\overset{\|}{O}}{R-C-H}, \quad ArH, \quad R-CH_2-Cl, \quad R-CH_3$$

24. (e) multiplicity of $H_b = n + 1$ ($n$ is the number of equivalent hydrogens on the adjacent carbon atoms) multiplicity of $H_a = 3$.
25. (a) The stretching vibration will be at a higher frequency as the bond order increases.

$$C\equiv C > C=C > C-C$$
$$C=O > C-O$$
$$C\equiv N > C=N > C-N$$

26. higher
27. less
28. greater
29. increased
30. upfield

31. shorter. The $\sigma \rightarrow \sigma^*$ transition requires more energy than a $n \rightarrow \sigma^*$ transition and thus occurs at a shorter wavelength.
32. longer. As the length of the conjugated system increases, the wavelength of absorption increases.
33. red

## Completion

34. spectrum or absorption spectrum
35. (a) $E = h\nu$ (b) $E = hc\tilde{\nu}$ (c) $E = \frac{hc}{\lambda}$
36. visible, ultraviolet
37. infrared
38. chemical shift. The term "chemical shift" can also be used to indicate the difference in resonance frequencies between peaks in the sample molecule itself.
39. 5.6. $\delta = 10 - \tau$
40. spin-spin coupling
41. 1-3-3-1. The area ratios of the peaks in a multiplet are given approximately by Pascal's triangle.
42. radiation source, monochromator, detector or photometer
43. potassium bromide
44. $cm^{-1}$, nanometer (nm) or millimicron (m$\mu$)
45. molecular ion
46. x-ray crystallography

## Matching

47. (i)
48. (f)
49. (d)
50. (g)
51. (c)
52. (b)
53. (h)
54. (e)
55. (a)

## Miscellaneous

56.

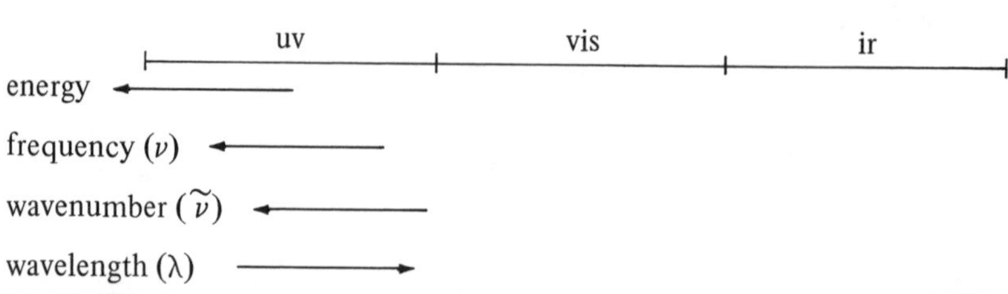


57. $\Delta E = hc\tilde{\nu}$

$$\Delta E = 2.8590 \times 10^{-3} \frac{\text{kcal cm}}{\text{mole}} \times \frac{1650}{\text{cm}}$$

$$\Delta E = 4.717 \frac{\text{kcal}}{\text{mole}} \text{ for C=C stretching vibration}$$

$$\Delta E = \frac{hc}{\lambda}$$

$$\Delta E = \frac{2.8590 \times 10^{-3} \dfrac{\text{kcal cm}}{\text{mole}} \times 10^{7} \dfrac{\text{nm}}{\text{cm}}}{175 \text{ nm}}$$

$$\Delta E = 163.4 \frac{\text{kcal}}{\text{mole}} \text{ for C=C } \pi \rightarrow \pi^* \text{ electronic transitions}$$

Note that these are very logical values. Electronic transitions in organic compounds occur in the uv and vis regions while stretching and bending vibrations occur in the ir region. The energy of radiation in the uv region > vis region > ir region. See Fig. 6.17 in the text for the energies associated with the electronic and infrared spectral regions.

58.

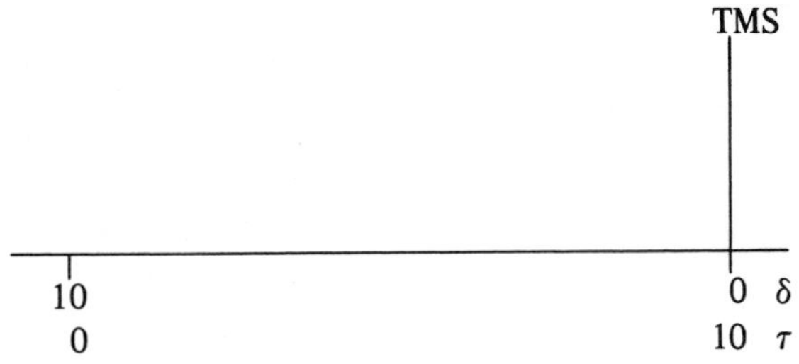


upfield ⟶

shielding ⟶

greater applied field ⟶

59. 0 to 16.7

$$\delta = \frac{\text{chemical shift in Hz}}{\text{spectrometer frequency in MHz}}$$

$$\delta = \frac{1000}{60}$$

$$\delta = 16.7$$

60.

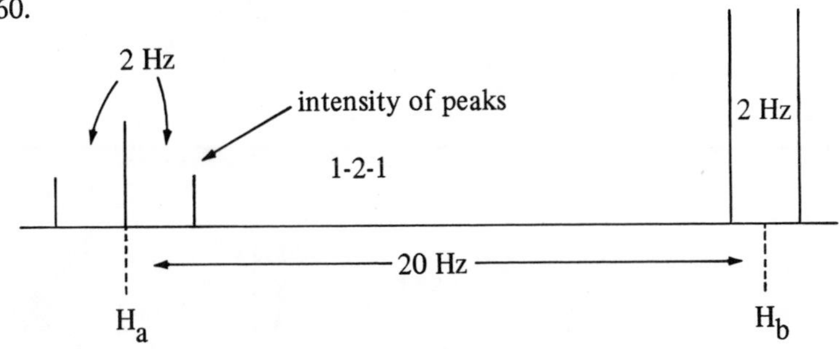

61. (a)

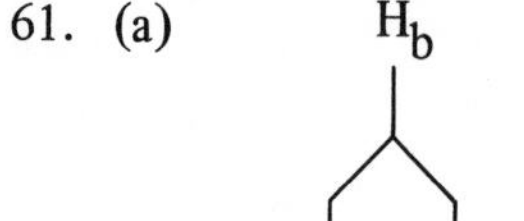


(b)

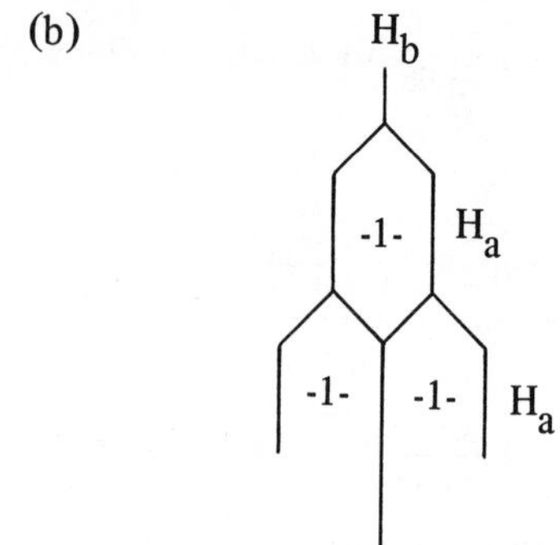


(c)

H
b
3
H
c
-1-
-1-
H
a

(d)

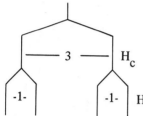

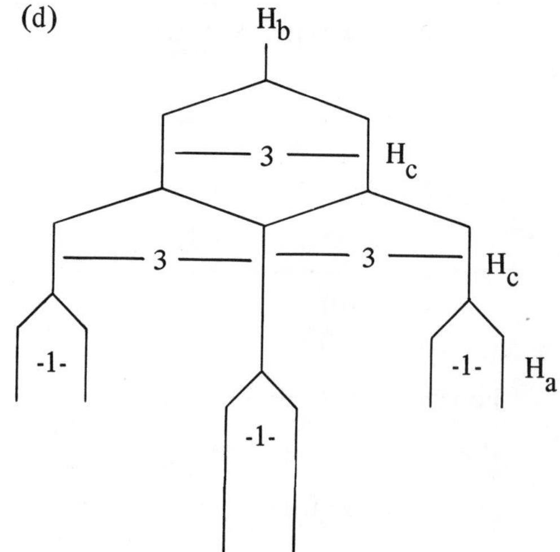


62. $\ominus = c - \frac{h}{2} + \frac{n}{2} - \frac{x}{2} + 1$

$\ominus = 7 - \frac{12}{2} + \frac{1}{2} - \frac{1}{2} + 1$

$\ominus = 2$

# 7

# Organic Halogen Compounds

## INTRODUCTION

Until now our study of organic compounds has been devoted primarily to compounds containing only carbon and hydrogen. In this chapter we shall study compounds which also contain carbon — halogen bonds. As in previous chapters we first must learn to name the compounds we are to deal with. Next, we shall study methods for the preparation of both aliphatic and aromatic halides, noticing that aliphatic or alkyl halides are reactive compounds and often are useful as intermediates in the synthesis of other compounds. The reactions of alkyl halides are principally of two types–**substitution** and **elimination**. The pathways **(reaction mechanisms)** of these reactions will be studied in some detail.

## PROBLEMS

**True–False**

1. T F Direct halogenation of alkanes produces isomeric and polysubstituted alkyl halides.
2. T F Thionyl chloride is an excellent chlorinating reagent because the by-products of the reaction, $SO_2$ and HCl, are gases.
3. T F Primary and secondary iodo- and fluoroalkanes are often prepared from bromo- or chloroalkanes *via* halogen exchange.
4. T F Chloroform and carbon tetrachloride are highly flammable.

5. T F Alkyl halides undergo mainly elimination and nucleophilic substitution reactions.
6. T F The $S_N1$ reaction of an optically active compound yields an optically active product.
7. T F Aryl halides are more reactive than alkyl halides.
8. T F Vinyl chloride is less reactive than most alkyl chlorides.
9. T F Allyl halides are very reactive.
10. T F Compounds that form the most stable carbonium ion intermediates undergo $S_N1$ reactions most readily.

**Multiple Choice**

11. Which alkyl halide forms the corresponding alkene most readily?
    (a) $CH_3-CH_2-Br$
    (b) $CH_3-\underset{\displaystyle CH_3}{\underset{|}{CH}}-Br$
    (c) $CH_3-\overset{\displaystyle CH_3}{\overset{|}{\underset{\displaystyle CH_3}{\underset{|}{C}}}}-Br$
12. Which alkyl halide reacts most rapidly by nucleophilic substitution?
    (a) $CH_3-CH_2-F$
    (b) $CH_3-CH_2-Cl$
    (c) $CH_3-CH_2-Br$
    (d) $CH_3-CH_2-I$
13. Which of the following factors influence whether a reaction will proceed by an $S_N1$, $S_N2$, E1, or E2 mechanism?
    (a) structure of the alkyl halide
    (b) solvent
    (c) concentration of reagents
    (d) nature of the nucleophile
    (e) all of the above
14. Which compound is benzal chloride?

    (a) $C_6H_5-CCl_3$

    (b) $C_6H_5-\overset{\displaystyle Cl}{\overset{|}{\underset{\displaystyle Cl}{\underset{|}{CH}}}}$

    (c) 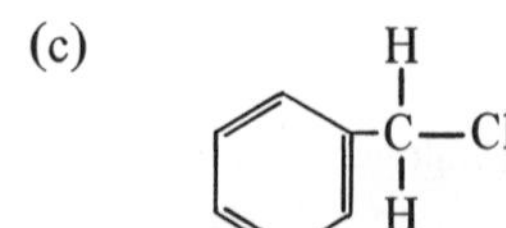


    (d) $Cl-C_6H_4-CH3$

15. Which compound is a tertiary alkyl halide?
    (a) $CH_3-Cl$
    (b) 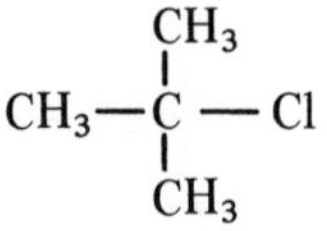 $CH_3-C(CH_3)_2-Cl$ (central C bearing $CH_3$ above and $CH_3$ below)
    (c) $CH_2=CH-Cl$ (Cl on the CH)
    (d) 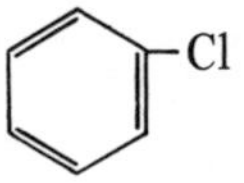 $C_6H_5-Cl$ (benzene ring bearing Cl)
    (e) 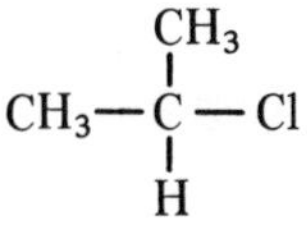 $CH_3-CH(CH_3)-Cl$ (central C bearing $CH_3$ above and H below)

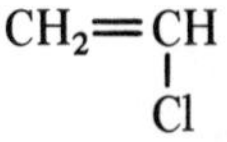

16. The structure of chloroform is
    (a) $CH_3Cl$
    (b) $CH_2Cl_2$
    (c) $CHCl_3$
    (d) $CCl_4$
    (e) $CH_3I$

17. Which reagent(s) cannot be used to prepare an alkyl halide from an alcohol?
    (a) $HCl + ZnCl_2$
    (b) NaCl
    (c) $PCl_5$
    (d) Concentrated HBr
    (e) $SOCl_2$

18. Which compound reacts most rapidly by an $S_N1$ mechanism?
    (a) $CH_3-Cl$
    (b) 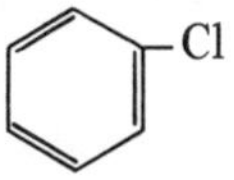 $C_6H_5-Cl$ (benzene ring bearing Cl)
    (c) $CH_3-CH=CH-Cl$
    (d) $Cl-CH_2-CH=CH_2$
    (e) 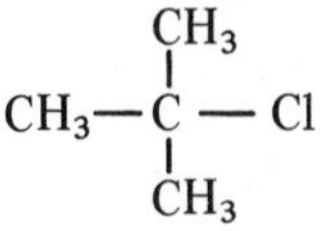 $CH_3-C(CH_3)_2-Cl$ (central C bearing $CH_3$ above and $CH_3$ below)

19. An example of the Friedel-Crafts reaction is
    (a) $2\ CH_3-CH_2-Cl + 2\ Na \longrightarrow CH_3-CH_2-CH_2-CH_3 + 2\ NaCl$
    (b) 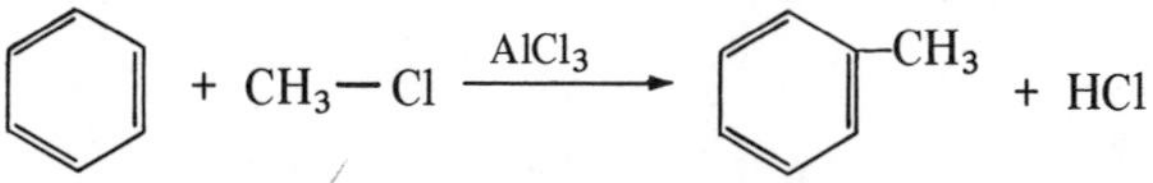 $C_6H_6 + CH_3-Cl \xrightarrow{AlCl_3} C_6H_5-CH_3 + HCl$ (benzene ring; benzene ring bearing $CH_3$)

*(cont'd)*

(c)

2 (iodobenzene) + 2 Cu $\xrightarrow{\text{heat}}$ (biphenyl) + 2 CuI

(d) $CH_3—CH_2—Cl$ + NaI ⟶ $CH_3—CH_2—I$ + NaCl

(e)

(bromobenzene) + Mg $\xrightarrow[\text{ether}]{\text{anhydrous}}$ (phenyl—MgBr) $\xrightarrow{H_2O}$ (benzene) + Mg(OH)Br

20. Which of the following is not a true statement about $S_N1$ reactions?
    - (a) The formation of a carbonium ion is the rate-determining step.
    - (b) Primary alkyl halides seldom, if ever, react by this mechanism.
    - (c) An increase in concentration of either of the reactants will increase the rate of the reaction.
    - (d) The designation $S_N1$ means substitution, nucleophilic, unimolecular.
    - (e) An elimination reaction leading to an alkene is a competing reaction.

**Completion**

21. The detailed pathway by which a reaction proceeds is the ________________.
22. The slowest step of a reaction is the ____________ - ____________ step.
23. An optically active compound which undergoes a reversal of configuration during a $S_N2$ reaction is said to have undergone a ____________ ____________.
24. An electron-rich reagent which seeks a positive center is called a ____________ reagent.
25. An optically active compound which undergoes an $S_N1$ reaction will give a ____________ ____________.
26. The ____________ of a reaction is determined by the number of molecules (or other species) involved in the transition state for the rate-determining step.

**Matching** (Match each compound to the appropriate description.)

(a) $(Cl—C_6H_4)_2CH—CCl_3$

(b) $CF_3—CHClBr$

(c) $Cl—C_6H_4—Cl$

(d) $Hg_2F_2$

(e) $CCl_2F_2$

(f) [hexachlorocyclohexane ring: each ring carbon bears one H and one Cl]

H Cl
Cl H
H— —Cl
Cl— —H
H Cl
Cl H

(g) $\left(-\mathrm{CH_2}-\mathrm{CCl_2}-\right)_x$

(h) $CH_2{=}CHCl$

(i) $SOCl_2$

(j) $\left(-\mathrm{CF_2}-\mathrm{CF_2}-\right)_x$

27. _____ a chlorinating agent

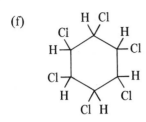

28. _____ an effective, relatively nontoxic inhalation anesthetic known as "halothane"

29. _____ a refrigerant and propellant in aerosol sprays which may cause depletion of the ozone layer in the atmosphere

30. _____ the organic halogen compound manufactured in greatest amount in the United States, used to prepare polyvinyl chloride

31. _____ a "hard" insecticide called DDT

32. _____ a waxy plastic often used as a liner in frying pans

33. _____ a "hard" insecticide called BHC or 666

34. _____ a moth-repellant

35. _____ a reagent used to prepare alkyl fluorides via halogen exchange

36. _____ an inert plastic of which "Saran Wrap" is an example

**Nomenclature** (Assign acceptable names to the following compounds.)

37. $SOCl_2$

____________________

38. 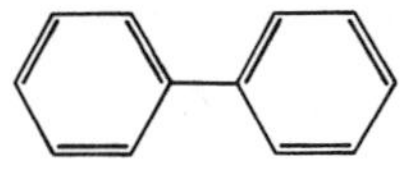

~~de~~ bis phenyl

39. $CCl_4$

tetrachloro carbon

40. $C_6H_5-CCl_3$

____________________

41. $CH_2{=}CH{-}CH_2{-}Cl$

____________________

42. $Cl{-}CH{=}CH{-}CH_3$

____________________

43. 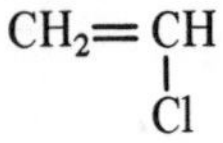

44. 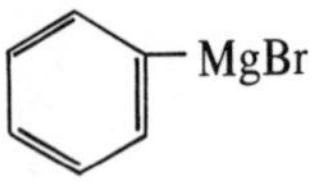

**Reactions** (Complete the following.)

45.

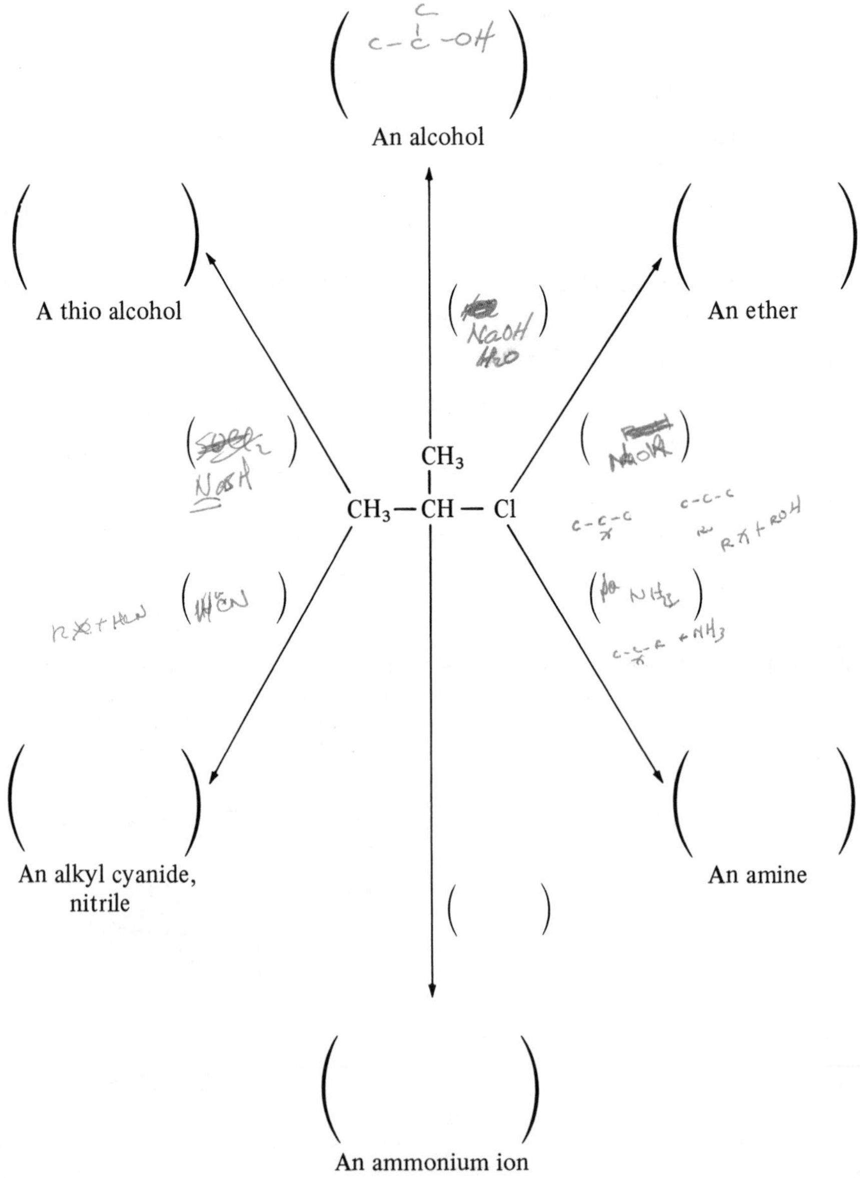

46.

$C_6H_5-CH_2-CH_3$ $\xrightarrow{Br_2,\ \text{sunlight}}$ ( ) + HBr

$C_6H_5-CH_2-CH_3$ $\xrightarrow{Br_2,\ Fe}$ ( ) + ( ) + HBr

47.

$CH_3-\underset{\substack{|\\CH_3}}{CH}-CH_2-CH_2-Br$ $\xrightarrow{OH^-,\ S_N2\ \text{mechanism}}$ ( )

$CH_3-\underset{\substack{|\\CH_3}}{CH}-CH_2-CH_2-Br$ $\xrightarrow{OH^-,\ E2\ \text{mechanism}}$ ( )

48.

$$CH_3-\overset{\overset{O}{\|}}{C}-CH_3 + (3\_\_\_\_\_\_ + 4\_\_\_\_\_\_) \rightarrow CH_3-\overset{\overset{O}{\|}}{C}-O^-\ Na^+ + CHCl_3 + 3\ H_2O + 3\ NaCl$$

## Syntheses

49. Using benzene and any other reagents you may require, devise a synthetic route for the preparation of styrene, $C_6H_5-CH=CH_2$.

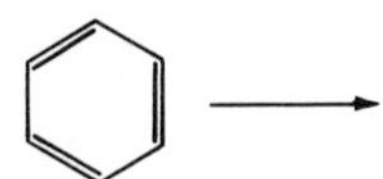

50. Starting with isopropyl alcohol, and any other reagents necessary, show how you could prepare propylamine, $CH_3{-}CH_2{-}CH_2{-}NH_2$.

$$\underset{\displaystyle OH}{CH_3{-}\underset{|}{C}H{-}CH_3} \quad \longrightarrow$$

51. Using isobutyl alcohol as your only carbon containing compound, devise a method for the preparation of 2,5-dimethylhexane.

$$\underset{\displaystyle CH_3}{CH_3{-}\underset{|}{C}H{-}CH_2{-}OH} \quad \longrightarrow$$

52. Starting with isopropyl alcohol, and any other reagents necessary, show how you could prepare propane.

$$CH_3{-}\overset{\displaystyle CH_3}{\underset{\displaystyle H}{\overset{|}{\underset{|}{C}}}}{-}OH \quad \longrightarrow$$

53. Starting with *tert*-butyl alcohol, and any other reagents necessary, devise a method for the preparation of 1,2-dibromo-2-methylpropane.

$$CH_3{-}\overset{\displaystyle CH_3}{\underset{\displaystyle CH_3}{\overset{|}{\underset{|}{C}}}}{-}OH \quad \longrightarrow$$

**Miscellaneous**

54. Using simple chemical tests, how could you distinguish between the following?

(a) [benzene ring]—Br

(b) [benzene ring]—$CH_2$—Br

(c) $CH_2{=}CH{-}Br$

55. A primary alkyl bromide (A), $C_4H_9Br$, reacted by an E2 mechanism with hot sodium hydroxide to give compound (B). Compound (B) reacted with hydrogen bromide to give an isomer of (A), (C). When (A) was reacted with sodium it gave compound (D), $C_8H_{18}$, which was different than the compound produced when *n*-butyl bromide was reacted with sodium. Draw the structure of (A) and write equations for all the reactions.

( ) $\xrightarrow[\text{NaOH, E2}]{\text{heat,}}$ ( ) $\xrightarrow{\text{HBr}}$ ( )

$C_4H_9Br$ (A) — (B) — $C_4H_9Br$ (C)

(A) ↓ Na

( )

$C_8H_{18}$ (D)

56. A secondary alcohol (A), $C_3H_8O$, reacted with thionyl chloride to give compound (B), $C_3H_7Cl$. Compound (B) reacted with benzene in the presence of aluminum chloride to give compound (C), $C_9H_{12}$. Draw structures for (A), (B), and (C) and write equations for all the reactions.

( ) $\xrightarrow{SOCl_2}$ ( ) $\xrightarrow[AlCl_3]{C_6H_6,}$ ( )

$C_3H_8O$ (A) — $C_3H_7Cl$ (B) — $C_9H_{12}$ (C)

**Questions 57, 58, 60, and 62** (Identify each reaction as $S_N1$, $S_N2$, E1, or E2.)

57. ______

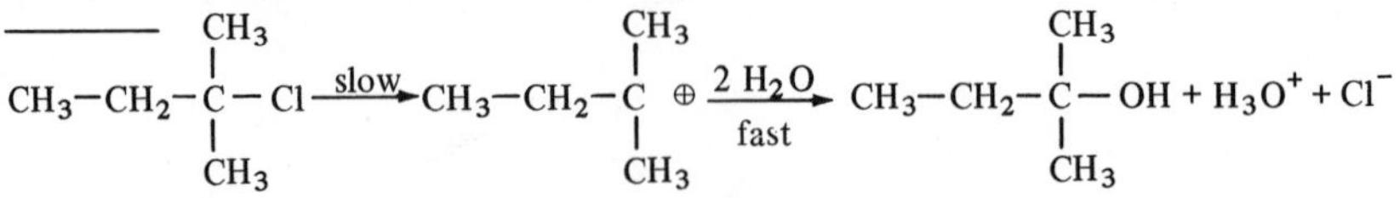

$$CH_3-CH_2-\underset{CH_3}{\overset{CH_3}{\underset{|}{\overset{|}{C}}}}-Cl \xrightarrow{\text{slow}} CH_3-CH_2-\underset{CH_3}{\overset{CH_3}{\underset{|}{\overset{|}{C}}}}\oplus \xrightarrow[\text{fast}]{2\,H_2O} CH_3-CH_2-\underset{CH_3}{\overset{CH_3}{\underset{|}{\overset{|}{C}}}}-OH + H_3O^+ + Cl^-$$

58. ______

$$CH_3-CH_2-Cl + OH^- \longrightarrow CH_2{=}CH_2 + H_2O + Cl^-$$

59. Draw the formula for the transition state for 58.

$$\left( \qquad\qquad\qquad \right)$$

60. ______

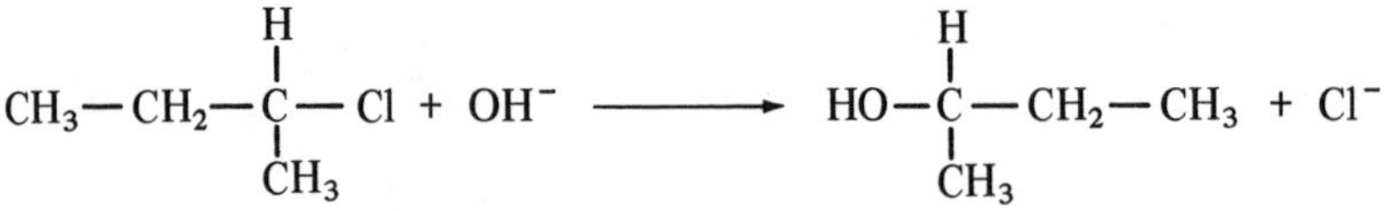

$$CH_3-CH_2-\underset{CH_3}{\overset{H}{\underset{|}{\overset{|}{C}}}}-Cl + OH^- \longrightarrow HO-\underset{CH_3}{\overset{H}{\underset{|}{\overset{|}{C}}}}-CH_2-CH_3 + Cl^-$$

(*S*) (*R*)

61. Draw the formula for the transition state for 60.

$$\left( \qquad\qquad\qquad \right)$$

62.

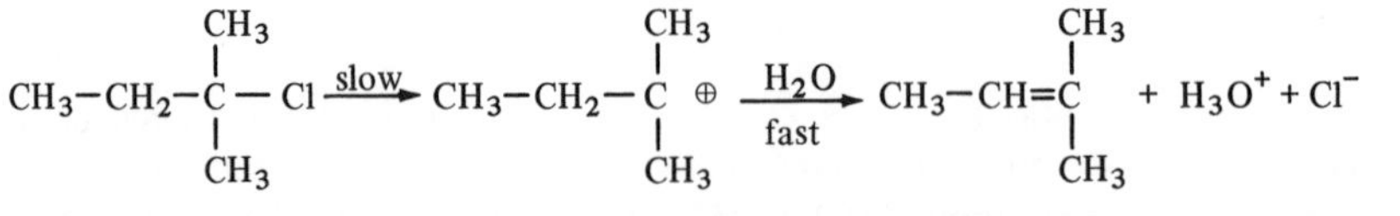

$$CH_3-CH_2-\underset{CH_3}{\overset{CH_3}{\underset{|}{\overset{|}{C}}}}-Cl \xrightarrow{\text{slow}} CH_3-CH_2-\underset{CH_3}{\overset{CH_3}{\underset{|}{\overset{|}{C}}}}\oplus \xrightarrow[\text{fast}]{H_2O} CH_3-CH{=}\underset{CH_3}{\overset{CH_3}{\underset{|}{\overset{|}{C}}}} + H_3O^+ + Cl^-$$

63. Give the structure of the compound formed in highest yield by the E2 elimination of HCl from:

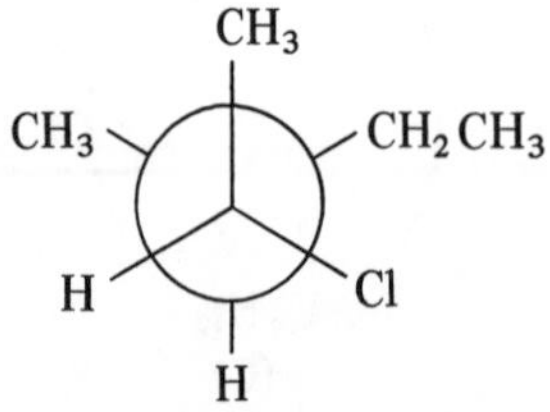

(Newman projection — front carbon: $CH_3$ (top), H (lower left), Cl (lower right); back carbon: $CH_3$ (upper left), $CH_2CH_3$ (upper right), H (bottom))

64. Give a structure consistent with each of the following sets of nmr data.

(a) $C_3H_7Br$

*a* δ 1.7, doublet, 6H
*b* δ 4.2, septet, 1H

(b) $C_3H_6Br_2$

*a* δ 2.4, quintet, 2H
*b* δ 3.7, triplet, 4H

(c) $C_4H_9Br$

*a* δ 1.0, doublet, 6H
*b* δ 2.0, multiplet, 1H
*c* δ 3.3, doublet, 2H

## ANSWERS

### True-False

1. True. Free radical halogenation of alkanes is not a specific reaction.
2. True
3. True. Reactions leading directly to the fluoroalkanes are usually rapid and difficult to handle while reactions leading to iodoalkanes are often too slow to be practical.
4. False
5. True
6. False. The intermediate in an $S_N1$ reaction is a planar carbonium ion. The nucleophile can attack from either side, leading to a racemate.
7. False. The electrons of chlorine interact with the $\pi$ electrons of the benzene ring causing the chlorine to be held more tightly to aryl groups than to alkyl groups.
8. True. The vinyl group also has $\pi$ electrons which can interact with the electrons of chlorine.
9. True. The $\pi$ electrons in the allyl group are in such a position that they stabilize the carbonium ion that is formed when the halide group is lost from allyl halide.

$$H_2C{=}CH{-}CH_2^{\oplus} \longleftrightarrow {}^{\oplus}CH_2{-}CH{=}CH_2$$

10. True

### Multiple Choice

11. (c) The order of reactivity in elimination reactions of the halides by class is tertiary > secondary > primary.
12. (d) The order of reactivity in nucleophilic substitution reactions of the halides is iodides > bromides > chlorides > fluorides.
13. (e)
14. (b)
15. (b)
16. (c)
17. (b)

18. (e) The ionization step is the slowest and thus is the rate-determining step in $S_N1$ reactions. The reaction that proceeds via the more stable intermediate is the most rapid reaction. The order of reactivity of halogen compounds in $S_N1$ reactions is:

    tertiary > allylic > benzylic > secondary > primary > vinyl

19. (b)
20. (c) The rate-determining step of a $S_N1$ reaction is the ionization step. Only an increase in the concentration of the material which is undergoing ionization will increase the speed of the reaction.

## Completion

21. reaction mechanism
22. rate-determining
23. Walden inversion
24. nucleophilic (Lewis base)
25. racemic mixture
26. molecularity

## Matching

27. (i)
28. (b)
29. (e)
30. (h)
31. (a)
32. (j)
33. (f)
34. (c)
35. (d)
36. (g)

## Nomenclature

37. thionyl chloride
38. biphenyl
39. carbon tetrachloride
40. benzotrichloride
41. allyl chloride, 3-chloro-1-propene
42. 1-chloro-1-propene
43. vinyl chloride, chloroethene
44. phenylmagnesium bromide

**Reactions**

45.

$CH_3-CH(CH_3)-Cl$

$\xrightarrow{(Na^+OH^-)}$ $CH_3-CH(CH_3)-OH$ An alcohol

$\xrightarrow{(Na^+SH^-)}$ $CH_3-CH(CH_3)-SH$ A thio alcohol

$\xrightarrow{(Na^+OR^-)}$ $CH_3-CH(CH_3)-O-R$ An ether

$\xrightarrow{(Na^+CN^-)}$ $CH_3-CH(CH_3)-CN$ An alkyl cyanide, nitrile

$\xrightarrow{(Na^+\,NH_2^-)}$ $CH_3-CH(CH_3)-NH_2$ An amine

$\xrightarrow{(:NH_3)}$ $CH_3-CH(CH_3)-\overset{+}{N}H_3$ An ammonium ion

46.

$C_6H_5-CH_2-CH_3 \xrightarrow[\text{sunlight}]{Br_2,} C_6H_5-CH(Br)-CH_3 + HBr$

$C_6H_5-CH_2-CH_3 \xrightarrow[\text{Fe}]{Br_2,} o\text{-}Br-C_6H_4-CH_2-CH_3 + p\text{-}Br-C_6H_4-CH_2-CH_3 + HBr$

47.

$$CH_3-\underset{\substack{|\\CH_3}}{CH}-CH_2-CH_2-Br \xrightarrow[\text{mechanism}]{OH^-,\ S_N2} \left(CH_3-\underset{\substack{|\\CH_3}}{CH}-CH_2-CH_2-OH\right)$$

$$CH_3-\underset{\substack{|\\CH_3}}{CH}-CH_2-CH_2-Br \xrightarrow[\text{E2 mechanism}]{OH^-,} \left(CH_3-\underset{\substack{|\\CH_3}}{CH}-CH=CH_2\right)$$

48.

$$CH_3-\overset{\substack{O\\\|}}{C}-CH_3 + \left(3\ Cl_2 + 4\ NaOH\right) \longrightarrow CH_3-\overset{\substack{O\\\|}}{C}-O^-\ Na^+ + CHCl_3 + 3\ H_2O + 3\ NaCl$$

**Syntheses**

49.

$$C_6H_6 + CH_3-CH_2-Cl \xrightarrow{AlCl_3} C_6H_5-CH_2-CH_3 \xrightarrow[uv]{Br_2} C_6H_5-\underset{\substack{|\\Br}}{CH}-CH_3 \xrightarrow[\text{KOH}]{\text{alcoholic}} C_6H_5-CH=CH_2$$

50.

$$CH_3-\underset{\substack{|\\OH}}{CH}-CH_3 \xrightarrow[\text{heat}]{H_2SO_4} CH_3-CH=CH_2 \xrightarrow[\text{peroxides}]{HBr,} CH_3-CH_2-CH_2-Br \xrightarrow{NaNH_2} CH_3-CH_2-CH_2-NH_2$$

51.

$$2\ CH_3-\underset{\substack{|\\CH_3}}{CH}-CH_2-OH \xrightarrow{SOCl_2} 2\ CH_3-\underset{\substack{|\\CH_3}}{CH}-CH_2-Cl \xrightarrow{2\ Na} CH_3-\underset{\substack{|\\CH_3}}{CH}-CH_2-CH_2-\underset{\substack{|\\CH_3}}{CH}-CH_3$$

(Wurtz Reaction)

52.

$$CH_3-\underset{H}{\overset{CH_3}{\underset{|}{\overset{|}{C}}}}-OH \xrightarrow{SOCl_2} CH_3-\underset{H}{\overset{CH_3}{\underset{|}{\overset{|}{C}}}}-Cl \xrightarrow{Mg} CH_3-\underset{H}{\overset{CH_3}{\underset{|}{\overset{|}{C}}}}-Mg-Cl \xrightarrow{H_2O} CH_3-CH_2-CH_3$$

or

$$CH_3-\underset{H}{\overset{CH_3}{\underset{|}{\overset{|}{C}}}}-OH \xrightarrow[\text{heat}]{H_2SO_4,} CH_3-CH{=}CH_2 \xrightarrow[\text{cat.}]{H_2,} CH_3-CH_2-CH_3$$

53.

$$CH_3-\underset{CH_3}{\overset{CH_3}{\underset{|}{\overset{|}{C}}}}-OH \xrightarrow[\text{heat}]{H_2SO_4,} CH_3-\underset{CH_3}{\overset{CH_2}{\underset{|}{\overset{\|}{C}}}} \xrightarrow{Br_2} CH_3-\underset{CH_3}{\overset{CH_2Br}{\underset{|}{\overset{|}{C}}}}-Br$$

**Miscellaneous**

54.

| Compound | dilute $KMnO_4$ or $Br_2$ | alcohol, $AgNO_3$ |
|---|---|---|
| (a) bromobenzene ($C_6H_5$–Br) | (a) no change | (a) no change |
| (b) benzyl bromide ($C_6H_5$–$CH_2$—Br) | (b) no change | (b) ppt. AgBr |
| (c) $CH_2{=}CH{-}Br$ | (c) decolorized | |

55. There can be only two primary alkyl bromides with molecular formula $C_4H_9Br$. They are

$$CH_3-CH_2-CH_2-CH_2-Br \quad \text{and} \quad CH_3-\overset{CH_3}{\overset{|}{C}H}-CH_2-Br$$

*n*-Butyl bromide          Isobutyl bromide

The problem states that when the alkyl bromide (A) was reacted with sodium, it gave a compound different than that produced by *n*-butyl bromide. (A) must therefore be isobutyl bromide.

$$\left(CH_3-\underset{}{\overset{CH_3}{\overset{|}{C}}}H-CH_2-Br\right) \xrightarrow[\text{NaOH, E2}]{\text{heat,}} \left(CH_3-\overset{CH_3}{\overset{|}{C}}=CH_2\right) \xrightarrow{HBr} \left(CH_3-\overset{CH_3}{\underset{Br}{\overset{|}{\underset{|}{C}}}}-CH_3\right)$$

(A) (B) (C)

(A) $\xrightarrow{\text{Na}}$

$$\left(CH_3-\overset{CH_3}{\overset{|}{C}}H-CH_2-CH_2-\overset{CH_3}{\overset{|}{C}}H-CH_3\right)$$

(D)

56. There can be only one secondary alcohol of formula $C_3H_8O$.

$$\left(CH_3-\overset{CH_3}{\underset{H}{\overset{|}{\underset{|}{C}}}}-OH\right) \xrightarrow{SOCl_2} \left(CH_3-\overset{CH_3}{\underset{H}{\overset{|}{\underset{|}{C}}}}-Cl\right) \xrightarrow[AlCl_3]{C_6H_6} \left(C_6H_5-\underset{CH_3}{\underset{|}{C}}H-CH_3\right)$$

(A) (B) (C)

57. $S_N1$. An —OH group is being substituted for the —Cl group and it is a unimolecular reaction. The rate of the reaction depends on the concentration of *tert*-pentyl chloride and how fast it ionizes.

58. E2. The hydrogen halide, HCl, has been eliminated from ethyl chloride. The rate of reaction depends on both the concentrations of ethyl chloride and the hydroxide ion.

59. $\left[ \begin{array}{c} OH \\ \vdots \\ H \\ H_2C \cdots C H_2 \\ \vdots \\ Cl \end{array} \right]^-$

···· indicates bonds being made or broken. The stereochemistry of the E2 reaction is anti-coplanar. The two groups that are leaving are anti to each other and in the same plane as the two carbons forming the double bond.

60. $S_N2$ bimolecular substitution, substrate undergoes a Walden inversion.

61. $\left[ HO \cdots \overset{H \quad CH_2CH_3}{C} \cdots Cl \; (\text{with } CH_3 \text{ below}) \right]^-$

62. E1 unimolecular elimination

63. Carbon 1 of the Newman projection formula must be rotated in respect to carbon 2 so that the elimination of HCl can occur in an anti-coplanar manner.

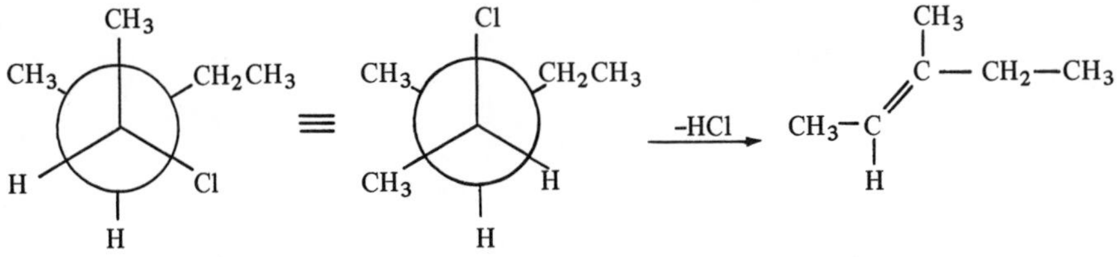
CH3
CH3
CH2CH3
H
Cl
H
≡
Cl
CH3
CH2CH3
CH3
H
H
-HCl
CH3
C — CH2—CH3
CH3—C
H

(*E*)-3-Methyl-2-pentene

64. (a)

$$\underset{a}{CH_3}-\underset{b}{\overset{Br}{\overset{|}{C}H}}-\underset{a}{CH_3}$$

(b) $Br-\underset{b}{CH_2}-\underset{a}{CH_2}-\underset{b}{CH_2}-Br$

(c) $\underset{a}{(CH_3)_2}\underset{b}{CH}-\underset{c}{CH_2}-Br$

# 8

# Alcohols, Phenols, and Ethers

## INTRODUCTION

In this chapter we shall study three families of oxygen-containing compounds in which the oxygen atom is singly bonded to two other atoms. The chemical behavior of a **hydroxyl group**

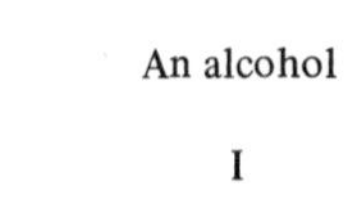

An alcohol

I

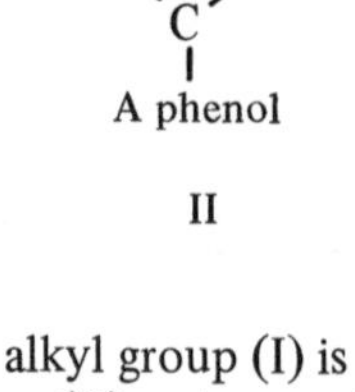

A phenol

II

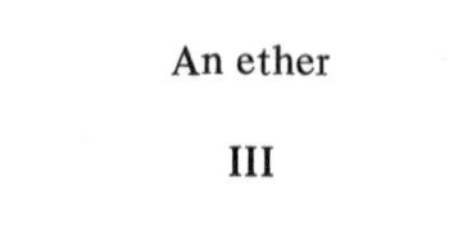

An ether

III

(—OH) attached to an alkyl group (I) is sufficiently different from that of one attached to an aryl group (II) to be considered separately as two different classes of compounds–namely, **alcohols** and **phenols**. We shall find that phenols are stronger acids than water, but alcohols are somewhat less acidic than water. The acidity of phenols can be attributed to the greater resonance stabilization of the phenoxide ion compared to that of phenol. There is no resonance stabilization shown by either alkoxide ions, nor by alcohols. The increased solvation of the hydroxide ion compared to that of the alkoxide ion is a principal factor in making water a stronger acid than alcohol.

$C_6H_5-\ddot{\underset{..}{O}}:^-$ $\qquad$ $R-\ddot{\underset{..}{O}}:^-$ $\qquad$ $H-\ddot{\underset{..}{O}}:^-$

Phenoxide ion $\qquad$ Alkoxide ion $\qquad$ Hydroxide ion

We shall study reactions in which the O—H bond is cleaved and others in which the C—O bond is cleaved. We also will see that the carbon atom bonded to oxygen may be easily oxidized further to other functional groups.

In the case of the phenols, we will deal mainly with reactions resulting in electrophilic substitution at the *ortho* and *para* ring positions.

The ethers represent an extremely stable group of compounds with few reactions other than those involving a C—O bond cleavage.

## PROBLEMS

### True–False

1. T F Another name for methyl alcohol is carbinol.
2. T F The suffix *-ol* indicates the presence of an —OH in the molecule.
3. T F Methanol is sometimes called grain alcohol.
4. T F The molecular formula for ethanol and methyl ether are the same, $C_2H_6O$; therefore their boiling points are very nearly the same.
5. T F Ethanol and methyl ether are not isomers because they have different functional groups.
6. T F Alcohols and phenols are more acidic than water.
7. T F Methyl alcohol is a very poisonous substance and must never be confused with ethyl alcohol in its physiological behavior.
8. T F Pure ethyl alcohol cannot be separated by fractional distillation from an aqueous mixture containing more than 5% water.
9. T F The phenolic group is an *ortho–para* director.
10. T F Halogen atoms attached to an aromatic ring are not easily displaced.
11. T F Ethers are very water soluble.
12. T F Ethers are relatively inert.
13. T F The boiling points of the ethers are closer to the boiling points of alkanes of nearly the same molecular weight than they are to the boiling points of the isomeric alcohols.
14. T F Ethyl ether, although often used as an organic solvent, is a potential fire hazard.
15. T F The hydroxyl of a phenol is not as easily displaced as the hydroxyl of an alcohol.

16. T F Tertiary alcohols are not easily oxidized.
17. T F Most reactions in which the hydroxyl group is being replaced are acid catalyzed.

acid & base

**Multiple Choice**

18. The compound with the lowest boiling point is

(a) $H_2O$
(b) $CH_3-CH_2-OH$
(c) $CH_3-O-CH_3$

19. The compound which reacts most readily with Lucas reagent is

(a) $CH_3-CH_2-Cl$
(b) $CH_3-CH_2-OH$
(c) $CH_2=CH_2$
(d) $CH_3-O-CH_3$
(e) $CH_3-C(CH_3)_2-OH$ (central C bearing $CH_3$ above and $CH_3$ below)

20. The structure that more nearly represents the correct bond angle between the carbon, oxygen, and hydrogen atoms in methyl alcohol is

(a) $H_3C-\ddot{\underset{..}{O}}-H$ (linear)
(b) $H_3C-\ddot{O}:$ with H bonded to O angled down to the right
(c) $H_3C-\ddot{O}:$ with H bonded to O straight down
(d) $H_3C-\ddot{O}\cdot$ with H bonded to O angled down to the left

21. A correct name for $CH_3-CH_2-CH(CH_3)-OH$ is

(a) *sec*-butyl alcohol
(b) ethylmethylcarbinol
(c) 2-butanol
(d) none of the above
(e) all of the above

22. Alcohols can be easily distinguished from phenols because:

(a) phenols are soluble in sodium hydroxide, but alcohols are not.
(b) alcohols are soluble in sodium hydroxide, but phenols are not.
(c) phenols are soluble in sodium bicarbonate, but alcohols are not.
(d) alcohols are soluble in sodium bicarbonate, but phenols are not.
(e) phenols react with metallic sodium, but alcohols do not.

23. The reaction

$$HO^- + CH_3-\underset{\underset{Br}{|}}{CH}-CH_2-CH_3 \longrightarrow \left[ HO\text{----}\overset{\overset{CH_3}{|}}{\underset{H\ \ \ C_2H_5}{C}}\text{---}Br \right]^- \longrightarrow Br^- + HO-\overset{\overset{CH_3}{|}}{\underset{\underset{H}{|}}{C}}-C_2H_5$$

(transition state)

is classified as a

(a) unimolecular nucleophilic substitution reaction.
(b) bimolecular nucleophilic substitution reaction.
(c) unimolecular electrophilic substitution reaction.
(d) bimolecular electrophilic substitution reaction.
(e) none of these.

24. The compound that reacts most readily with metallic sodium is

(a) $CH_3-CH_2-OH$

(b)
$$CH_3-\underset{\displaystyle CH_3}{\underset{|}{CH}}-OH$$

(c)
$$CH_3-\underset{\displaystyle CH_3}{\underset{|}{\overset{\displaystyle CH_3}{\overset{|}{C}}}}-OH$$

(d) none of them will react
(e) they all react at the same rate

25. The structure that most nearly represents the correct bond angle between the C—O—C bonds in methyl ether is

(a) $CH_3-\ddot{\underset{..}{O}}-CH_3$

(b) $CH_3-\ddot{O}:$ with $CH_3$ bonded diagonally down from O

(c)
$$CH_3-\underset{\displaystyle CH_3}{\underset{|}{\ddot{O}:}}$$

26. An unknown compound is soluble in cold concentrated sulfuric acid, but insoluble in sodium hydroxide. It does not decolorize bromine and does not react with metallic sodium. The class of compounds to which the unknown belongs is

(a) alkanes
(b) alkenes
(c) alcohols
(d) phenols
(e) ethers

27. The compound in which hydrogen bonding is not possible is

(a) $CH_3-O-CH_3$
(b) $CH_3-CH_2-OH$
(c) $H_2O$
(d) HF
(e)
$$CH_3-\underset{\displaystyle O}{\underset{\|}{C}}-OH$$

## Completion

28. A common method of purification of liquids is fractional distillation. Sometimes a mixture of liquids will form a constant-boiling mixture which cannot be further separated by fractional distillation. Such a constant–boiling mixture is called an ____________________.

29. Ethyl alcohol rendered unfit for use in beverages by the addition of substances repugnant and difficult to remove is said to be____________________.

30. Alcohols can be prepared by the hydration of alkenes with sulfuric acid and water. The only primary alcohol that can be prepared by this method is ____________________.

31. An indirect method of hydration of alkenes which can be used to prepare other primary alcohols is ____________________.

32. The name given to the reaction indicated by the equation below is ____________ ____________ ____________________.

$$CH_3-\underset{\displaystyle H}{\overset{\displaystyle CH_3}{\underset{|}{\overset{|}{C}}}}-O^-Na^+ + CH_3-CH_2-I \longrightarrow CH_3-\underset{\displaystyle H}{\overset{\displaystyle CH_3}{\underset{|}{\overset{|}{C}}}}-O-CH_2-CH_3 + NaI$$

33. Compounds secreted by animals and insects for the purpose of intraspecies communication are called ____________________.

**Matching** (Match each compound to the appropriate description.)

(a) benzene ring–OH

(b) $\underset{OH}{CH_2}-\underset{OH}{CH_2}$

(c) $CH_3-OH$

(d) Cl, Cl-substituted benzene ring–$O-CH_2-\overset{O}{\overset{\|}{C}}-OH$

(e) $CH_2$——$CH_2$ bridged by O

(f) $\underset{OH}{CH_2}-\underset{OH}{CH}-\underset{OH}{CH_2}$

(g) benzene ring with OH, Cl, Cl

(h) $\underset{ONO_2}{CH_2}-\underset{ONO_2}{CH}-\underset{ONO_2}{CH_2}$

(i) $CH_3-(CH_2)_{11}-OSO_3^-\ Na^+$

(j) benzene ring with OH, Cl, Cl, Cl, Cl, Cl

34. ____ The principal component in "permanent" types of antifreeze.
35. ____ The high explosive component of dynamite.
36. ____ The triol which is a by-product in the manufacture of soap from animal fats.
37. ____ The compound commonly called "wood alcohol".
38. ____ The compound used as a comparison standard for measuring the germicidal efficiency of antiseptics.
39. ____ A compound commonly used as wood preservative.
40. ____ A sex attractant (lonestar tick).
41. ____ A detergent.
42. ____ A herbicide (2,4-D).
43. ____ A cyclic ether.

**Structural Formulas** (Write the correct structural formula for each of the following compounds.)

44. Ethylene oxide

45. Dimethylisopropylcarbinol

46. Phenol

47. Benzyl alcohol

48. 1,3-Butanediol

49. Anisole

50. 2-Ethoxypropane

51. Cyclohexyl phenyl ether

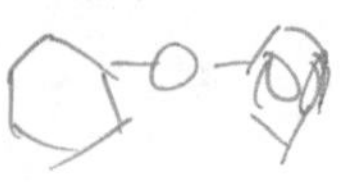

52. 3-Penten-2-ol

53. Resorcinol

54. *p*-Cresol

55. Ethylene glycol

56. Benzyne

57. Hydroquinone

58. Potassium *tert*-butoxide

59. *o*-Benzoquinone

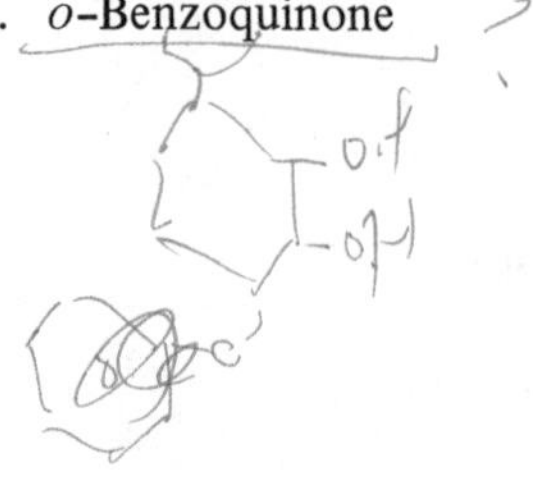

**Reactions** (Complete the following.)

60. $CH_3-CH_2-O-CH_2-CH_3 + 2HI \longrightarrow \quad + H_2O$

61.

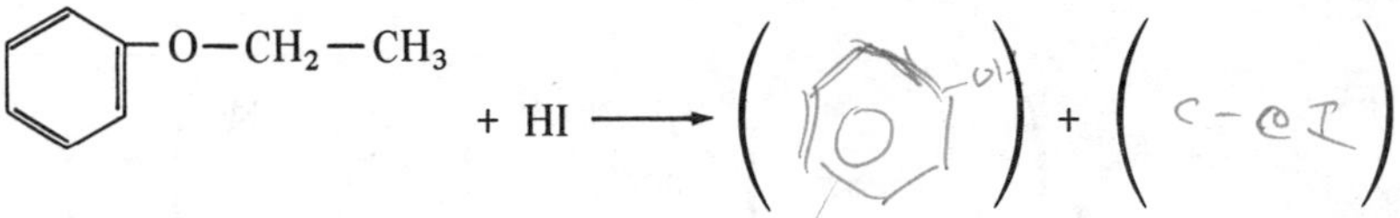

62. $$CH_3-\underset{CH_3}{\overset{CH_3}{\underset{|}{\overset{|}{C}}}}-O^-Na^+ + CH_3-CH_2-Cl \longrightarrow (\quad) + (\quad)$$

63. $$CH_3-O^-Na^+ + CH_3-\underset{CH_3}{\overset{CH_3}{\underset{|}{\overset{|}{C}}}}-Cl \longrightarrow (\quad) + (\quad) + (\quad)$$

64. $$CH_3-CH_2-\overset{OH}{\overset{|}{CH}}-CH_2-CH_3 \xrightarrow[KMnO_4]{hot} (\quad)$$

65. $$CH_3-CH_2-CH_2-CH_2-OH \xrightarrow[H_2SO_4]{K_2Cr_2O_7,} (\quad)$$

66. $$CH_3-CH{=}CH_2 + O_2 \xrightarrow{Ag} (\quad) \xrightarrow[H_2O]{HCl,} (\quad)$$

67. $$CH_3-CH{=}CH_2 + HOCl \longrightarrow (\quad) \xrightarrow[Na_2CO_3]{H_2O,} (\quad)$$

68.

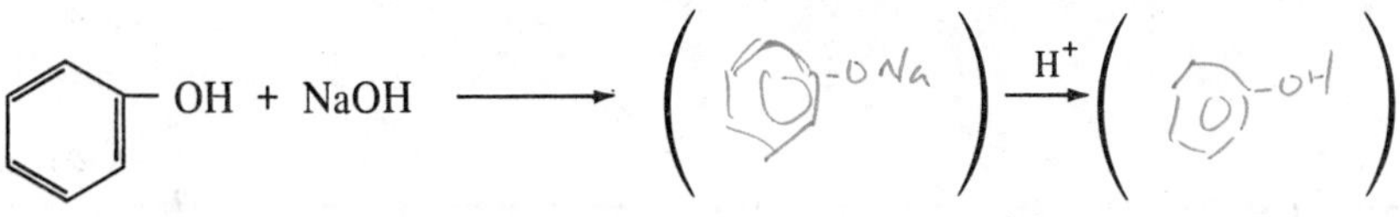

69. $$CH_3-\underset{CH_3}{\overset{CH_3}{\underset{|}{\overset{|}{C}}}}-OH + HCl \xrightarrow{ZnCl_2} (\quad)$$

70.

Benzene $\xrightarrow{Cl_2,\ FeCl_3}$ ( ) $\xrightarrow[300^\circ C,\ 200\ atm]{H_2O,\ NaOH,}$ ( ) $\xrightarrow{H^+}$ ( )

Benzene $\xrightarrow[CH_3-CH=CH_2]{AlCl_3,}$ ( ) $\xrightarrow{O_2}$ ( ) $\xrightarrow{H^+}$ ( )

Benzene $\xrightarrow{H_2SO_4}$ ( ) $\xrightarrow{NaOH}$ ( ) $\xrightarrow[fusion]{NaOH}$ ( ) $\xrightarrow{H^+}$ ( )

71. $CH_3-CH=CH_2 \xrightarrow[\text{then } H_2O]{H_2SO_4,}$ ( )

72. $CH_3-CH=CH_2 \xrightarrow[H_2O_2 \text{ and } OH^-]{B_2H_6,\ \text{then}}$ ( )

**Syntheses**

73. Starting with 2-methyl-1-propanol, and any other reagents necessary, show by equations how you would prepare 3-methyl-1-butanol.

74. Starting with 2-propanol, and any other reagents necessary, show how you could prepare 3-methyl-1-butanol.

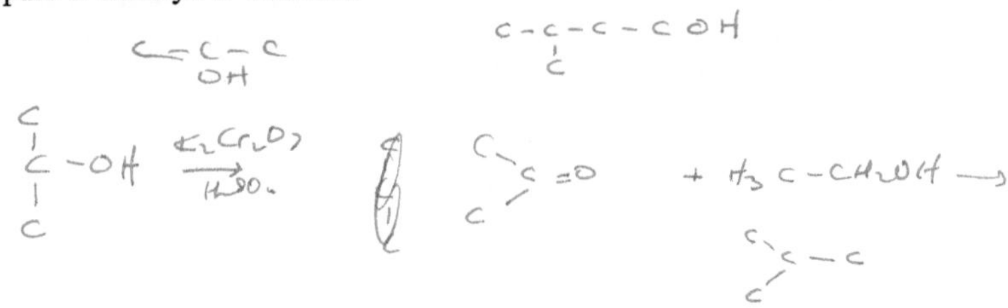

75. Using methyl alcohol and ethyl alcohol as the only carbon-containing compounds, show how to prepare methyl acetate, $CH_3—\overset{\displaystyle O}{\overset{\|}{C}}—O—CH_3$.

76. Using isopropyl alcohol as your only carbon-containing compound, show how to prepare 2,3-dimethyl-2-butanol,

$$CH_3—\underset{HO}{\overset{CH_3}{C}}—\underset{CH_3}{CH}—CH_3.$$

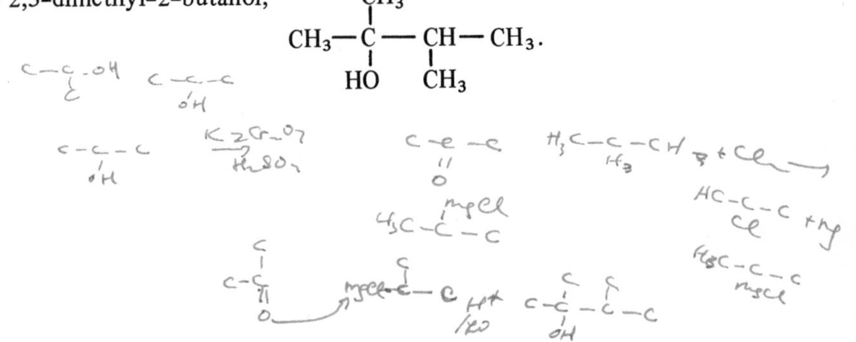

77. Using benzene as a starting material, and any other reagents necessary, show how you could prepare anisole, $C_6H_5—OCH_3$.

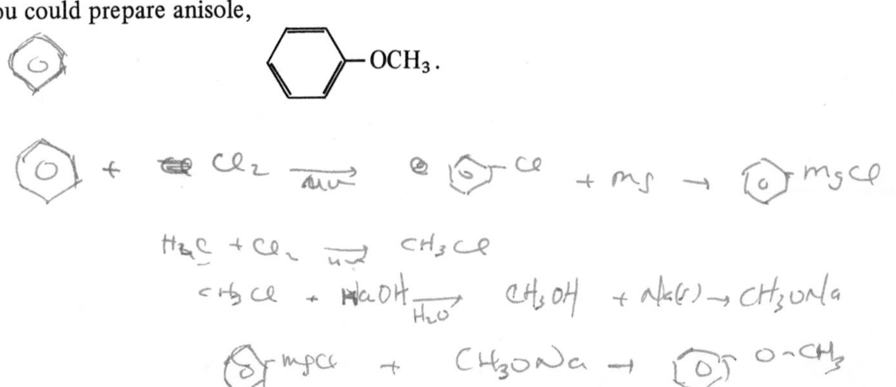

## Miscellaneous

78. Before a flask of anhydrous ether was distilled, it was tested with an acidified solution of potassium iodide. The characteristic brown color of iodine appeared. Was it safe to begin the distillation? Why?

79. The addition of a proton to ammonia or to an amine results in the formation of an amm*onium* ion.

$$R\ddot{N}H_2 + H^+ \longrightarrow R\overset{+}{N}H_3 \text{ (amm}\textit{onium}\text{ ion)}$$

What is the name and structure of the ion which results from the addition of a proton to an ether or to an alcohol?

80. Circle the hydroxyl group in glycerol that corresponds to that of a secondary alcohol.

$$\underset{\displaystyle OH}{\underset{|}{CH_2}}-\underset{\displaystyle OH}{\underset{|}{CH}}-\underset{\displaystyle OH}{\underset{|}{CH_2}}$$

(Give the polar number of the carbon atom in each of the following.)

81. ____ $CH_3OH$

82. ____ $R-CH_2-OH$

83. ____ $R_2CH-OH$

84. ____ $R_3C-OH$

85. ____ $R_4C$

86. ____ $R_2CH_2$

87. ____ $CH_4$

88. ____ $H_2C{=}O$

89. ____ $R-\underset{\displaystyle O}{\underset{\|}{CH}}$

90. ____ $R_2C{=}O$

(Balance the following two equations.)

91. $CH_3-\underset{\displaystyle OH}{\underset{|}{CH}}-CH_3 + KMnO_4 \rightarrow CH_3-\underset{\displaystyle O}{\underset{\|}{C}}-CH_3 + MnO_2 + KOH + H_2O$

92. $C_6H_5CH_3 + KMnO_4 \longrightarrow C_6H_5CO_2K + MnO_2 + KOH + H_2O$

93. When 2-methyl-1-butanol is heated in concentrated sulfuric acid, 2-methyl-2-butene is the chief product. Write a mechanism that explains this.

94. Give a structure consistent with each of the following sets of nmr data.

(a) $C_7H_8O$

*a* δ 2.2, singlet, 3H
*b* δ 6.4, singlet, 1H
*c* δ 6.8, multiplet, 4H

(b) $C_7H_8O$

*a* δ 2.4, singlet, 1H
*b* δ 4.6, singlet, 2H
*c* δ 7.3, multiplet, 5H

(c) $C_7H_8O$

*a* δ 3.8, singlet, 3H
*b* δ 7.1, multiplet, 5H

## ANSWERS

### True–False

1. True
2. True
3. False. Methanol is called **wood alcohol**. Ethanol is **grain alcohol**.

4. False. Ethanol is capable of hydrogen bonding, but methyl ether is not. Thus, ethanol (b.p. 78) boils higher than methyl ether (b.p. -24).
5. False. Different compounds that have the same molecular formula are called **isomers**.
6. False. (acidity) phenols > water > alcohols (See introduction)
7. True
8. True. Ethyl alcohol and water form an azeotrope of composition 95% alcohol, 5% water.
9. True
10. True. Aryl halides do not easily undergo nucleophilic substitution reactions.
11. False
12. True
13. True. Neither alkanes nor ethers are capable of hydrogen bonding, but alcohols do hydrogen bond and thus boil higher.
14. True
15. True
16. True
17. True. The hydroxide ion is a poor leaving group and is not easily removed in either substitution or elimination reactions. Protonation of the hydroxyl group changes the leaving group from hydroxide ion to water, a good leaving group.

$$R-\ddot{\underset{..}{O}}-H \longrightarrow R^+ + :\ddot{\underset{..}{O}}H^- \quad \text{(not likely)}$$

$$R-\ddot{\underset{..}{O}}-H + H^+ \longrightarrow R-\overset{+}{\ddot{O}}H_2 \longrightarrow R^+ + H_2O \quad \text{(possible)}$$

**Multiple Choice**

18. (c) Water and alcohols can form hydrogen bonds, but ethers cannot. If compounds are intermolecularly hydrogen bonded, it takes more energy to separate the molecules from each other so they can enter the vapor phase. Thus, the boiling point is higher.
19. (e) Lucas reagent is used as a test to distinguish the classes of alcohols. Tertiary alcohols react the fastest.
20. (b)
21. (e)
22. (a) Most phenols are acidic enough to dissolve in sodium hydroxide, but not in sodium bicarbonate. Alcohols are not acidic and do not dissolve in either.
23. (b) This is a $S_N2$ reaction. The hydroxide ion is a nucleophilic reagent and the rate of the reaction would depend upon the concentration of both hydroxide ion and 2-bromobutane since both are involved in the rate-determining step.
24. (a) Alkyl groups are inductively electron releasing.

$$R-CH_2-OH$$

This has the effect of holding the $H^+$ more tightly and decreasing the reactivity. The order of reactivity of alcohols with sodium is

primary > secondary > tertiary

25. (b)
26. (e) Alkanes are not soluble in cold concentrated sulfuric acid. Alkenes decolorize bromine. Alcohols and phenols react with metallic sodium.
27. (a)

**Completion**

28. azeotrope
29. denatured
30. ethanol
31. hydroboration
32. Williamson ether synthesis
33. pheromones

**Matching**

34. (b)
35. (h)
36. (f)
37. (c)
38. (a)
39. (j)
40. (g)
41. (i)
42. (d)
43. (e)

**Structural Formulas**

44. $CH_2-CH_2$ (ring closed through O)

45. $CH_3-CH(CH_3)-C(CH_3)_2-OH$ (H₃C on CH; CH₃ above and below C)

46. $C_6H_5-OH$

47. $C_6H_5-CH_2-OH$

48. $CH_3-CH(OH)-CH_2-CH_2-OH$

49. $C_6H_5-O-CH_3$

50. $CH_3-CH(-O-CH_2-CH_3)-CH_3$

51. $C_6H_5-O-C_6H_{11}$ (benzene ring–O–cyclohexane ring)

52. $CH_3-CH{=}CH-CH(OH)-CH_3$

53. (benzene ring with OH at positions 1 and 3)

57. (benzene ring with HO and OH at positions 1 and 4)

54. (benzene ring with $CH_3$ and OH)

58.
$$CH_3-\underset{CH_3}{\overset{CH_3}{\overset{|}{\underset{|}{C}}}}-O^-\,K^+$$

55.
$$\underset{OH}{\underset{|}{CH_2}}-\underset{OH}{\underset{|}{CH_2}}$$

59. (ring with two =O groups)

56. (benzene ring)

**Reactions**

60. $CH_3-CH_2-O-CH_2-CH_3 + 2\,HI \rightarrow \left(2\,CH_3-CH_2-I\right) + H_2O$

61. (benzene ring)$-O-CH_2-CH_3$ + HI → ((benzene ring)$-OH$) + $\left(CH_3-CH_2-I\right)$

Exercises 60 and 61 illustrate the difference in reactivity between an alkyl-oxygen linkage and an aryl-oxygen linkage.

62.
$$CH_3-\underset{CH_3}{\overset{CH_3}{\overset{|}{\underset{|}{C}}}}-O^-\,Na^+ + CH_3-CH_2-Cl \rightarrow \left(CH_3-\underset{CH_3}{\overset{CH_3}{\overset{|}{\underset{|}{C}}}}-O-CH_2-CH_3\right) + \left(NaCl\right)$$

If a primary alkyl halide is reacted with an alkoxide, substitution occurs and the product is an ether.

63.
$$CH_3-O^-\,Na^+ + CH_3-\underset{CH_3}{\overset{CH_3}{\overset{|}{\underset{|}{C}}}}-Cl \rightarrow \left(CH_3-\underset{CH_2}{\underset{\|}{C}}-CH_3\right) + \left(CH_3-OH\right) + \left(NaCl\right)$$

If a tertiary alkyl halide is reacted with an alkoxide, elimination occurs and the product is an alkene.

64. $CH_3-CH_2-\underset{|}{\overset{OH}{CH}}-CH_2-CH_3 \xrightarrow{KMnO_4} \left(CH_3-CH_2-\overset{O}{\overset{\|}{C}}-CH_2-CH_3\right)$

Oxidation of a secondary alcohol produces a ketone which is stable toward most oxidizing reagents.

65. $CH_3-CH_2-CH_2-CH_2-OH \xrightarrow[H_2SO_4]{K_2Cr_2O_7,} \left(CH_3-CH_2-CH_2-\overset{O}{\overset{\|}{C}}-OH\right)$

Oxidation of a primary alcohol produces an aldehyde ($R-\overset{O}{\overset{\|}{C}}-H$) which is easily oxidized to a carboxylic acid.

66. $CH_3-CH{=}CH_2 + O_2 \xrightarrow{Ag} \left(CH_3-CH\text{—}CH_2 \text{ (epoxide, bridged by O)}\right) \xrightarrow[HCl]{H_2O,} \left(CH_3-\underset{OH}{\underset{|}{CH}}-CH_2-OH\right)$

67. $CH_3-CH{=}CH_2 + HOCl \rightarrow \left(CH_3-\underset{OH}{\underset{|}{CH}}-CH_2Cl\right) \xrightarrow[Na_2CO_3]{H_2O,} \left(CH_3-\underset{OH}{\underset{|}{CH}}-CH_2OH\right)$

68. $C_6H_5-OH + NaOH \rightarrow \left(C_6H_5-O^-\,Na^+\right) \xrightarrow{H^+} \left(C_6H_5-OH\right)$

69. $CH_3-\underset{CH_3}{\underset{|}{\overset{CH_3}{\overset{|}{C}}}}-OH + HCl \xrightarrow{ZnCl_2} \left(CH_3-\underset{CH_3}{\underset{|}{\overset{CH_3}{\overset{|}{C}}}}-Cl\right)$

70.

$C_6H_6 \xrightarrow[FeCl_3]{Cl_2} \left(C_6H_5-Cl\right) \xrightarrow[300^\circ C,\ 200\ atm]{\text{dilute NaOH,}} \left(C_6H_5-O^-\,Na^+\right) \xrightarrow{H^+} \left(C_6H_5-OH\right)$

$C_6H_6 \xrightarrow[CH_3-CH=CH_2]{AlCl_3,} \left(C_6H_5-\overset{CH_3}{\overset{|}{CH}}-CH_3\right) \xrightarrow{O_2} \left(C_6H_5-\underset{CH_3}{\underset{|}{\overset{CH_3}{\overset{|}{C}}}}-OOH\right) \xrightarrow{H^+} \left(C_6H_5-OH\right)$

$C_6H_6 \xrightarrow{H_2SO_4} \left(C_6H_5-SO_3H\right) \xrightarrow{NaOH} \left(C_6H_5-SO_3^-\,Na^+\right) \xrightarrow[\text{fusion}]{NaOH} \left(C_6H_5-O^-\,Na^+\right) \xrightarrow{H^+} \left(C_6H_5-OH\right)$

71.
$$CH_3-CH=CH_2 \xrightarrow[\text{then } H_2O]{H_2SO_4,} CH_3-\underset{\displaystyle OH}{\underset{|}{CH}}-CH_3$$

72.
$$CH_3-CH=CH_2 \xrightarrow[\text{then } H_2O_2 \text{ and } OH^-]{B_2H_6,} CH_3-CH_2-CH_2OH$$

Note that the reactions in exercises 71 and 72 compliment each other. The addition of ionic reagents ($H^+$ and $HSO_4^-$) proceeds according to Markovnikov's Rule, whereas the hydroboration reaction gives the "anti" Markovnikov product.

### Syntheses

73. This synthesis calls for the extension of the alkyl chain by a methylene unit ($-CH_2-$). One of the easiest ways to accomplish this is by the addition of a Grignard reagent to formaldehyde.

$$CH_3-\underset{\displaystyle CH_3}{\underset{|}{CH}}-CH_2-OH \xrightarrow{PCl_3} CH_3-\underset{\displaystyle CH_3}{\underset{|}{CH}}-CH_2-Cl \xrightarrow[\text{ether}]{Mg,} CH_3-\underset{\displaystyle CH_3}{\underset{|}{CH}}-CH_2-Mg-Cl$$

$$\downarrow CH_2O$$

$$CH_3-\underset{\displaystyle CH_3}{\underset{|}{CH}}-CH_2-CH_2-OH \xleftarrow[H_2O]{H^+,} CH_3-\underset{\displaystyle CH_3}{\underset{|}{CH}}-CH_2-CH_2-O-Mg-Cl$$

74. This synthesis requires the extension of the alkyl chain by two carbons ($-CH_2-CH_2-$). An easy way to accomplish this is by the addition of a Grignard reagent to ethylene oxide.

$$CH_3-\underset{\displaystyle CH_3}{\underset{|}{CH}}-OH \xrightarrow{PCl_3} CH_3-\underset{\displaystyle CH_3}{\underset{|}{CH}}-Cl \xrightarrow[\text{ether}]{Mg,} CH_3-\underset{\displaystyle CH_3}{\underset{|}{CH}}-Mg-Cl$$

$$\downarrow \; \underset{\diagdown \; O \; \diagup}{CH_2 - CH_2}$$

$$CH_3-\underset{\displaystyle CH_3}{\underset{|}{CH}}-CH_2-CH_2-OH \xleftarrow[H_2O]{H^+,} CH_3-\underset{\displaystyle CH_3}{\underset{|}{CH}}-CH_2-CH_2-O-Mg-Cl$$

75. Esters can be prepared by the reaction of carboxylic acids and alcohols. The carboxylic acid portion of the ester contains two carbon atoms and the alcohol portion one.

$$CH_3-CH_2-OH \xrightarrow[H_2SO_4]{K_2Cr_2O_7,} CH_3-\overset{\displaystyle O}{\overset{\|}{C}}-OH \xrightarrow[H^+]{CH_3OH,} CH_3-\overset{\displaystyle O}{\overset{\|}{C}}-O-CH_3$$

76. Two isopropyl groups are apparently reacted in some fashion to prepare the 6-carbon final product. Again the addition of a Grignard reagent to a ketone would seem to be a good method of preparation.

$$CH_3-\underset{}{\overset{OH}{\overset{|}{C}H}}-CH_3 \xrightarrow{KMnO_4} CH_3-\overset{O}{\overset{\|}{C}}-CH_3$$

$$CH_3-\overset{OH}{\overset{|}{C}H}-CH_3 \xrightarrow{PCl_3} CH_3-\overset{Cl}{\overset{|}{C}H}-CH_3 \xrightarrow[\text{ether}]{Mg,} CH_3-\overset{CH_3}{\overset{|}{C}H}-Mg-Cl$$

$$\downarrow CH_3-\overset{O}{\overset{\|}{C}}-CH_3$$

$$CH_3-\underset{H_3C}{\underset{|}{\overset{OH}{\overset{|}{C}}}}-\underset{CH_3}{\underset{|}{C}H}-CH_3 \xleftarrow[H_2O]{H^+,} CH_3-\underset{CH_3}{\underset{|}{\overset{O-Mg-Cl}{\overset{|}{C}}}}-\underset{CH_3}{\underset{|}{C}H}-CH_3$$

77. One of the best ways to prepare an ether is by the Williamson ether synthesis.

$$C_6H_6 \xrightarrow{H_2SO_4} C_6H_5-SO_3H \xrightarrow{NaOH} C_6H_5-SO_3^-\ Na^+$$

$$\downarrow \text{NaOH fusion}$$

$$C_6H_5-OCH_3 \xleftarrow{CH_3I} C_6H_5-O^-\ Na^+$$

**Miscellaneous**

78. No. The formation of iodine indicates peroxides are present in the ether. They may explode upon heating.

79. $$R-\ddot{O}{:}(-R) + H^+ \longrightarrow R-\overset{+}{\ddot{O}}(-R)-H \qquad \text{Ox}\textit{onium}\text{ ion}$$

$$R-\ddot{O}{:}(-H) + H^+ \longrightarrow R-\overset{+}{\ddot{O}}(-H)-H \qquad \text{Ox}\textit{onium}\text{ ion}$$

80. $\underset{\text{OH}}{\underset{|}{CH_2}}-\underset{\text{(OH)}}{\underset{|}{CH}}-\underset{\text{OH}}{\underset{|}{CH_2}}$

81.

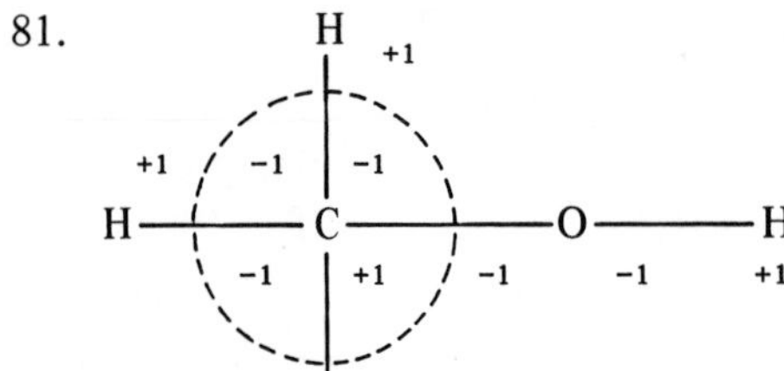


Polar number = -2

82.

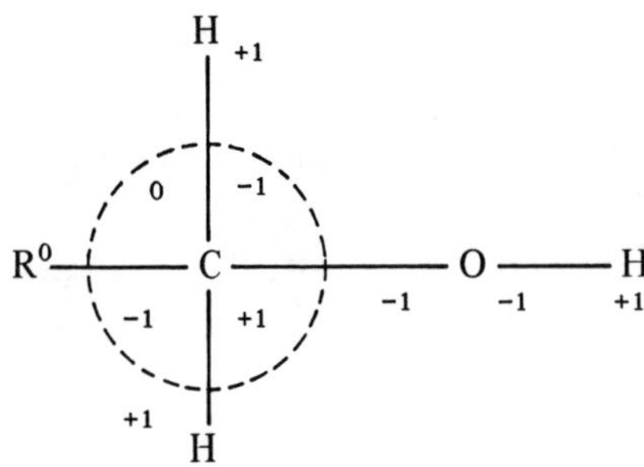


Polar number = -1

83. 0
84. +1
85. 0
86. -2
87. -4
88. 0
89. +1
90. +2

91.

oxidation (loss of 2 in polar number): (0) → (+2)

$$CH_3-\underset{\text{OH}}{\underset{|}{CH}}-CH_3 + KMnO_4 \longrightarrow CH_3-\underset{\text{O}}{\underset{\|}{C}}-CH_3 + MnO_2 + KOH + H_2O$$

(+7) → (+4): reduction (gain of 3 in polar number)

Equate oxidation and reduction:

$$3\ CH_3{-}\underset{\displaystyle OH}{\underset{|}{CH}}{-}CH_3 + 2\ KMnO_4 \rightarrow 3\ CH_3{-}\underset{\displaystyle O}{\underset{\|}{C}}{-}CH_3 + 2\ MnO_2 + KOH + H_2O$$

Balance remainder of equation:

$$3\ CH_3{-}\underset{\displaystyle OH}{\underset{|}{CH}}{-}CH_3 + 2\ KMnO_4 \rightarrow 3\ CH_3{-}\underset{\displaystyle O}{\underset{\|}{C}}{-}CH_3 + 2\ MnO_2 + 2\ KOH + 2\ H_2O$$

92.

oxidation (loss of 6 in polar number): (-3) → (+3)

$$C_6H_5CH_3 + KMnO_4 \longrightarrow C_6H_5CO_2K + MnO_2 + KOH + H_2O$$

(+7) → (+4)
reduction (gain of 3 in polar number)

Equate oxidation and reduction using simpliest ratio which is 2:1, not 6:3.

$$C_6H_5CH_3 + 2\ KMnO_4 \longrightarrow C_6H_5CO_2K + 2\ MnO_2 + KOH + H_2O$$

The remainder of the equation is also now balanced.

93.

$$HO{-}CH_2{-}\overset{\displaystyle CH_3}{\overset{|}{\underset{\displaystyle H}{\underset{|}{C}}}}{-}CH_2{-}CH_3 \overset{H^+}{\rightleftharpoons} H_2\overset{+}{O}{-}CH_2{-}\overset{\displaystyle CH_3}{\overset{|}{\underset{\displaystyle H}{\underset{|}{C}}}}{-}CH_2{-}CH_3$$

$\Big\updownarrow$ $-H_2O$

$$CH_3{-}\overset{\displaystyle CH_3}{\overset{|}{\underset{+}{C}}}{-}CH_2{-}CH_3 \rightleftharpoons {+}CH_2{-}\overset{\displaystyle CH_3}{\overset{|}{\underset{\displaystyle H}{\underset{|}{C}}}}{-}CH_2{-}CH_3$$

a tertiary carbonium ion (more stable) — a primary carbonium ion

$\Big\updownarrow$ $-H^+$

$$CH_3{-}\overset{\displaystyle CH_3}{\overset{|}{C}}{=}CH{-}CH_3$$

94. (a) $CH_3$–$C_6H_4$–OH (para-methylphenol)

$CH_3$: *a*; OH: *b*; *c* (aromatic hydrogens)

(b) $C_6H_5$–$CH_2$–OH

$CH_2$: *b*; OH: *a*; *c* (aromatic hydrogens)

(c) $C_6H_5$–O–$CH_3$

$CH_3$: *a*; *b* (aromatic hydrogens)

# 9

# *The Aldehydes and Ketones*

## INTRODUCTION

In the chapter on the alcohols and ethers we studied oxygen-containing compounds in which the oxygen atom was singly bonded to carbon. In this chapter we shall study oxygen-containing compounds in which the oxygen atom is doubly bonded to carbon.

The $\gt C{=}O$ structure, called a **carbonyl group**, is the functional group of aldehydes and ketones. Carbonyl compounds are very important in synthetic organic chemistry because of their great number and variety of reactions. The many reactions of the carbonyl compounds may be more easily understood and certainly better remembered if we think of the carbon–oxygen double bond in somewhat the same perspective as we did the carbon-carbon double bond. However, we must remember that the oxygen atom is more electronegative than carbon and has a greater attraction for the $\pi$ electrons. Because of this difference in electronegativity, the saturated aldehydes and ketones for the most part will give reactions in which the nucleophilic part of the reagent will add to the carbonyl carbon and the electrophilic part of the reagent will add to the oxygen atom of the carbonyl group. This is considered to be a normal or 1,2-addition. $\alpha,\beta$-Unsaturated aldehydes and ketones may undergo 1,2- (normal) or 1,4- (conjugate) addition or both.

Another point to remember: the attraction that the oxygen atom has for the $\pi$ electrons of the carbonyl group is induced through the carbonyl carbon to the adjacent (alpha) carbon atom. This effect weakens the C—H bonds of the alpha carbon and confers upon the attached hydrogen atoms a slight acidity. As a result, a number of reactions of carbonyl compounds, such as the aldol condensation and haloform reaction, involve the alpha hydrogens.

Finally, the chemical behavior that distinguishes the aldehydes from the ketones is largely due to structure. The aldehyde function is a terminal one. That is, it is always carbon number 1, and it still has one hydrogen atom bonded to it. As you study this chapter you will notice this makes an aldehyde a much more reactive species than a ketone.

## PROBLEMS

### True–False

1. T F Aldehydes and ketones are not capable of hydrogen bonding to themselves and therefore have lower boiling points than alcohols of approximately the same molecular weight.
2. T F The higher molecular weight aldehydes and ketones are very water soluble.
3. T F The higher molecular weight aldehydes and most ketones have fragrant odors.
4. T F Ketones, like aldehydes, are easily oxidized to carboxylic acids.
5. T F A reagent which is specific for the reduction of carbonyl groups to alcohols is sodium borohydride, $NaBH_4$.
6. T F Formaldehyde is a gas but is not usually handled in that form.
7. T F Bakelite is a phenol–formaldehyde copolymer.
8. T F Tollens' reagent is a strong oxidizing agent and will oxidize aldehydes and ketones to carboxylic acids.
9. T F Fehling's solution will oxidize aliphatic aldehydes, but not aromatic aldehydes.
10. T F Hydrogen atoms on carbon atoms alpha to the carbonyl group are acidic.

### Multiple Choice

11. Which product is formed by the oxidation of propionaldehyde?
    (a) acetic acid
    (b) formic and acetic acid
    (c) propionic acid
    (d) *n*-propyl alcohol

12. Which of the following reactions could be used for the synthesis of acetophenone,

$$C_6H_5-\overset{O}{\overset{\|}{C}}-CH_3?$$

(a)

$$C_6H_6 + CH_3-\overset{O}{\overset{\|}{C}}-O-\overset{O}{\overset{\|}{C}}-CH_3 \xrightarrow{AlCl_3}$$

(b) $C_6H_5-\overset{O}{\overset{\|}{C}}-Cl + CH_4 \xrightarrow{AlCl_3}$

(c) $C_6H_5-CH{=}CH_2 + O_2 \longrightarrow$

(d) $C_6H_5-\overset{O}{\overset{\|}{C}}-Cl + CH_3Cl \xrightarrow{AlCl_3}$

13. Which of the following, on oxidation, yields a compound of formula $C_4H_8O$ that gives a negative result with Fehling's or Tollens' reagent?

(a) $CH_3-CH_2-CH_2-CH_2-OH$

(b) $CH_3-CH_2-\underset{OH}{\underset{|}{CH}}-CH_3$

(c) $(CH_3)_2-CH-CH_2-OH$

(d) $CH_3-CH_2-O-CH_2-CH_3$

14. Which structure is most nearly correct for formaldehyde?

(a) $H-\overset{H}{\overset{|}{C}}{=}O$

(b) $H-\overset{O}{\overset{\|}{C}}-H$

(c) $\overset{O}{\overset{\|}{C}}$ with H and H bonded at angles below (planar)

(d) $O{=}C$ with H---- and H bonds (not planar)

15. Which of the following reagents would you use to convert acetophenone to isopropylmethylphenylcarbinol,

$$C_6H_5-\underset{OH}{\underset{|}{\overset{CH_3}{\overset{|}{C}}}}-\underset{H}{\underset{|}{\overset{CH_3}{\overset{|}{C}}}}-CH_3?$$

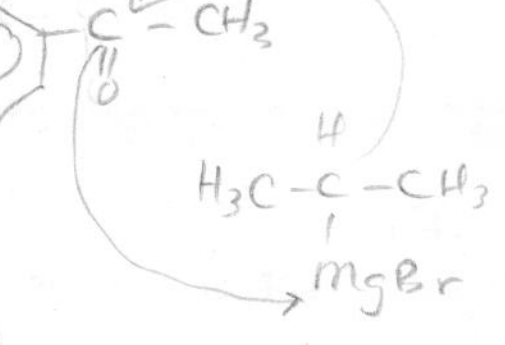

(a) $CH_3-CH_2-CH_2-Mg-Br$, followed by HX hydrolysis

(b) $CH_3-\underset{Br}{\underset{|}{CH}}-CH_3$, $AlCl_3$

(c) $CH_3-\underset{CH_3}{\underset{|}{CH}}-Mg-Br$, followed by HX hydrolysis

(d) $CH_3-\underset{\displaystyle OH}{\underset{|}{C}H}-CH_3$, Zn

16. An unknown compound gave a positive Tollens' test. Treatment of the unknown with iodine and sodium hydroxide gave a solid which was identified as iodoform ($CHI_3$). The unknown was which of the following compounds?

(a) $CH_3-CH_2-OH$

(b) $C_6H_5-\overset{\displaystyle O}{\overset{\|}{C}}-CH_3$

(c) $CH_3-CH_2-\underset{\displaystyle H}{\underset{|}{C}}=O$

(d) $CH_3-\underset{\displaystyle H}{\underset{|}{C}}=O$

(e) $CH_3-\underset{\displaystyle O}{\underset{\|}{C}}-CH_3$

**Completion**

17. Aldehydes and ketones are characterized by a carbon-oxygen double bond, $-\underset{|}{C}=O$. This group is called a ________________ group.

18. The only aldehyde that can be prepared by hydration of alkynes is ________________. Hydration of alkyl-substituted acetylenes give ________________.

19. An acyl group has the structure ________________ and is ________________ (activating, deactivating) when attached to the benzene nucleus.

20. A carbonyl group attached directly to the benzene nucleus directs electrophilic reagents to the ________________ position.

21. The molecular formula of lithium aluminum hydride is ________________.

22. Paraldehyde is a cyclic trimer of ________________.

23. A reaction in which two or more unlike monomers polymerize with each other is called ________________.

24. The formula for Tollens' reagent is ________________.

25. Reactions in which two reactants combine with the loss of a molecule of water (or alcohol) are called ________________.

26. Isomers of the following type are known as ________________.

$$R-\overset{\displaystyle OH}{\overset{|}{C}}=CH_2 \rightleftharpoons R-\overset{\displaystyle O}{\overset{\|}{C}}-CH_3$$

27. A tear-producing substance is called a ________________.

28. Biosynthesis of cell constitutents is called ____________________ and degradation is called ____________________.

**Matching** (Match each structure with its class of compound.)

(a) $R-\underset{\substack{|\\R'}}{C}=N-\underset{\substack{|\\H}}{N}-C_6H_5$ (phenyl ring)

(b) $R-\overset{\substack{H\\|}}{\underset{\substack{|\\OR'}}{C}}-OH$

(c) $R-\overset{\substack{R'\\|}}{\underset{\substack{|\\CN}}{C}}-OH$

(d) $R-\overset{\substack{R'\\|}}{\underset{\substack{|\\OR''}}{C}}-OR''$

(e) $R-\overset{\substack{H\\|}}{C}=N-\overset{\substack{H\\|}}{N}-C_6H_3(NO_2)_2$ (2,4-dinitrophenyl ring)

(f) $R-\underset{\substack{|\\R'}}{C}=N-\underset{\substack{|\\H}}{N}-\underset{\substack{\|\\O}}{C}-NH_2$

(g) $R-\underset{\substack{|\\R'}}{C}=N-OH$

(h) $R-\overset{\substack{R'\\|}}{\underset{\substack{|\\OH}}{C}}-SO_3^-\,Na^+$

(i) $R-\overset{\substack{H\\|}}{\underset{\substack{|\\OR'}}{C}}-OR'$

(j) $R-\underset{\substack{|\\H}}{C}=N-OH$

(k) $R-\overset{\substack{Cl\\|}}{\underset{\substack{|\\Cl}}{C}}-R'$

(l) $R-\overset{\substack{OH\\|}}{\underset{\substack{|\\H}}{C}}-OH$

(m) $R-\overset{\substack{OH\\|}}{C}=CH_2$

(n) $R-\overset{-}{C}H-\overset{+}{P}(C_6H_5)_3 \longleftrightarrow R-CH=P(C_6H_5)_3$

(o) $R-\overset{\substack{H\\|}}{C}=\overset{\substack{H\\|}}{C}-\overset{\substack{O\\\|}}{C}-R'$

(p) $\begin{array}{ccccc} & & H & & R' \\ & & | & & | \\ & R- & C & - & C-R'' \\ & & | & & | \\ (C_6H_5)_3 & - & P & - & O \end{array}$

29. ______ Cyanohydrin

30. ______ Enol

31. ______ Bisulfite addition product

32. ____ Aldehyde hydrate
33. ____ Acetal
34. ____ Ketal
35. ____ *gem*-Dihalide
36. ____ Hemiacetal
37. ____ Oxaphosphetane
38. ____ Phenylhydrazone
39. ____ Aldoxime
40. ____ Ylid
41. ____ Ketoxime
42. ____ 2,4-Dinitrophenylhydrazone
43. ____ $\alpha, \beta$-Unsaturated carbonyl
44. ____ Semicarbazone

(Identify each of the following as a "name" reaction.)

(a)

$$C_6H_6 + CH_3-CH=CH_2 \xrightarrow{AlCl_3} C_6H_5-\underset{CH_3}{\overset{H}{C}}-CH_3$$

(b)

$$2\ C_6H_5-\overset{H}{C}=O + NaOH \xrightarrow{heat} C_6H_5-\overset{O}{\overset{\|}{C}}-O^-Na^+ + C_6H_5-CH_2OH$$

(c)

$$C_6H_5-CH_3 + \underbrace{Cl-\underset{H}{C}=O}_{HCl\ +\ CO} \xrightarrow{AlCl_3} H-\overset{O}{\overset{\|}{C}}-C_6H_4-CH_3 + HCl$$

(d)

$$C_6H_6 + (CH_3-\overset{O}{\overset{\|}{C}})_2O \xrightarrow{AlCl_3} C_6H_5-\overset{O}{\overset{\|}{C}}-CH_3 + CH_3-\overset{O}{\overset{\|}{C}}-OH$$

(e) $2\ CH_3-CH_2-Cl + 2\ Na \longrightarrow CH_3-CH_2-CH_2-CH_3 + 2\ NaCl$

(f)

$$C_6H_5-\overset{O}{\overset{\|}{C}}-CH_3 + 3\,Cl_2 + 4\,NaOH \rightarrow C_6H_5-\overset{O}{\overset{\|}{C}}-O^-Na^+ + 3\,H_2O + 3\,NaCl + HCCl_3$$

(g)

$$C_6H_5-O^-\,Na^+ + CH_3I \rightarrow C_6H_5-O-CH_3 + NaI$$

(h)

$$(HOCH_2)_3-C-\underset{H}{\underset{|}{C}}=O + CH_2O + NaOH \rightarrow (HOCH_2)_4C + H-\underset{O}{\underset{\|}{C}}-O^-\,Na^+$$

(i)

$$C_6H_5-\overset{O}{\overset{\|}{C}}-CH_3 \xrightarrow[HCl]{Zn(Hg),} C_6H_5-CH_2-CH_3$$

(j)

$$2\,CH_3-CH_2-\overset{O}{\overset{\|}{C}}-H \xrightarrow{NaOH} CH_3-CH_2-\overset{H}{\underset{HO}{C}}-\overset{CH_3}{\underset{H}{C}}-\underset{H}{\underset{|}{C}}=O$$

(k)

$$CH_3-CH_2-\overset{O}{\overset{\|}{C}}-CH_2-CH_3 + (C_6H_5)_3P=CH-CH_2-CH_3 \longrightarrow$$

$$CH_3-CH_2-\underset{CH_3}{\underset{|}{\underset{CH_2}{\underset{|}{C}}}}=CH-CH_2-CH_3 + (C_6H_5)_3P=O$$

45. ______ Gatterman-Koch reaction
46. ______ Haloform reaction
47. ______ Aldol condensation
48. ______ Wittig reaction
49. ______ Cannizzaro reaction
50. ______ Crossed Cannizzaro reaction
51. ______ Clemmensen reduction
52. ______ Friedel-Crafts acylation
53. ______ Williamson ether synthesis

54. ______ Friedel-Crafts alkylation a

55. __e__ Wurtz reaction

The IUPAC system of nomenclature uses certain suffixes to designate specific functional groups. Give the class and characteristic suffix of each compound and name the compound by the IUPAC system (as done for the example).

| | Class | Suffix | Compound | Name |
|---|---|---|---|---|
| *Example* | Alkane | -ane | $CH_3-CH_3$ | Ethane |
| 56. | Alkene | -ene | $CH_2=CH_2$ | ethene |
| 57. | Alkynes | -yne | $HC\equiv CH$ | ethyne |
| 58. | Alcohol | -ol | $CH_3-CH_2-OH$ | ethanol |
| 59. | Alkanal | al | $CH_3-\overset{H}{\overset{\vert}{C}}=O$ | ethanal |
| 60. | Alkanone | one | $CH_3-\overset{O}{\overset{\Vert}{C}}-CH_3$ | propanone |

**Nomenclature** (Assign an acceptable name to each of the following.)

61. $C_6H_5-\overset{H}{\overset{\vert}{C}}=O$

benzaldehyde

62. $CH_3-CH=CH-\overset{H}{\overset{\vert}{C}}=O$

2-butenal

63. $CH_3-\overset{O}{\overset{\Vert}{C}}-CH_3$

dimethyl ketone

propanone

64. $C_6H_5-\overset{O}{\overset{\Vert}{C}}-C_6H_5$

diphenyl ketone

benzophenone

65. $H-\overset{H}{\overset{\vert}{C}}=O$

formaldehyde

methanal

66. $CH_3-\overset{H}{\overset{\vert}{C}}=O$

ethanal

acetaldehyde

67. $CH_3-\overset{O}{\overset{\Vert}{C}}-CH_2-\overset{CH_3}{\overset{\vert}{CH}}-CH_3$

4 methyl 2-pentanone

isobutyl methyl ketone

68. $H_2C-CH_2$, $H_2C$, $CH_2$, $C=O$ (ring)

cyclopentanone

69.

$$\text{C}_6\text{H}_5-\overset{\text{H}}{\underset{}{\text{C}}}=\overset{\text{H}}{\text{C}}-\overset{\text{H}}{\text{C}}=\text{O}$$

____________________

70.

$$\text{Cl}-\overset{\text{Cl}}{\underset{\text{Cl}}{\text{C}}}-\overset{\text{H}}{\text{C}}=\text{O}$$

____________________

**Reactions** (Complete the following:)

71. $\text{H}-\text{C}\equiv\text{C}-\text{H} + \text{Na} \xrightarrow[\text{NH}_3]{\text{liquid}}$ ( )

$C_2HNa$ → ($CH_3-CH_2-I$) → ( ) $C_4H_6$

$C_4H_6$ → ($H_2O$, $HgSO_4$, $H_2SO_4$) → ( ) $C_4H_8O$

$C_4H_8O$ → ($CH_3-MgBr$) → ( ) $C_5H_{11}OMgBr$

$C_4H_8O$ → (NaOH, $I_2$) → ( ) $C_3H_5O_2Na$ + ( ) $CHI_3$

$C_5H_{11}OMgBr$ → ($H_2O$, $H^+$) → ( ) $C_5H_{12}O$

$C_5H_{12}O$ → ($H_2SO_4$, heat) → ( ) $C_5H_{10}$

72.

$C_{10}H_{13}N_3O$

$H^+$, $H_2N-\overset{O}{\overset{\|}{C}}-NH-NH_2$

$C_9H_{12}O$

NaBH$_4$

Benzene + $CH_3-CH_2-\overset{O}{\overset{\|}{C}}-Cl$ $\xrightarrow{AlCl_3}$

$C_9H_{10}O$

Zn(Hg), HCl

CO, HCl, $AlCl_3$

$C_{10}H_{12}O$

$C_9H_{12}$

1. $CH_3-CH_2-\overset{O}{\overset{\|}{C}}-H$, $OH^-$
2. heat

Phenyl$-NH-NH_2$

$C_{16}H_{18}N_2$

$C_{13}H_{16}O$

$Ag(NH_3)_2OH$

$C_{13}H_{19}NO_2$

73.

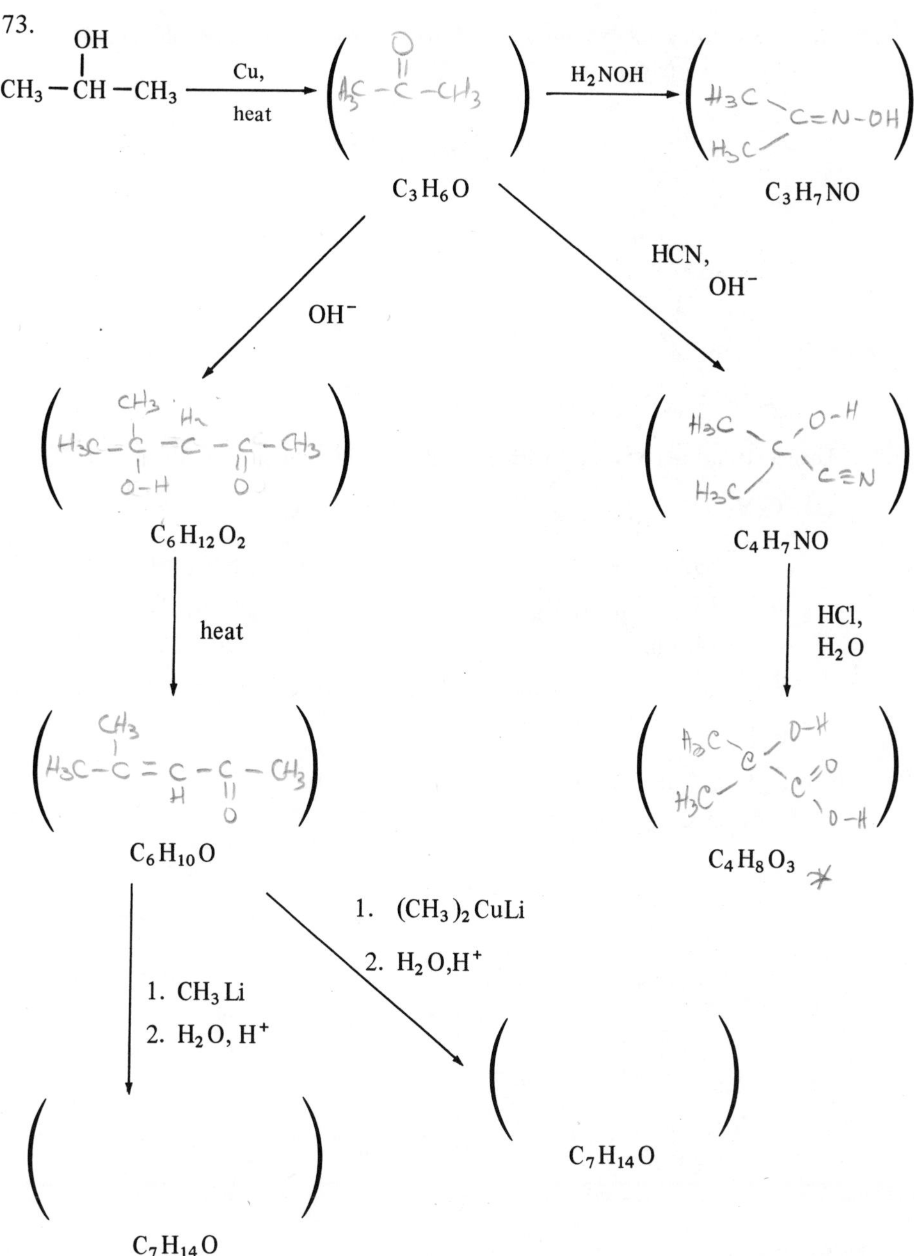
OH
$CH_3-CH-CH_3$
Cu,
heat
$C_3H_6O$
$H_2NOH$
$C_3H_7NO$
HCN,
$OH^-$
$OH^-$
$C_6H_{12}O_2$
$C_4H_7NO$
heat
HCl,
$H_2O$
$C_6H_{10}O$
$C_4H_8O_3$
1. $(CH_3)_2CuLi$
2. $H_2O,H^+$
1. $CH_3Li$
2. $H_2O, H^+$
$C_7H_{14}O$
$C_7H_{14}O$

74. Starting with cyclohexanone, show how the Wittig reaction could be used to prepare methylenecyclohexane

$=CH_2$

75. Give a structure consistent with each of the following sets of nmr data.

(a) $C_8H_8O_2$

*a* δ 3.9, singlet, 3 H
*b* δ 7.3, multiplet, 4H
*c* δ 9.8, singlet, 1H

(b) $C_5H_{10}O$

*a* δ 0.9, triplet, 3H
*b* δ 1.6, sextet, 2H
*c* δ 2.1, singlet, 3H
*d* δ 2.4, triplet, 2H

(c) $C_5H_{10}O$

*a* δ 1.1, triplet, 6H
*b* δ 2.5, quartet, 4H

## ANSWERS

### True–False

1. True. Aldehydes and ketones do not have hydrogen atoms attached to highly electronegative atoms (i.e., fluorine, oxygen, nitrogen) and therefore are not capable of hydrogen bonding to themselves.
2. False. Those containing more than four carbon atoms are practically insoluble in water.

3. True
4. False. Aldehydes can be oxidized to carboxylic acids by mild oxidizing agents, but ketones are fairly resistant to oxidation because carbon-carbon bonds must be ruptured.
5. True
6. True. Formaldehyde is marketed as formalin, an aqueous solution, or the solid polymer, paraformaldehyde.
7. True
8. False. Tollens' reagent, $Ag(NH_3)_2OH$, is a weak oxidizing agent and will only oxidize aldehydes to carboxylic acids. This test is used to distinguish between aldehydes and ketones.
9. True. This is a method of distinguishing aliphatic aldehydes from aromatic aldehydes.
10. True. This is the basis of such reactions as the aldol condensation and the haloform reaction. Acetone is not as strong an acid as water, but is considerably stronger than acetylene.

**Multiple Choice**

11. (c)

$$CH_3-CH_2-\overset{\overset{\Large H}{|}}{C}=O \xrightarrow{\text{oxidation}} CH_3-CH_2-\overset{\overset{\Large O}{\|}}{C}-OH$$

12. (a)
13. (b)
14. (c) The carbon atom is $sp^2$ hybridized resulting in a planar molecule with bond angles of approximately 120°.
15. (c)

$$C_6H_5-\overset{\overset{\Large O}{\|}}{C}-CH_3 + CH_3-\overset{\overset{\Large CH_3}{|}}{CH}-Mg-Br \rightarrow C_6H_5-\underset{\underset{\underset{\Large MgBr}{|}}{\underset{O}{|}}}{\overset{\overset{\Large CH_3}{|}}{C}}-\underset{\underset{\Large CH_3}{|}}{\overset{\overset{\Large H}{|}}{C}}-CH_3$$

$$\downarrow H_2O,\ HX$$

$$C_6H_5-\underset{\underset{\Large OH}{|}}{\overset{\overset{\Large CH_3}{|}}{C}}-\underset{\underset{\Large H}{|}}{\overset{\overset{\Large CH_3}{|}}{C}}-CH_3$$

16. (d) All except propionaldehyde, (c), would produce iodoform on reaction with iodine and sodium hydroxide, the haloform reaction. Only the aldehydes would also give a positive Tollens' test. The only compound that would give both reactions is acetaldehyde.

**Completion**

17. carbonyl
18. acetaldehyde, ketones
19. $R-\overset{\overset{\large O}{\|}}{C}-$ , (deactivating)
20. *meta*
21. $LiAlH_4$
22. acetaldehyde
23. copolymerization
24. $Ag(NH_3)_2OH$
25. condensations
26. tautomers
27. lachrymator
28. anabolism, catabolism

**Matching**

29. (c)
30. (m)
31. (h)
32. (l)
33. (i)
34. (d)
35. (k)
36. (b)
37. (p)
38. (a)
39. (j)
40. (n)
41. (g)
42. (e)
43. (o)
44. (f)
45. (c)
46. (f)
47. (j)
48. (k)
49. (b)
50. (h)
51. (i)
52. (d)
53. (g)
54. (a)
55. (e)

| | Class | Suffix | Compound | Name |
|---|---|---|---|---|
| *Example* | Alkane | -ane | $CH_3-CH_3$ | Ethane |
| 56. | Alkene | -ene | $CH_2=CH_2$ | Ethene |
| 57. | Alkyne | -yne | $HC\equiv CH$ | Ethyne |
| 58. | Alcohol | -ol | $CH_3-CH_2-OH$ | Ethanol |
| 59. | Aldehyde | -al | $CH_3-\overset{\overset{\large H}{\vert}}{C}=O$ | Ethanal |
| 60. | Ketone | -one | $CH_3-\overset{\overset{\large O}{\|}}{C}-CH_3$ | Propanone |

**Nomenclature**

61. benzaldehyde
62. crotonaldehyde or 2-butenal
63. acetone, dimethyl ketone, or propanone
64. benzophenone or diphenyl ketone
65. formaldehyde or methanal
66. acetaldehyde or ethanal

67. 4-methyl-2-pentanone or isobutyl methyl ketone
68. cyclopentanone
69. cinnamaldehyde or 3-phenyl-2-propenal
70. trichloroacetaldehyde or chloral

## Reactions

71.

$$H-C\equiv C-H + Na \xrightarrow{\text{liq. } NH_3} \left( H-C\equiv C\colon^- Na^+ \right) \quad C_2HNa$$

$$\xrightarrow{CH_3-CH_2-I} \left( H-C\equiv C-CH_2-CH_3 \right) \quad C_4H_6$$

$$\xrightarrow{H_2O,\ HgSO_4,\ H_2SO_4} \left( CH_3-\underset{\underset{O}{\|}}{C}-CH_2-CH_3 \right) \quad C_4H_8O$$

$$C_4H_8O \xrightarrow{CH_3-MgBr} \left( CH_3-\underset{\underset{OMgBr}{|}}{\overset{\overset{CH_3}{|}}{C}}-CH_2-CH_3 \right) \quad C_5H_{11}OMgBr$$

$$C_4H_8O \xrightarrow{NaOH,\ I_2} \left( CH_3-CH_2-\overset{\overset{O}{\|}}{C}-O^- Na^+ \right) + \left( CHI_3 \right) \quad C_3H_5O_2Na \quad CHI_3$$

$$C_5H_{11}OMgBr \xrightarrow{H_2O,\ H^+} \left( CH_3-\underset{\underset{OH}{|}}{\overset{\overset{CH_3}{|}}{C}}-CH_2-CH_3 \right) \quad C_5H_{12}O$$

$$\xrightarrow[\text{heat}]{H_2SO_4,} \left( CH_3-\overset{\overset{CH_3}{|}}{C}=CH-CH_3 \right) \quad C_5H_{10}$$

72.

$C_6H_6 + CH_3-CH_2-C(=O)-Cl \xrightarrow{AlCl_3} C_6H_5-C(=O)-CH_2-CH_3$ ($C_9H_{10}O$)

$C_9H_{10}O \xrightarrow{H_2N-C(=O)-NH-NH_2,\ H^+} C_6H_5-C(=N-NH-C(=O)-NH_2)-CH_2-CH_3$ ($C_{10}H_{13}N_3O$)

$C_9H_{10}O \xrightarrow{NaBH_4} C_6H_5-CH(OH)-CH_2-CH_3$ ($C_9H_{12}O$)

$C_9H_{10}O \xrightarrow{Zn(Hg),\ HCl} C_6H_5-CH_2-CH_2-CH_3$ ($C_9H_{12}$)

$C_9H_{12} \xrightarrow{CO,\ HCl,\ AlCl_3} O=CH-C_6H_4-CH_2-CH_2-CH_3$ ($C_{10}H_{12}O$)

$C_{10}H_{12}O \xrightarrow{C_6H_5-NH-NH_2} C_6H_5-NH-N=CH-C_6H_4-CH_2-CH_2-CH_3$ ($C_{16}H_{18}N_2$)

$C_{10}H_{12}O \xrightarrow{1.\ CH_3-CH_2-C(=O)-H,\ OH^-;\ 2.\ heat} O=CH-C(CH_3)=CH-C_6H_4-CH_2-CH_2-CH_3$ ($C_{13}H_{16}O$)

$C_{13}H_{16}O \xrightarrow{Ag(NH_3)_2OH} NH_4^+\ {}^-O-C(=O)-C(CH_3)=CH-C_6H_4-CH_2-CH_2-CH_3$ ($C_{13}H_{19}NO_2$)

73. $CH_3-CH(OH)-CH_3 \xrightarrow[\text{heat}]{\text{Cu,}} \left( CH_3-\overset{O}{\overset{\|}{C}}-CH_3 \right) \xrightarrow{H_2NOH} \left( CH_3-\overset{N-OH}{\overset{\|}{C}}-CH_3 \right)$

$C_3H_6O$ $\qquad$ $C_3H_7NO$

$C_3H_6O \xrightarrow{OH^-} \left( CH_3-C(CH_3)(OH)-CH_2-\overset{O}{\overset{\|}{C}}-CH_3 \right)$ $C_6H_{12}O_2$

$C_3H_6O \xrightarrow{HCN,\ OH^-} \left( CH_3-C(OH)(C\equiv N)-CH_3 \right)$ $C_4H_7NO$

$C_6H_{12}O_2 \xrightarrow{\text{heat}} \left( CH_3-C(CH_3)=CH-\overset{O}{\overset{\|}{C}}-CH_3 \right)$ $C_6H_{10}O$

$C_4H_7NO \xrightarrow{HCl,\ H_2O} \left( CH_3-C(CH_3)(OH)-CO_2H \right)$ $C_4H_8O_3$

$C_6H_{10}O \xrightarrow{1.\ CH_3Li;\ 2.\ H_2O,\ H^+} \left( CH_3-C(CH_3)=CH-C(OH)(CH_3)-CH_3 \right)$

$C_7H_{14}O$

1,2-addition

$C_6H_{10}O \xrightarrow{1.\ (CH_3)_2CuLi;\ 2.\ H_2O,\ H^+} \left( CH_3-C(CH_3)(CH_3)-CH_2-\overset{O}{\overset{\|}{C}}-CH_3 \right)$

$C_7H_{14}O$

1,4-addition

74. $(C_6H_5)_3P + CH_3Br \longrightarrow CH_3{-}\overset{+}{P}(C_6H_5)_3\ Br^-$

$$CH_3{-}\overset{+}{P}(C_6H_5)_3\ Br^- + CH_3CH_2CH_2CH_2Li \longrightarrow CH_3CH_2CH_2CH_3 + LiBr +$$

$$^-CH_2{-}\overset{+}{P}(C_6H_5)_3 \longleftrightarrow CH_2{=}P(C_6H_5)_3$$

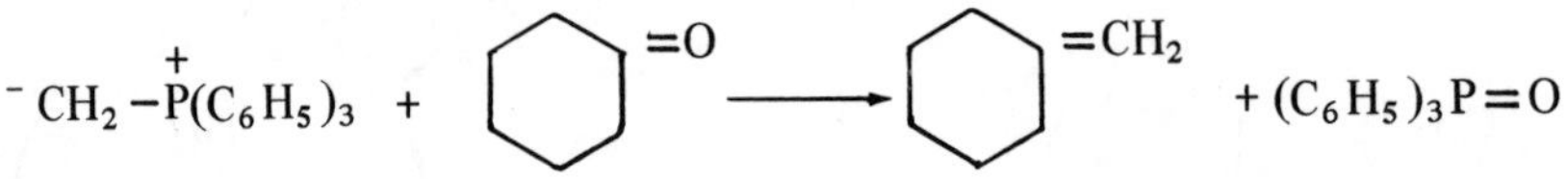

$$^-CH_2{-}\overset{+}{P}(C_6H_5)_3 + \text{cyclohexanone } (=O) \longrightarrow \text{methylenecyclohexane } (=CH_2) + (C_6H_5)_3P{=}O$$

75. (a)

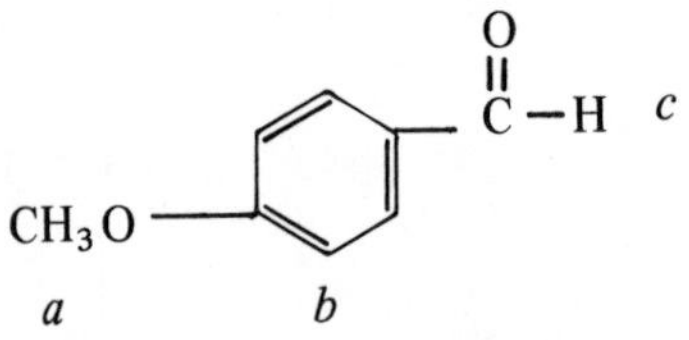

$$CH_3O{-}C_6H_4{-}\overset{O}{\overset{\|}{C}}{-}H$$

$CH_3O$: *a*; ring: *b* (aromatic hydrogens); CHO: *c*

(b)

$$CH_3{-}\overset{O}{\overset{\|}{C}}{-}CH_2{-}CH_2{-}CH_3$$

*c*    *d*    *b*    *a*

(c)

$$CH_3{-}CH_2{-}\overset{O}{\overset{\|}{C}}{-}CH_2{-}CH_3$$

*a*    *b*    *b*    *a*

# 10

# The Carboxylic Acids and Their Derivatives

## INTRODUCTION

In this chapter we shall study the properties of a functional group which contains two oxygen atoms. The **carboxyl** group, $-C\begin{smallmatrix}\nearrow O \\ \searrow OH\end{smallmatrix}$, is a combination of a hydroxyl and a carbonyl group. We will find, however, that the chemical properties of the carboxyl group are unique and not typical of alcohols, aldehydes, or ketones. Inasmuch as the preparations and reactions of the carboxylic acid derivatives are closely interrelated, they also will be discussed in this chapter. The carboxylic acid derivatives that we shall consider in this chapter are: **acyl halides, esters, amides,** and **anhydrides.**

## PROBLEMS

**True–False**

1. T F "Fatty" acids are carboxylic acids of the aliphatic series.
2. T F Carboxylic acids are also known as mineral acids.
3. T F Acetic acid is a stronger acid than hydrochloric acid.
4. T F The presence of an electron withdrawing group on the $\alpha$-carbon atom of a carboxylic acid increases its acid strength.

5. T F 2-Chlorobutanoic acid and 3-chlorobutanoic acid have ionization constants which are practically identical.
6. T F Lower molecular weight carboxylic acids have sweet, perfume-like odors.
7. T F Carboxylic acids have high boiling points because of hydrogen bond formation between acid molecules.
8. T F The neutralization equivalent of a monoprotic acid is the same as its molecular weight.
9. T F Acid halides are extremely reactive compounds.
10. T F Acetic anhydride is by far the most important acid anhydride.
11. T F A number of esters are responsible for the odor of certain fruits and are used in artificial flavorings.
12. T F Sodium benzoate is more soluble in water than is benzoic acid.

**Multiple Choice**

13. Which compound represented by the following is not acidic and does not yield an acid when hydrolyzed?

(a) $CH_3-C(=O)-O-CH_3$

(b) $CH_3-C(=O)-OH$

(c) $CH_3-C(CH_3)_2-OH$

(d) $C_6H_5-OH$

(e) $CH_3-\overset{O}{\overset{\|}{C}}-O-\overset{O}{\overset{\|}{C}}-CH_3$

14. The Grignard reagent, $CH_3-CH_2-MgBr$, can be used in the preparation of
   (a) ethane
   (b) propionic acid
   (c) 3-ethyl-3-pentanol
   (d) all of these
   (e) none of these

15. Which is the strongest acid?
   (a) $CH_3-CO_2H$
   (b) $ClCH_2-CO_2H$
   (c) $Cl_2CH-CO_2H$
   (d) $Cl_3C-CO_2H$

16. Which reaction does not yield an ester as one of the products?

*(cont'd)*

(a) a carboxylic acid is heated with an alcohol in the presence of a mineral acid
(b) a Grignard reagent is added to a carboxylic acid
(c) an acyl halide is treated with an alcohol
(d) an acid anhydride is treated with an alcohol
(e) an alkyl halide is heated with the salt of a carboxylic acid

17. An acceptable name for $CH_3{-}CH_2{-}\underset{\displaystyle Br}{\underset{|}{C}H}{-}CH_2{-}CO_2H$ is

(a) γ-bromovaleric acid
(b) β-bromopentanoic acid
(c) 3-bromopentanoic acid
(d) 2-bromopentanoic acid

18. The product obtained when acetic acid is treated with phosphorus trichloride is

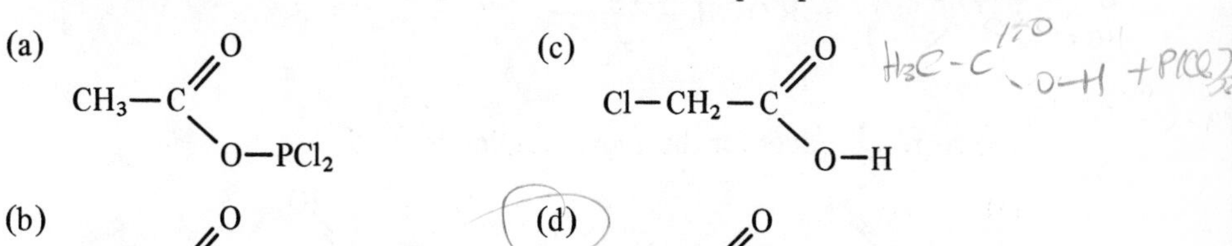

(a) $CH_3{-}C({=}O){-}O{-}PCl_2$
(c) $Cl{-}CH_2{-}C({=}O){-}O{-}H$
(b) $CH_3{-}C({=}O){-}O{-}Cl$
(d) $CH_3{-}C({=}O){-}Cl$

19. Which compound has the highest boiling point?

(a) $CH_3{-}CH_3$
(b) $CH_3{-}CH_2{-}OH$
(c) $CH_3{-}O{-}CH_3$
(d) $CH_3{-}C({=}O){-}OH$

20. The neutralization equivalent of trimesic acid, $HO_2C{-}C_6H_3(CO_2H){-}CO_2H$, is

(a) 210
(b) 105
(c) 70
(d) 630

21. Which reaction is an example of a Fischer esterification?

(a)

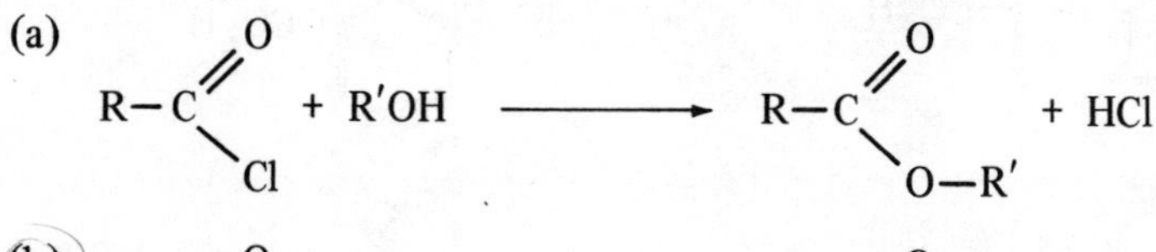

$$R{-}C({=}O){-}Cl + R'OH \longrightarrow R{-}C({=}O){-}O{-}R' + HCl$$

(b)

$$R{-}C({=}O){-}OH + R'OH \underset{}{\overset{H^+}{\rightleftharpoons}} R{-}C({=}O){-}OR' + H_2O$$

(c)

$$\begin{array}{l}R-C(=O)\\ \quad\;\; O\\ R-C(=O)\end{array} + R'OH \longrightarrow R-C(=O)-OR' + R-C(=O)-OH$$

(d)

$$R-C(=O)-O^- Ag^+ + R'I \longrightarrow R-C(=O)-OR' + AgI \downarrow$$

22.

$$R-C(=O)-OH + R'O^{18}H \xrightarrow{H^+} ?$$

The correct product for the above reaction is

(a) $R-C(=O)-OR' + H_2O^{18}$

(b) $R-C(=O)-O^{18}R' + H_2O$

(c) $R'-C(=O)-OR + H_2O^{18}$

(d) $R'-C(=O)-O^{18}R + H_2O$

(e) $R-C(=O^{18})-OR' + H_2O$

23. Which acid is weaker than benzoic?

(a) $CH_3-C_6H_4-CO_2H$ (para)

(b) $Cl-C_6H_4-CO_2H$ (para)

(c) $O_2N-C_6H_4-CO_2H$ (para)

(d) $C_6H_4(CO_2H)(Cl)$ (ortho)

**Completion**

24. The functional group $-C(=O)-OH$ is known as a carboxyl group.

25. The stronger the acid, the ______________ (larger, smaller) the $K_a$ and the ______________ (larger, smaller) the $pK_a$.
26. The weight of an acid in grams that is required to neutralize one gram equivalent weight of base is the ______________ ______________ of the acid.
27. Alkali metal salts of long chain fatty acids are called ______________.
28. The reaction of an acyl halide with an alcohol produces an ______________.
29. The reaction of an acyl halide with ammonia produces an ______________.
30. Acyl halides react with salts of acids to yield ______________.
31. Alkaline hydrolysis of an ester is called ______________.
32. The sour principle of vinegar is ______________.
33. Structural isomers in rapid equilibrium, such as the keto and enol forms below, are known as ______________.

$$CH_3-\overset{O}{\overset{\|}{C}}-CH_2-\overset{O}{\overset{\|}{C}}-OCH_3 \rightleftharpoons H_3C-\overset{O-H\cdots}{\overset{|}{C}}=\underset{H}{C}-\overset{\cdots O}{\overset{\|}{C}}-OCH_3$$

**Matching** (Match the common name and the IUPAC name with the proper formula.)

| Common Names | IUPAC Names |
|---|---|
| (a) Acetic acid | (s) Octanoic acid |
| (b) Caproic acid | (t) Pentanoic acid |
| (c) Propionic acid | (u) Methanoic acid |
| (d) Capric acid | (v) Hexanoic acid |
| (e) Formic acid | (w) Decanoic acid |
| (f) *n*-Butyric acid | (x) Ethanoic acid |
| (g) Caprylic acid | (y) Butanoic acid |
| (h) *n*-Valeric acid | (z) Propanoic acid |

| | Common | IUPAC |
|---|---|---|
| 34. $CH_3-CH_2-CH_2-CO_2H$ | ____ | ____ |
| 35. $CH_3-(CH_2)_8-CO_2H$ | ____ | ____ |
| 36. $CH_3-CO_2H$ | ____ | ____ |
| 37. $CH_3-(CH_2)_4-CO_2H$ | ____ | ____ |
| 38. $CH_3-(CH_2)_3-CO_2H$ | ____ | ____ |
| 39. $H-CO_2H$ | ____ | ____ |

| | Common | IUPAC |
|---|---|---|
| 40. $CH_3—CH_2—CO_2H$ | c | |
| 41. $CH_3—(CH_2)_6—CO_2H$ | g | |

**Structural Formulas** (Write the correct structural formula for each of the following compounds.)

42. Salicylic Acid

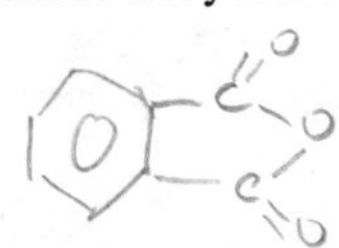

43. Mandelic acid

44. Phthalic anhydride

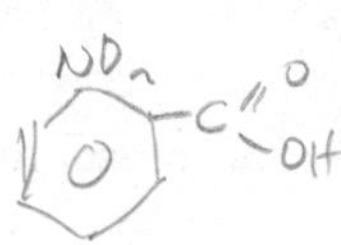

45. Ethyl formate

46. 2-Nitrobenzoic acid

47. α-Phenylpropionic acid

48. 2-Bromobutanoic acid

49. Ammonium valerate

50. Benzonitrile

51. Butyryl chloride

52. 2-Methylpropanamide

**Reactions** (Complete the following.)

53.

$$CH_3-CH_2-CH_2-C(=O)-O-MgCl$$

$$CH_3-CH_2-CH_2-CH_2-OH$$

HCl

$$CH_3-CH_3-CH_2-C(=O)OH$$

$H_3O^+$, heat

$$CH_3-CH_2-CH_2-C\equiv N$$

KCN

54.

$$C_6H_5-CO_2H \xrightarrow{NaOH} (\quad) \xrightarrow{HCl} (\quad)$$

55.

$CH_3-C(=O)-O-C(=O)-CH_3$ —( $H_2O$ )→ $CH_3-C(=O)OH$ ←( $H_2O$ )— $CH_3-C(=O)Cl$

$CH_3-C(=O)-O-C(=O)-CH_3$ —( $H_3COH$ )→ $CH_3-C(=O)OCH_3$ ←( $H_3COH$ )— $CH_3-C(=O)Cl$

$CH_3-C(=O)-O-C(=O)-CH_3$ —( $NH_3$ )→ $CH_3-C(=O)NH_2$ ←( $2NH_3$ )— $CH_3-C(=O)Cl$

$CH_3-C(=O)-O-C(=O)-CH_3$ —( benzene ), $AlCl_3$ → $C_6H_5-C(=O)-CH_3$ ← ( benzene ), $AlCl_3$ — $CH_3-C(=O)Cl$

56.

$CH_3-CH_2-C(=O)-O-C(=O)-CH_2-CH_3$ —$NH_3$→ ( $H_3C-CH_2-C(=O)-NH_2$ ) $C_3H_7NO$

$C_3H_7NO$ —NaOH→ ( $H_3C-CH_2-C(=O)ONa$ ) $C_3H_5O_2Na$

$C_3H_7NO$ —$P_4O_{10}$, heat→ ( $H_3C-CH_2-C\equiv N$ ) $C_3H_5N$

$CH_3-CH_2-C(=O)-O-C(=O)-CH_2-CH_3$ —$H_2O$→ ( $H_3C-CH_2-C(=O)OH$ ) $C_3H_6O_2$

$C_3H_5O_2Na$ —HCl→ $C_3H_6O_2$

$C_3H_5N$ —$H_3O^+$, heat→ $C_3H_6O_2$

57.

$CH_3-CH(CH_3)-OH$ → $CH_3-CH(CH_3)-Cl$ $\xrightarrow{KCN}$ ( ) $C_4H_7N$

$C_4H_7N$ $\xrightarrow{H_3O^+,\ heat}$ ( ) $C_4H_8O_2$

$C_4H_8O_2$ $\xrightarrow{SOCl_2}$ $CH_3-CH(CH_3)-C(=O)Cl$

$CH_3-CH(CH_3)-C(=O)Cl$ $\xrightarrow{CH_3-CH_2-OH}$ ( ) $C_6H_{12}O_2$

$C_6H_{12}O_2$ $\xrightarrow{NaOH}$ ( ) $C_4H_7O_2Na$

$C_6H_{12}O_2$ $\xrightarrow{NH_3}$ ( ) $C_4H_9NO$

$C_6H_{12}O_2$ $\xrightarrow{1.\ 2\ CH_3MgCl\quad 2.\ HCl}$ ( ) $C_6H_{14}O$

$C_6H_{12}O_2$ $\xrightarrow{1.\ LiAlH_4\quad 2.\ HCl}$ ( ) $C_4H_{10}O$ + ( ) $C_2H_6O$

58.

( )

$C_6H_{12}O_2$

$CH_3-\underset{CH_3}{CH}-OH$, $H^+$ (from $CH_3-CH_2-COOH$)

$CH_3-CH-OH$ with $CH_3$ on CH (from $C_3H_5OCl$)

( )

$C_3H_5O_2Na$

NaOH

$CH_3-CH_2-C(=O)-OH$

PCl₃ → ( )

$C_3H_5OCl$

$CH_3-CH_2-C(=O)-Cl$

NH₃

( )

$C_6H_{10}O_3$

$NH_4OH$

( )

$C_3H_7NO$

heat

( )

$C_3H_9NO_2$

59. $H{-}\overset{\overset{\displaystyle O}{\|}}{C}{-}H + CH_3MgBr \xrightarrow[\text{hydrolysis}]{\text{then}}$

60. $CH_3{-}CH_2{-}\overset{\overset{\displaystyle O}{\|}}{C}{-}H + CH_3MgBr \xrightarrow[\text{hydrolysis}]{\text{then}}$

61. $CH_3{-}CH_2{-}\overset{\overset{\displaystyle O}{\|}}{C}{-}CH_2{-}CH_3 + CH_3MgBr \xrightarrow[\text{hydrolysis}]{\text{then}}$

62. $CH_3{-}CH_2{-}\overset{\overset{\displaystyle O}{\|}}{C}{-}CH_2{-}CH_3 + (CH_3)_2Cd \xrightarrow[\text{hydrolysis}]{\text{then}}$

63. $CH_3{-}CH_2{-}\overset{\overset{\displaystyle O}{\|}}{C}{-}CH_2{-}CH_3 + CH_3Li \xrightarrow[\text{hydrolysis}]{\text{then}}$

64. $CH_3{-}CH_2{-}\overset{\overset{\displaystyle O}{\|}}{C}{-}OH + CH_3MgBr \xrightarrow[\text{hydrolysis}]{\text{then}}$

65. $CH_3{-}CH_2{-}\overset{\overset{\displaystyle O}{\|}}{C}{-}OH + CH_3Li \xrightarrow[\text{hydrolysis}]{\text{then}}$

66. $CH_3{-}CH_2{-}\overset{\overset{\displaystyle O}{\|}}{C}{-}Cl + (CH_3)_2Cd \xrightarrow[\text{hydrolysis}]{\text{then}}$

67. $CH_3{-}CH_2{-}\overset{\overset{\displaystyle O}{\|}}{C}{-}O{-}CH_2{-}CH_3 + CH_3MgBr \xrightarrow[\text{hydrolysis}]{\text{then}}$

**Miscellaneous**

68. The $K_a$ of butyric acid in $H_2O$ at 25° is $1.5 \times 10^{-5}$. What is the p$K_a$? What is the pH of the solution when $[HA] = [A^-]$?

69. Using simple chemical tests, how would you distinguish between the following?

cyclohexanol (cyclohexane ring–OH) — phenol (benzene ring–OH) — benzoic acid (benzene ring–$CO_2H$)

**Sets 70-72** (Draw a circle around the strongest acid in each pair.)

70. $C_6H_5-CO_2H$ $\qquad$ $Cl-C_6H_4-CO_2H$

71. $ClCH_2-CO_2H$ $\qquad$ $I-CH_2-CO_2H$

72. $Cl-CH_2-CH_2-CO_2H$ $\qquad$ $CH_3-CH(Cl)-CO_2H$

(Give the class of compound and an acceptable name for each of the following.)

| | Compound | Class of Compound | Name |
|---|---|---|---|
| 73. | $CH_3-C(=O)-OH$ | acetic acid | |
| 74. | $CH_3-C(=O)-O^-Na^+$ | | |
| 75. | $CH_3-C\equiv N$ | | |
| 76. | $CH_3-C(=O)-Cl$ | | |
| 77. | $CH_3-C(=O)-O-C(=O)-CH_3$ | | |
| 78. | $CH_3-C(=O)-O-CH_3$ | | |
| 79. | $CH_3-C(=O)-NH_2$ | | |

80. A neutral organic compound (A) of molecular formula $C_6H_{12}O_2$ was saponified to give compounds (B) and (C). After acidification of the reaction mixture, an acid (B) was isolated and found to have a neutralization equivalent of 88. (C) gave a positive iodoform test and reacted with Lucas reagent to produce an alkyl halide after heating for several hours. What two structures are possible for (A)?

**Questions 81-84** (Write an acceptable mechanism.)

81.

$$CH_3-CH_2-\overset{\overset{\displaystyle O}{\|}}{C}-H + CH_3NH_2 \xrightarrow[\text{acidic}]{\text{slightly}} CH_3-CH_2-\overset{\overset{\displaystyle H}{|}}{C}=N-CH_3 + H_2O$$

82.

$$CH_3-CH_2-\overset{\overset{\displaystyle O}{\|}}{C}-O-CH_2-CH_3 + CH_3NH_2 \xrightarrow[\text{acidic}]{\text{slightly}} CH_3-CH_2-\overset{\overset{\displaystyle O}{\|}}{C}-\overset{\overset{\displaystyle H}{|}}{N}-CH_3 + CH_3-CH_2-OH$$

83.

$$2\,CH_3-CH_2-\overset{\overset{\displaystyle O}{\|}}{C}-H \xrightarrow{\text{NaOH}} CH_3-CH_2-\overset{\overset{\displaystyle H}{|}}{C}=\overset{\overset{\displaystyle CH_3}{|}}{C}-\underset{\underset{\displaystyle O}{\|}}{C}-H + H_2O$$

84.

$$2\,CH_3-CH_2-\overset{\overset{\large O}{\|}}{C}-O-CH_2-CH_3 \xrightarrow{NaOCH_2CH_3} CH_3-CH_2-\overset{\overset{\large O}{\|}}{C}-\overset{\overset{\large CH_3}{|}}{CH}-\underset{\underset{\large O}{\|}}{C}-O-CH_2-CH_3 + CH_3CH_2OH$$

85. Give a structure consistent with each of the following sets of nmr data.

(a) $C_8H_8O_2$

*a* δ 3.6, singlet, 2H
*b* δ 7.2, singlet, 5H
*c* δ 11.9, singlet, 1H

(b) $C_8H_9NO$

*a* δ 2.3, singlet, 3H
*b* δ 6.5, broad, 2H
*c* δ 7.4, multiplet, 4H

(c) $C_2H_3OCl$

*a* δ 2.7, singlet, 3H

(d) $C_6H_{10}O_3$

*a* δ 1.2, triplet, 6H
*b* δ 2.4, quartet, 4H

(e) $C_9H_{10}O_2$

*a* δ 2.1, singlet, 3H
*b* δ 5.1, singlet, 2H
*c* δ 7.3, singlet, 5H

## ANSWERS

### True–False

1. True
2. False. Mineral acids are inorganic acids such as sulfuric, hydrochloric, and nitric.
3. False. All of the carboxylic acids are weaker than hydrochloric acid.
4. True
5. False. The inductive effect of the chlorine atom is lessened as more $-CH_2-$ groups are interposed between the chlorine atom and the carboxyl group.

   Examples:

| Acid | $K_a$ |
|---|---|
| 2-chlorobutanoic acid | $1.45 \times 10^{-3}$ |
| 3-chlorobutanoic acid | $8.8 \times 10^{-5}$ |
| 4-chlorobutanoic acid | $3.0 \times 10^{-5}$ |
| butanoic acid | $1.51 \times 10^{-5}$ |

6. False. With the exception of acetic acid the odors of the lower molecular weight carboxylic acids are very disagreeable.
7. True
8. True
9. True
10. True
11. True. Most esters have sweet, pleasant odors.
12. True. Sodium benzoate is a salt and ionizes in solution to give sodium cations and benzoate anions.

### Multiple Choice

13. (c) An alcohol is a weaker acid than water.
14. (d) $CH_3-CH_2-MgBr + H_2O \longrightarrow$ ethane

$$CH_3-CH_2-MgBr \xrightarrow[2.\ H_2O,\ H^+]{1.\ CO_2} \text{propionic acid}$$

$$CH_3-CH_2-MgBr \xrightarrow[2.\ H_2O,\ H^+]{1.\ CH_3-CH_2-\overset{O}{\overset{\|}{C}}-CH_2-CH_3} \text{3-ethyl-3-pentanol}$$

15. (d)
16. (b) Carboxylic acids react with Grignard reagents to liberate the hydrocarbon corresponding to the alkyl group of the reagent and to form the magnesium halide salt of the acid.

17. (c) The common name would be $\beta$-bromovaleric acid. The $\alpha$-carbon atom is the carbon atom adjacent to the carboxyl group in the common naming system. The carbon atom of the carboxyl group is carbon number 1 in systematic nomenclature.
18. (d)
19. (d) The hydrogen bonding between carboxylic acid molecules is more intense than that between alcohol molecules.
20. (c) The neutralization equivalent of an acid is the same as its equivalent weight. The molecular weight of the acid is 210 and, since there are three acid functions, the equivalent weight is 70.
21. (b) All of the reactions are valid preparations of esters, however.
22. (b) Cleavage occurs between the carbon and the oxygen of the carboxyl group.
23. (a) Ring-activating substituents, when in the *para* position, decrease the acidity of benzoic acid.

**Completion**

24. carboxyl
25. larger, smaller
26. neutralization equivalent
27. soaps
28. ester
29. amide
30. acid anhydrides
31. saponification
32. acetic acid
33. tautomers

**Matching**

34. (f,y)
35. (d,w)
36. (a,x)
37. (b,v)
38. (h,t)
39. (e,u)
40. (c,z)
41. (g,s)

**Structural Formulas**

42. $CO_2H$ / OH (on benzene ring)

43. OH / $CH-CO_2H$ (on benzene ring)

44. O, C, O, C, O (anhydride ring fused to benzene ring)

45. $H-\overset{O}{\overset{\|}{C}}-O-CH_2-CH_3$

46. (benzene ring with $CO_2H$ and $NO_2$ substituents)

47. (benzene ring with $\overset{CH_3}{|}CH-CO_2H$ substituent)

48. $CH_3-CH_2-\underset{Br}{\underset{|}{CH}}-CO_2H$

49. $CH_3-CH_2-CH_2-CH_2-\overset{O}{\overset{||}{C}}-O^- NH_4^+$

50. (benzene ring with $C\equiv N$ substituent)

51. $CH_3-CH_2-CH_2-\overset{O}{\overset{||}{C}}-Cl$

52. $CH_3-\underset{CH_3}{\underset{|}{CH}}-\overset{O}{\overset{||}{C}}-NH_2$

**Reactions**

53.

$\left(CH_3-CH_2-CH_2-MgCl\right)$

$\xrightarrow{(CO_2)}$ $CH_3-CH_2-CH_2-C(=O)-O-MgCl$

$\xrightarrow{HCl}$ $CH_3-CH_2-CH_2-C(=O)OH$

$CH_3-CH_2-CH_2-CH_2-OH$ $\xrightarrow{\left(K_2Cr_2O_7,\ H_2SO_4,\ heat\right)}$ $CH_3-CH_2-CH_2-C(=O)OH$

$\left(CH_3-CH_2-CH_2-Cl\right)$ $\xrightarrow{KCN}$ $CH_3-CH_2-CH_2-C\equiv N$ $\xrightarrow{H_3O^+,\ heat}$ $CH_3-CH_2-CH_2-C(=O)OH$

54.

$C_6H_5CO_2H \xrightarrow{NaOH} (C_6H_5CO_2^-\,Na^+) \xrightarrow{HCl} (C_6H_5CO_2H)$

55.

$(CH_3CO)_2O \xrightarrow{(H_2O)} CH_3COOH \xleftarrow{(H_2O)} CH_3COCl$

$(CH_3CO)_2O \xrightarrow{(CH_3OH)} CH_3COOCH_3 \xleftarrow{(CH_3OH)} CH_3COCl$

$(CH_3CO)_2O \xrightarrow{(NH_3)} CH_3CONH_2 \xleftarrow{(NH_3)} CH_3COCl$

$(CH_3CO)_2O \xrightarrow[AlCl_3]{(C_6H_6)} C_6H_5COCH_3 \xleftarrow[AlCl_3]{(C_6H_6)} CH_3COCl$

56.

$(CH_3CH_2CO)_2O \xrightarrow{NH_3} (CH_3-CH_2-CONH_2)$, $C_3H_7NO$

$C_3H_7NO \xrightarrow{NaOH} (CH_3-CH_2-COO^-\,Na^+)$, $C_3H_5O_2Na$

$C_3H_7NO \xrightarrow{P_4O_{10},\ heat} (CH_3-CH_2-C{\equiv}N)$, $C_3H_5N$

$(CH_3CH_2CO)_2O \xrightarrow{H_2O} (CH_3-CH_2-COOH)$, $C_3H_6O_2$

$C_3H_5O_2Na \xrightarrow{HCl} (CH_3-CH_2-COOH)$, $C_3H_6O_2$

$C_3H_5N \xrightarrow{H_3O^+,\ heat} (CH_3-CH_2-COOH)$, $C_3H_6O_2$

57.

$CH_3-CH(CH_3)-OH$ $\xrightarrow{(PCl_3,\ PCl_5,\ \text{or}\ SOCl_2)}$ $CH_3-CH(CH_3)-Cl$ $\xrightarrow{KCN}$ $(CH_3-CH(CH_3)-C{\equiv}N)$ $C_4H_7N$

$\xrightarrow{H_3O^+,\ \text{heat}}$ $(CH_3-CH(CH_3)-CO_2H)$ $C_4H_8O_2$

$\xrightarrow{SOCl_2}$ $CH_3-CH(CH_3)-C(=O)Cl$

$\xrightarrow{CH_3-CH_2-OH}$ $(CH_3-CH(CH_3)-C(=O)-O-CH_2-CH_3)$ $C_6H_{12}O_2$

$\xrightarrow{NaOH}$ $(CH_3-CH(CH_3)-C(=O)-O^-\ Na^+)$ $C_4H_7O_2Na$

$\xrightarrow{NH_3}$ $(CH_3-CH(CH_3)-C(=O)-NH_2)$ $C_4H_9NO$

$\xrightarrow{1.\ 2\ CH_3MgCl;\ 2.\ HCl}$ $(CH_3-CH(CH_3)-C(OH)(CH_3)-CH_3)$ $C_6H_{14}O$

$\xrightarrow{1.\ LiAlH_4;\ 2.\ HCl}$ $(CH_3-CH(CH_3)-CH_2-OH)$ $C_4H_{10}O$ + $(CH_3-CH_2-OH)$ $C_2H_6O$

58.

Starting compound (center): $CH_3-CH_2-C(=O)-OH$

- NaOH → $CH_3-CH_2-C(=O)-O^- Na^+$ ($C_3H_5O_2Na$)
- $C_3H_5O_2Na$ + $CH_3-CH_2-C(=O)-Cl$ → $CH_3-CH_2-C(=O)-O-C(=O)-CH_2-CH_3$ ($C_6H_{10}O_3$)
- $CH_3-CH(CH_3)-OH$, $H^+$ → $CH_3-CH_2-C(=O)-O-CH(CH_3)-CH_3$ ($C_6H_{12}O_2$)
- $PCl_3$ → $CH_3-CH_2-C(=O)-Cl$ ($C_3H_5OCl$)
- $C_3H_5OCl$ + $CH_3-CH(CH_3)-OH$ → $CH_3-CH_2-C(=O)-O-CH(CH_3)-CH_3$ ($C_6H_{12}O_2$)
- $C_3H_5OCl$ + $NH_3$ → $CH_3-CH_2-C(=O)-NH_2$ ($C_3H_7NO$)
- $NH_4OH$ → $CH_3-CH_2-C(=O)-O^- NH_4^+$ ($C_3H_9NO_2$)
- $C_3H_9NO_2$, heat → $CH_3-CH_2-C(=O)-NH_2$ ($C_3H_7NO$)

59. $$H-C(=O)-H + CH_3MgBr \xrightarrow[\text{hydrolysis}]{\text{then}} CH_3-CH_2-OH$$

60. $$CH_3-CH_2-C(=O)-H + CH_3MgBr \xrightarrow[\text{hydrolysis}]{\text{then}} CH_3-CH_2-CH(CH_3)-OH$$

61. $$CH_3-CH_2-C(=O)-CH_2-CH_3 + CH_3MgBr \xrightarrow[\text{hydrolysis}]{\text{then}} CH_3-CH_2-C(OH)(CH_3)-CH_2-CH_3$$

62. $CH_3-CH_2-\overset{O}{\overset{\|}{C}}-CH_2-CH_3 + (CH_3)_2Cd \xrightarrow[\text{hydrolysis}]{\text{then}}$ no reaction

63. $CH_3-CH_2-\overset{O}{\overset{\|}{C}}-CH_2-CH_3 + CH_3Li \xrightarrow[\text{hydrolysis}]{\text{then}} CH_3-CH_2-\underset{CH_3}{\overset{OH}{C}}-CH_2-CH_3$

64. $CH_3-CH_2-\overset{O}{\overset{\|}{C}}-OH + CH_3MgBr \xrightarrow[\text{hydrolysis}]{\text{then}} CH_3-CH_2-\overset{O}{\overset{\|}{C}}-OMgBr + CH_4$

65. $CH_3-CH_2-\overset{O}{\overset{\|}{C}}-OH + CH_3Li \xrightarrow[\text{hydrolysis}]{\text{then}} CH_3-CH_2-\overset{O}{\overset{\|}{C}}-CH_3$

66. $CH_3-CH_2-\overset{O}{\overset{\|}{C}}-Cl + (CH_3)_2Cd \xrightarrow[\text{hydrolysis}]{\text{then}} CH_3-CH_2-\overset{O}{\overset{\|}{C}}-CH_3$

67. $CH_3-CH_2-\overset{O}{\overset{\|}{C}}-O-CH_2-CH_3 + CH_3MgBr \xrightarrow[\text{hydrolysis}]{\text{then}} CH_3-CH_2-\underset{CH_3}{\overset{OH}{C}}-CH_3$

**Miscellaneous**

68. $pK_a = -\log K_a$

$pK_a = -\log(1.5 \times 10^{-5})$

$pK_a = -(\log 1.5)-(\log 10^{-5})$

$pK_a = -0.18-(-5)$

$pK_a = 4.82$

$pK_a = pH_{half\text{-}neutralization}$

At half-neutralization, $[HA] = [A^-]$. Thus, the pH = 4.82.

69. $C_6H_5-CO_2H + \text{aqueous } NaHCO_3 \longrightarrow C_6H_5-CO_2^-\ Na^+$

(soluble in $H_2O$)

$C_6H_5-OH + \text{aqueous } NaOH \longrightarrow C_6H_5-O^-\ Na^+$

(soluble in $H_2O$)

Benzoic acid is a stronger acid than phenol. Phenol is not acidic enough to react with sodium bicarbonate, which is a weaker base than sodium hydroxide. Cyclohexanol is not acidic enough to react with either base.

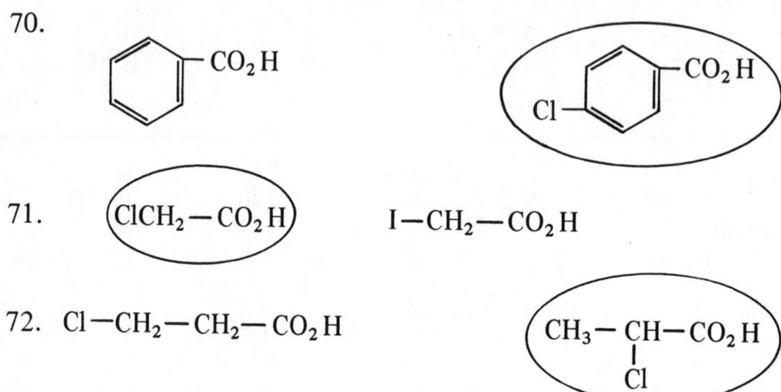


| | Compound | Class of Compound | Name |
|---|---|---|---|
| 73. | $CH_3-C(=O)-OH$ | carboxylic acid | acetic acid or ethanoic acid |
| 74. | $CH_3-C(=O)-O^-\ Na^+$ | salt of a carboxylic acid | sodium acetate or sodium ethanoate |
| 75. | $CH_3-C\equiv N$ | nitrile or alkyl cyanide | acetonitrile or ethanonitrile or methyl cyanide |
| 76. | $CH_3-C(=O)-Cl$ | acid halide or acyl halide | acetyl chloride or ethanoyl chloride |
| 77. | $CH_3-C(=O)-O-C(=O)-CH_3$ | acid anhydride | acetic anhydride or ethanoic anhydride |
| 78. | $CH_3-C(=O)-O-CH_3$ | ester | methyl acetate or methyl ethanoate |
| 79. | $CH_3-C(=O)-NH_2$ | amide | acetamide or ethanamide |

80. (B) is an acid with a neutralization equivalent of 88. The weight of a $—CO_2H$ group is 45. Therefore 88-45=43 or the weight of the alkyl group, R, attached to the carboxyl group.

$$\underbrace{\underset{15}{CH_3}—\underset{14}{CH_2}—\underset{14}{CH_2}}_{43} \quad + \quad \underbrace{—C(=O)OH}_{45} \quad = \quad 88$$

or

$$\underbrace{CH_3—\overset{H}{\underset{CH_3}{C}}}_{43} \; + \; \underbrace{—C(=O)OH}_{45} \; = \; 88$$

(B) must therefore be *n*-butyric acid, $CH_3—CH_2—CH_2—CO_2H$, or isobutyric acid,

$$CH_3—\overset{CH_3}{\overset{|}{CH}}—CO_2H$$

(C) must be primary alcohol since it reacts very slowly with Lucas reagent. The iodoform test is positive. The only primary alcohol that is capable of a positive iodoform reaction is ethyl alcohol.

The ester (A) could be either ethyl butyrate,

$$CH_3—CH_2—CH_2—\overset{O}{\overset{\|}{C}}—O—CH_2—CH_3$$

or ethyl isobutyrate,

$$CH_3—\overset{CH_3}{\overset{|}{CH}}—\underset{O}{\underset{\|}{C}}—O—CH_2—CH_3$$

Both compounds agree with the molecular formula.

81.

$$CH_3-CH_2-\overset{O}{\overset{\|}{C}}-H \underset{}{\overset{+H^+}{\rightleftharpoons}} CH_3-CH_2-\overset{+OH}{\overset{\|}{C}}-H \overset{CH_3NH_2}{\rightleftharpoons} CH_3-CH_2-\underset{+NH_2CH_3}{\overset{OH}{C}}-H$$

$$\overset{-H^+}{\rightleftharpoons}\ CH_3-CH_2-\underset{NHCH_3}{\overset{OH}{C}}-H \overset{+H^+}{\rightleftharpoons} CH_3-CH_2-\underset{NHCH_3}{\overset{+OH_2}{C}}-H \overset{-H_2O}{\rightleftharpoons} CH_3-CH_2-\underset{+NHCH_3}{\underset{\|}{C}}-H$$

$$\overset{-H^+}{\rightleftharpoons}\ CH_3-CH_2-\underset{NCH_3}{\underset{\|}{C}}-H$$

82.

$$CH_3-CH_2-\overset{O}{\overset{\|}{C}}-O-CH_2-CH_3 \overset{+H^+}{\rightleftharpoons} CH_3-CH_2-\overset{+OH}{\overset{\|}{C}}-O-CH_2-CH_3$$

$$\overset{CH_3NH_2}{\rightleftharpoons}\ CH_3-CH_2-\underset{+NH_2CH_3}{\overset{OH}{C}}-O-CH_2-CH_3 \overset{-H^+}{\rightleftharpoons} CH_3-CH_2-\underset{NHCH_3}{\overset{OH}{C}}-O-CH_2-CH_3$$

$$\overset{+H^+}{\rightleftharpoons}\ CH_3-CH_2-\underset{CH_3HN}{\overset{HO}{C}}-\underset{+}{\overset{H}{O}}-CH_2-CH_3 \overset{-CH_3CH_2OH}{\rightleftharpoons} CH_3-CH_2-\overset{+OH}{\overset{\|}{C}}-NHCH_3$$

$$\overset{-H^+}{\rightleftharpoons}\ CH_3-CH_2-\overset{O}{\overset{\|}{C}}-\overset{H}{N}-CH_3$$

83.

$$CH_3-CH_2-\overset{O}{\overset{\|}{C}}-H + OH^- \overset{-H_2O}{\rightleftharpoons} \left[CH_3-\underline{C}H-\overset{O}{\overset{\|}{C}}-H \longleftrightarrow CH_3-CH=\overset{O^-}{C}-H\right]$$

$$\overset{CH_3-CH_3-\overset{O}{\overset{\|}{C}}-H}{\rightleftharpoons}\ CH_3-CH_2-\underset{H}{\overset{O^-}{C}}-\underset{CH_3}{CH}-\overset{O}{\overset{\|}{C}}-H \overset{+H_2O}{\rightleftharpoons} OH^- + CH_3-CH_2-\underset{H}{\overset{OH}{C}}-\underset{CH_3}{CH}-\overset{O}{\overset{\|}{C}}-H$$

Aldols readily lose water when heated

$$\longrightarrow CH_3-CH_2-CH=\underset{CH_3}{C}-\overset{O}{\overset{\|}{C}}-H$$

84.

$$CH_3-CH_2-\overset{O}{\overset{\|}{C}}-OCH_2CH_3 + {}^{-}OCH_2CH_3 \xrightleftharpoons{-CH_3CH_2OH} \left[ CH_3-\underset{-}{C}H-\overset{O}{\overset{\|}{C}}-OCH_2CH_3 \longleftrightarrow CH_3-CH=\overset{O^-}{\overset{|}{C}}-OCH_2CH_3 \right]$$

$$\xrightleftharpoons{CH_3-CH_2-\overset{O}{\overset{\|}{C}}-OCH_2CH_3} CH_3-CH_2-\underset{OCH_2CH_3}{\underset{|}{\overset{^{-}O}{\overset{|}{C}}}}-\overset{CH_3}{\overset{|}{C}H}-\overset{O}{\overset{/\!/}{C}}-OCH_2CH_3 \xrightleftharpoons{-CH_3CH_2O^-} CH_3-CH_2-\overset{O}{\overset{\|}{C}}-\underset{CH_3}{\underset{|}{C}H}-\overset{O}{\overset{\|}{C}}-OCH_2CH_3$$

85. (a)

$C_6H_5-CH_2-CO_2H$, with $CH_2$ labelled *a*, $CO_2H$ labelled *c*, and ring labelled *b* (aromatic hydrogens)

(b)

$m$-$CH_3C_6H_4-\overset{O}{\overset{\|}{C}}-NH_2$, with $NH_2$ labelled *b*, ring labelled *c* (aromatic hydrogens), and $CH_3$ labelled *a*

(c)

$$CH_3-\overset{O}{\overset{\|}{C}}-Cl$$

*a*

(d)

$$(CH_3-CH_2-\overset{O}{\overset{\|}{C}})_2O$$

*a* *b*

(e)

$C_6H_5-CH_2-O-\overset{O}{\overset{\|}{C}}-CH_3$, with $CH_2$ labelled *b*, $CH_3$ labelled *a*, and ring labelled *c* (aromatic hydrogens)

# 11

# *Fats, Oils, Waxes; Soap and Detergents; Prostaglandins*

## INTRODUCTION

The organic matter of living cells is made up largely of **lipids**, **carbohydrates** (Chapter 14), and **proteins** (Chapter 15). This chapter is devoted primarily to the saponifiable lipids derived from the fatty acids: the fats, oils, and waxes. Waxes are esters of long-chain, unbranched fatty acids and long-chain alcohols. Fats and oils also are esters of fatty acids, but the alcohol in both fats and oils is glycerol (1,2,3-trihydroxypropane). Because of this common feature fats and oils may be called **glycerides**. Fats are solid at room temperature and oils are liquid. The reason for these differences in physical properties is due largely to differences in the structures of the fatty acids that make up the glyceride. The fatty acids in oils are largely unsaturated; those in fats are, for the most part, saturated.

We shall study the chemical reactions of fats and oils and see how such useful products as soaps, margarines, and detergents are prepared from them. We shall learn that the digestion of fats in the animal system involves a series of remarkable enzyme catalyzed chemical changes, each of which proceeds either by a 2-carbon degradation or a 2-carbon increase in the carbon chain length. These biochemical changes answer the question: Why do the natural long-chain fatty acids contain an even number of carbon atoms?

**Prostaglandins**, important biochemical regulators derived from the fatty acids, also will be studied briefly.

## PROBLEMS

### True–False

1. T F Lipids are plant or animal products, characterized by their insolubility in water and their ready solubility in water-immiscible organic solvents such as ether.
2. T F The number of calories liberated in the oxidation of a fat or oil is more than twice the amount liberated by an equal weight of protein or carbohydrate.
3. T F Fats and oils differ from waxes in that fats and oils are esters of glycerol.
4. T F Natural fats and oils are usually mixed glycerides.
5. T F The presence of unsaturation in the acid component of a fat tends to lower its melting point.
6. T F Fats are solids, but oils are liquids.
7. T F Oils are largely of vegetable origin.
8. T F The most important unsaturated fatty acids obtainable by the hydrolysis of oils are the $C_{18}$ acids.
9. T F The iodine value for triolein is zero.
10. T F Heavy metal salts of long chain fatty acids are insoluble in water.
11. T F Vegetable oils may be converted into semisolid cooking fats by a hydrogenation process known as *hardening*.
12. T F The mechanics of cleansing with soaps or with syndets is the same.
13. T F The addition of antioxidants to fats and oils retards rancidity.
14. T F Synthetic detergents are insoluble in hard water.
15. T F Fats, when oxidized in the body to carbon dioxide and water, liberate large amounts of energy that are used to perform body functions.
16. T F The body can synthesize fatty acids by modifying dietary fats or synthesizing them from proteins or carbohydrates. However, the "essential" fatty acids cannot be made and must be supplied in the diet.

### Multiple Choice

17. Which of the following compounds would not be classified as lipids?
    (a) Fats
    (b) Soaps
    (c) Waxes
    (d) Oils

18. Which compound would be classified as a wax?
    (a)
    $$CH_3-CH_2-\overset{\overset{\displaystyle O}{\|}}{C}-O-CH_2-CH_2-CH_3$$

(b)

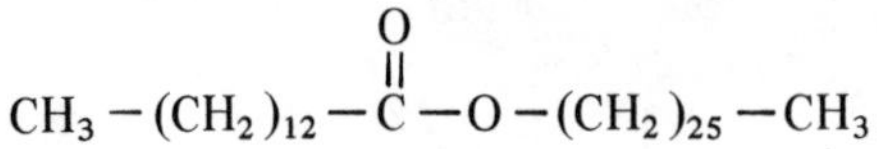

$$CH_3-(CH_2)_{12}-\overset{\overset{\displaystyle O}{\|}}{C}-O-(CH_2)_{25}-CH_3$$

(c)

$$\begin{array}{l} CH_3-(CH_2)_{14}-\overset{\overset{\displaystyle O}{\|}}{C}-O-CH_2 \\ CH_3-(CH_2)_{14}-\overset{\overset{\displaystyle O}{\|}}{C}-O-CH \\ CH_3-(CH_2)_{14}-\overset{\overset{\displaystyle O}{\|}}{C}-O-CH_2 \end{array}$$

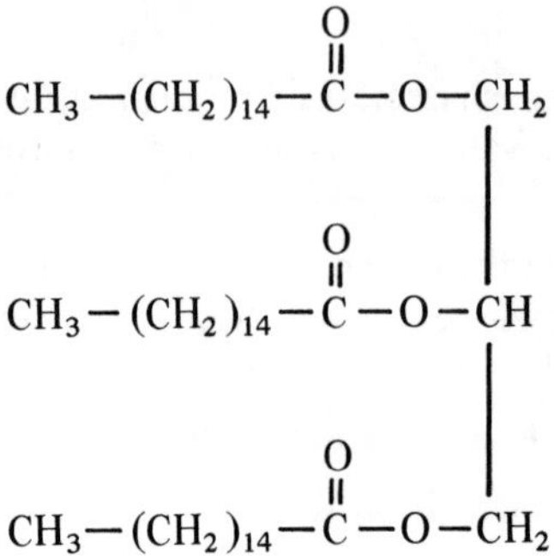

(d)

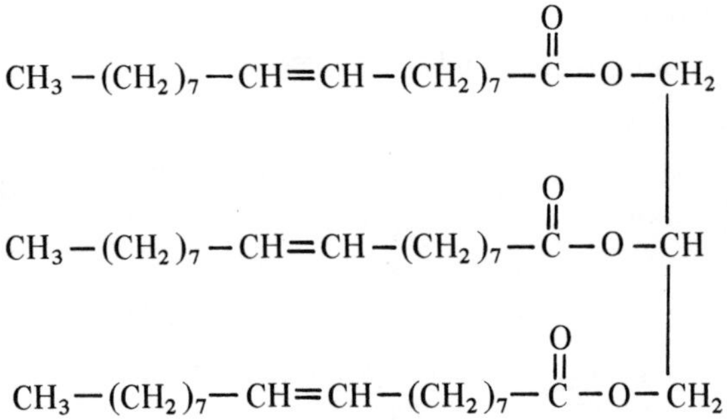

$$\begin{array}{l} CH_3-(CH_2)_7-CH{=}CH-(CH_2)_7-\overset{\overset{\displaystyle O}{\|}}{C}-O-CH_2 \\ CH_3-(CH_2)_7-CH{=}CH-(CH_2)_7-\overset{\overset{\displaystyle O}{\|}}{C}-O-CH \\ CH_3-(CH_2)_7-CH{=}CH-(CH_2)_7-\overset{\overset{\displaystyle O}{\|}}{C}-O-CH_2 \end{array}$$

(e)

$$\begin{array}{r} CH_3-(CH_2)_7-CH{=}CH-(CH_2)_7-\overset{\overset{\displaystyle O}{\|}}{C}-O-CH_2 \\ CH_3-(CH_2)_{14}-\overset{\overset{\displaystyle O}{\|}}{C}-O-CH \\ CH_3-(CH_2)_{16}-\overset{\overset{\displaystyle O}{\|}}{C}-O-CH_2 \end{array}$$

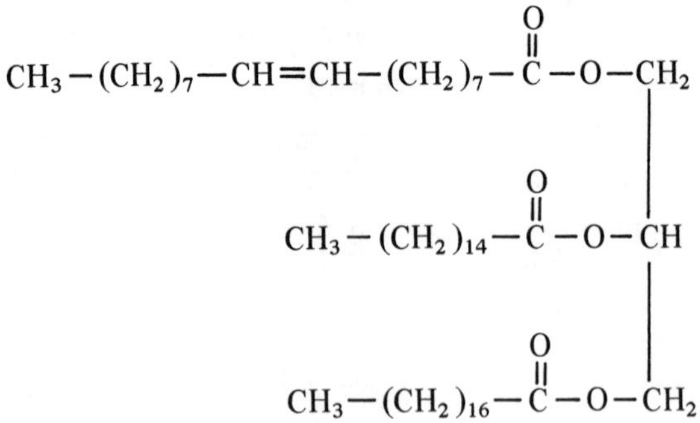

19. Which compound is a simple glyceride?

(a)

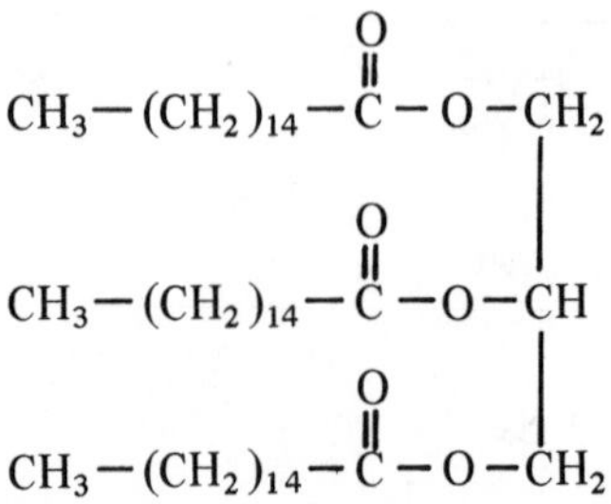

$$\begin{array}{l} CH_3-(CH_2)_{14}-\overset{\overset{\displaystyle O}{\|}}{C}-O-CH_2 \\ CH_3-(CH_2)_{14}-\overset{\overset{\displaystyle O}{\|}}{C}-O-CH \\ CH_3-(CH_2)_{14}-\overset{\overset{\displaystyle O}{\|}}{C}-O-CH_2 \end{array}$$

(b)

$$\begin{array}{l} CH_3-(CH_2)_{14}-\overset{\overset{\displaystyle O}{\|}}{C}-O-CH_2 \\ CH_3-(CH_2)_{16}-\overset{\overset{\displaystyle O}{\|}}{C}-O-CH \\ CH_3-(CH_2)_{14}-\overset{\overset{\displaystyle O}{\|}}{C}-O-CH_2 \end{array}$$

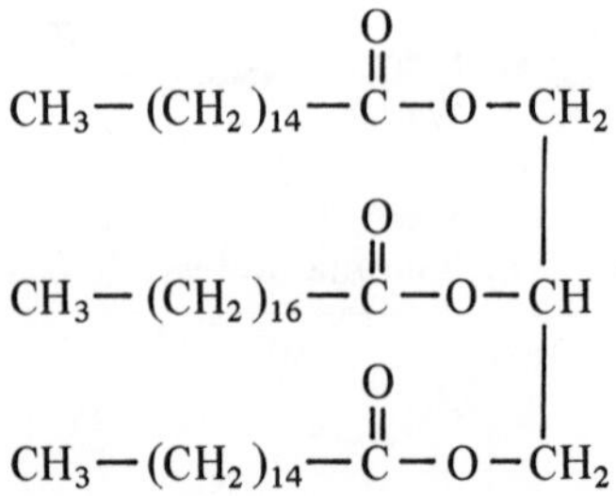

(c)

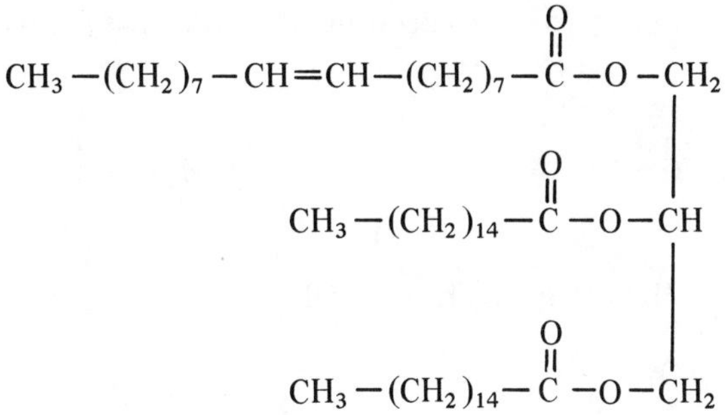

$$\begin{array}{l} CH_3-(CH_2)_7-CH{=}CH-(CH_2)_7-\overset{O}{\overset{\|}{C}}-O-CH_2 \\ \qquad\qquad\qquad\qquad CH_3-(CH_2)_{14}-\overset{O}{\overset{\|}{C}}-O-CH \\ \qquad\qquad\qquad\qquad CH_3-(CH_2)_{14}-\overset{O}{\overset{\|}{C}}-O-CH_2 \end{array}$$

(d)

$$\begin{array}{l} CH_3-(CH_2)_{14}-O-\overset{O}{\overset{\|}{C}}-CH_2 \\ CH_3-(CH_2)_{14}-O-\overset{O}{\overset{\|}{C}}-CH \\ CH_3-(CH_2)_{14}-O-\overset{O}{\overset{\|}{C}}-CH_2 \end{array}$$

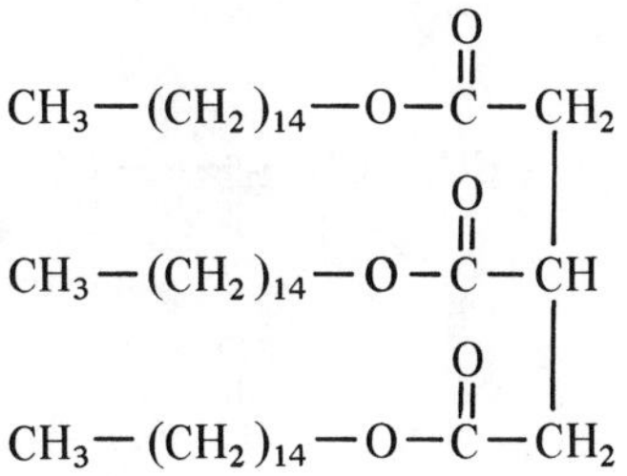

20. Which lipid would be classified as an oil?

(a)

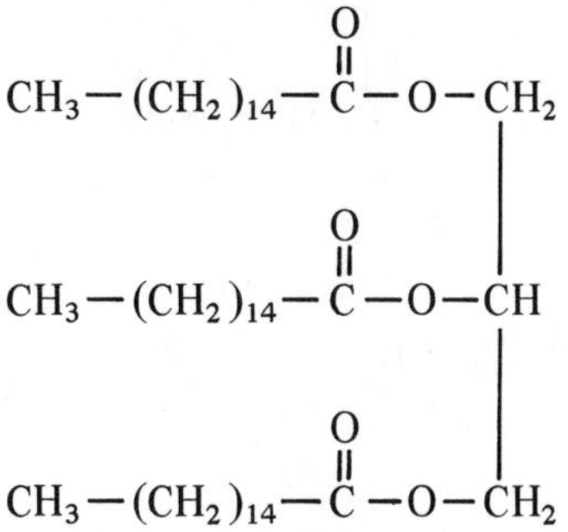

$$\begin{array}{l} CH_3-(CH_2)_{14}-\overset{O}{\overset{\|}{C}}-O-CH_2 \\ CH_3-(CH_2)_{14}-\overset{O}{\overset{\|}{C}}-O-CH \\ CH_3-(CH_2)_{14}-\overset{O}{\overset{\|}{C}}-O-CH_2 \end{array}$$

(b)

$$\begin{array}{l} CH_3-(CH_2)_{14}-\overset{O}{\overset{\|}{C}}-O-CH_2 \\ CH_3-(CH_2)_{10}-\overset{O}{\overset{\|}{C}}-O-CH \\ CH_3-(CH_2)_{16}-\overset{O}{\overset{\|}{C}}-O-CH_2 \end{array}$$

(c)

$$\begin{array}{l} CH_3-(CH_2)_7-CH{=}CH-(CH_2)_7-\overset{O}{\overset{\|}{C}}-O-CH_2 \\ CH_3-(CH_2)_7-CH{=}CH-(CH_2)_7-\overset{O}{\overset{\|}{C}}-O-CH \\ CH_3-(CH_2)_7-CH{=}CH-(CH_2)_7-\overset{O}{\overset{\|}{C}}-O-CH_2 \end{array}$$

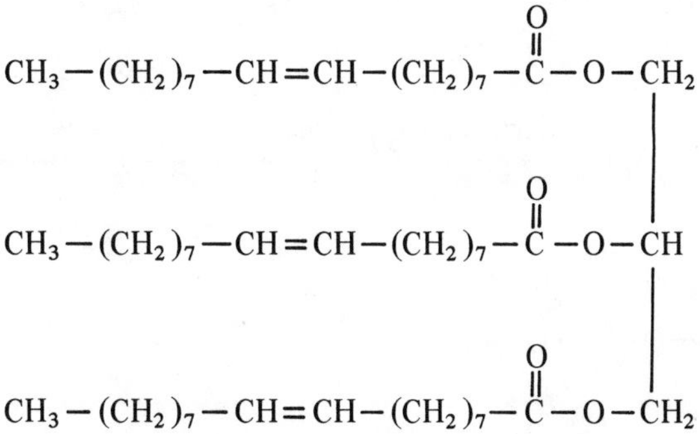

(d)

$$CH_3-(CH_2)_7-CH{=}CH-(CH_2)_7-\overset{O}{\overset{\|}{C}}-O-(CH_2)_{15}-CH_3$$

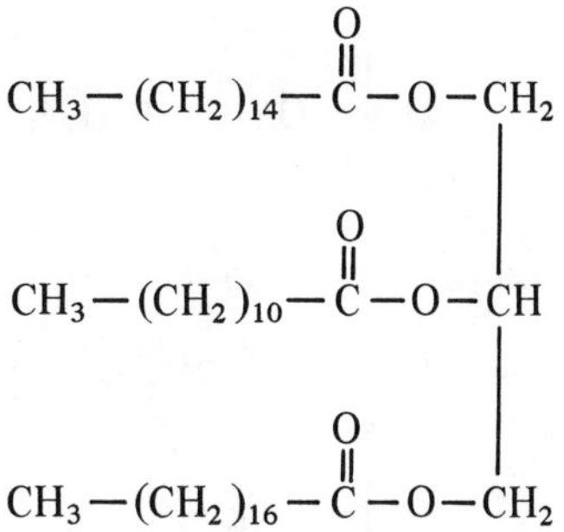

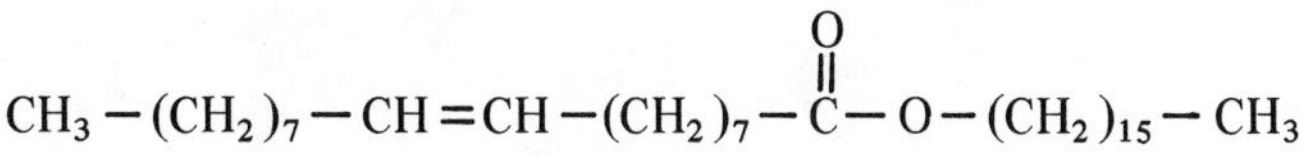

21. Which fatty acid is least likely to be isolated from the hydrolysis product of a fat?

(a) $CH_3-(CH_2)_{10}-\overset{O}{\overset{\|}{C}}-OH$

(b) $CH_3-(CH_2)_{14}-\overset{O}{\overset{\|}{C}}-OH$

(c) $CH_3-(CH_2)_7-CH=CH-(CH_2)_7-\overset{O}{\overset{\|}{C}}-OH$

(d) $CH_3-(CH_2)_{15}-\overset{O}{\overset{\|}{C}}-OH$

(e) $CH_3-(CH_2)_4-CH=CH-CH_2-CH=CH-(CH_2)_7-\overset{O}{\overset{\|}{C}}-OH$

22. Which compound is classified as a lecithin?

(a)

$$\begin{array}{l} H_2C-O-\overset{O}{\overset{\|}{C}}-(CH_2)_{14}-CH_3 \\ | \\ HC-O-\overset{O}{\overset{\|}{C}}-(CH_2)_{14}-CH_3 \\ | \\ H_2C-O-\overset{O}{\overset{\|}{C}}-(CH_2)_{14}-CH_3 \end{array}$$

(b)

$$\begin{array}{l} H_2C-O-\overset{O}{\overset{\|}{C}}-(CH_2)_{14}-CH_3 \\ | \\ HC-O-\overset{O}{\overset{\|}{C}}-(CH_2)_{14}-CH_3 \\ | \\ H_2C-O-\underset{O^-}{\underset{|}{\overset{O}{\overset{\|}{P}}}}-O-CH_2-CH_2-\overset{+}{N}(CH_3)_3 \end{array}$$

(c)

$$\begin{array}{l} H_2C-O-\overset{O}{\overset{\|}{C}}-(CH_2)_7-CH=CH-(CH_2)_7-CH_3 \\ | \\ HC-O-\overset{O}{\overset{\|}{C}}-(CH_2)_7-CH=CH-(CH_2)_7-CH_3 \\ | \\ H_2C-O-\overset{O}{\overset{\|}{C}}-(CH_2)_7-CH=CH-(CH_2)_7-CH_3 \end{array}$$

(d) R (on benzene ring, para to) $SO_2O^-\ Na^+$

(e)

$$CH_3-(CH_2)_{17}-O-\underset{\underset{O}{\downarrow}}{\overset{\overset{O}{\uparrow}}{S}}-O^-\ Na^+$$

23. Which compound is neither a soap nor a detergent?

(a) $CH_3-(CH_2)_{14}-\overset{\overset{O}{||}}{C}-O^- Na^+$

(b) $CH_3-(CH_2)_{15}-O-\underset{\underset{O}{\downarrow}}{\overset{\overset{O}{\uparrow}}{S}}-O^- Na^+$

(c) Benzene ring with $OCH_3$ (top), $C(CH_3)_3$, and $OH$ (bottom)

(d) Benzene ring with $(CH_2)_9-CH_3$ (top) and $O \leftarrow S \rightarrow O$ with $O^- Na^+$ (bottom)

24. Which compound is not easily biodegraded?

(a) Benzene ring with $(CH_2)_9-CH_3$ (top) and $O \leftarrow S \rightarrow O$ with $O^- Na^+$ (bottom)

(b) $CH_3-(CH_2)_{15}-O-\underset{\underset{O}{\downarrow}}{\overset{\overset{O}{\uparrow}}{S}}-O^- Na^+$

(c) ${}^+Na^- O-\underset{\underset{O}{\downarrow}}{\overset{\overset{O}{\uparrow}}{S}}-$ benzene ring — (highly branched alkyl group)

25. Which synthetic detergent is a sulfate?

(a) Benzene ring with $(CH_2)_9CH_3$ (top) and $O \leftarrow S \rightarrow O$ with $O^- Na^+$ (bottom)

(b) $CH_3-(CH_2)_{15}-O-\underset{\underset{O}{\downarrow}}{\overset{\overset{O}{\uparrow}}{S}}-O^- Na^+$

26. Which compound is a prostaglandin?

(a) $OCH_3$, $C(CH_3)_3$, OH

(b) $CH_3$, O, HO, $CH_3$, $(CH_2)_3-\overset{CH_3}{CH}-(CH_2)_3-\overset{CH_3}{CH}-(CH_2)_3-\overset{CH_3}{CH}-CH_3$

(c) $CO_2H$, OH

(d) $CH_3-\overset{O}{\overset{\|}{C}}-CH_2-\overset{O}{\overset{\|}{C}}-SCoA$

**Completion**

27. Esters of long-chain, unbranched fatty acids and long-chain alcohols are called ____________________.

28. Esters of the trihydroxy alcohol, glycerol, are called ____________________.

29. Glyceryl esters in which long-chain saturated acids predominate are ____________________.

30. Glyceryl esters in which unsaturated fatty acids predominate are ____________________.

31. The extent of unsaturation in a fat or oil is expressed in terms of its ____________________ ____________________.

32. Alkali metal salts of long-chain fatty acids are called ____________________.

33. Synthetic detergents are known as ____________________.

34. Hydrolysis and oxidation of edible fats and oils results in the formation of lower molecular weight volatile acids which have a disagreeable odor. This condition is known as ____________________.

35. Water which contains calcium, magnesium, or iron salts is said to be ____________________.

36. Enzymes which are active in the hydrolysis of fats or lipids are called ____________________.

37. The degradation, or destructive metabolism, of fats (and other foods) is called ____________________; biosynthesis, or constructive metabolism, is called ____________________.

38. The suffix ____________ is used in the naming of enzymes.

39. An enzyme which removes hydrogen is called a ____________________.

40. An enzyme which catalyzes chemical combination with oxygen is called an

    ____________________.

41. ____________________ are cyclic derivatives of 20-carbon atom fatty acids which function as biological regulators.

**Structural Formulas** (Write the correct structural formula for each of the following compounds.)

42. Glycerol

43. Palmitic acid

44. Triolein

45. Sodium stearate

46. Glyceryl lauropalmitooleate

47. Sodium lauryl sulfate

48. Sodium *p*-ethylbenzenesulfonate

**Reactions** (Complete the following.)

49.

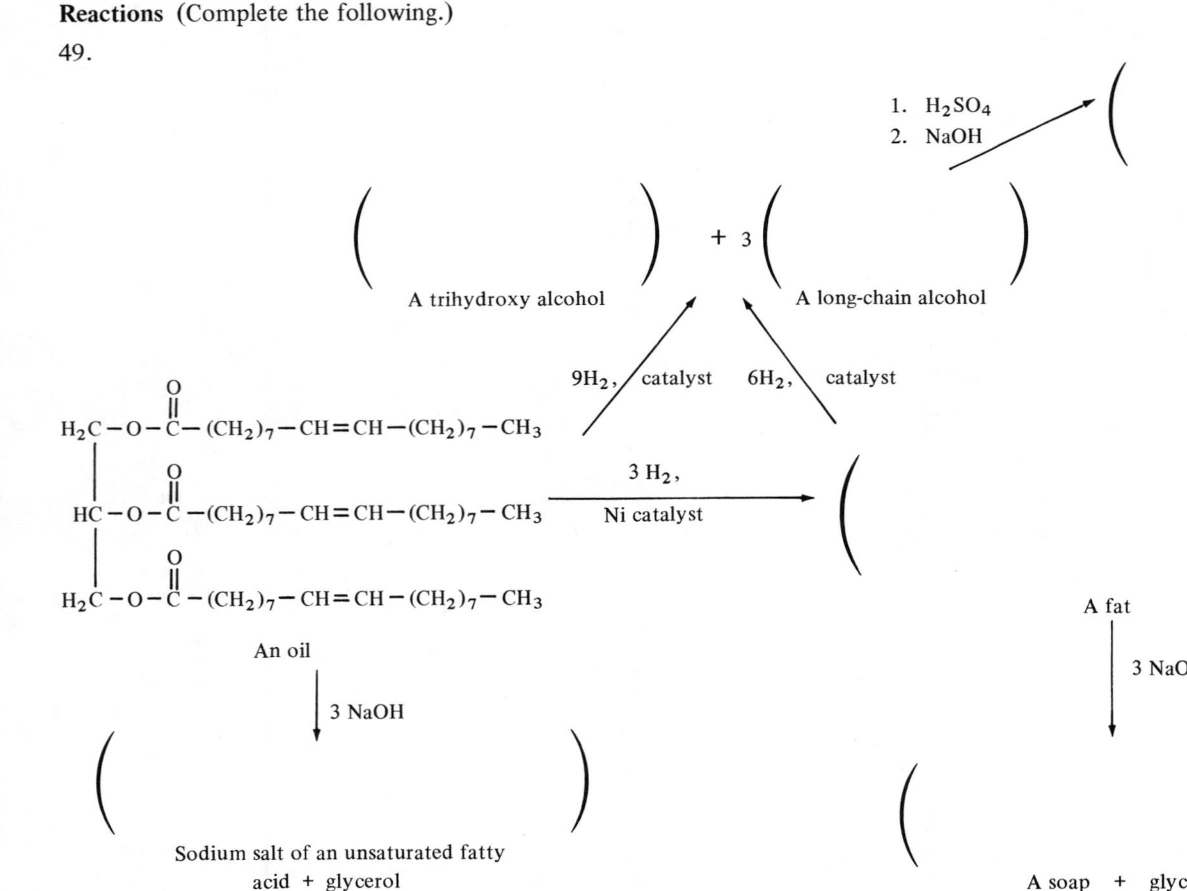

**Miscellaneous**

50. Determine the saponification value and the iodine value of trilinolein.

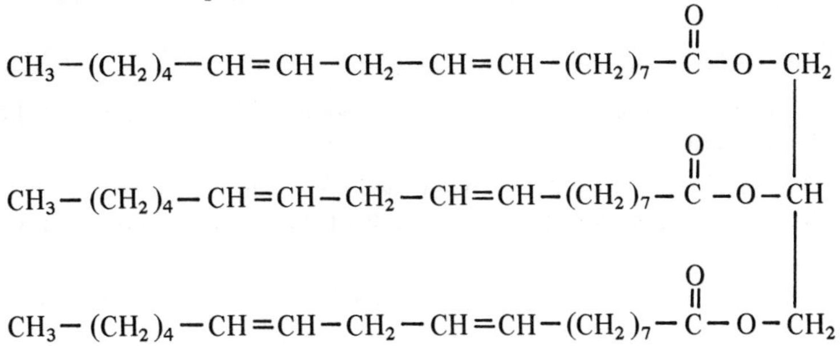

$$\begin{array}{l} CH_3-(CH_2)_4-CH{=}CH-CH_2-CH{=}CH-(CH_2)_7-\overset{\displaystyle O}{\overset{\|}{C}}-O-CH_2 \\ \qquad\qquad\qquad\qquad\qquad\qquad\qquad\qquad\qquad\qquad\qquad\qquad\quad | \\ CH_3-(CH_2)_4-CH{=}CH-CH_2-CH{=}CH-(CH_2)_7-\overset{\displaystyle O}{\overset{\|}{C}}-O-CH \\ \qquad\qquad\qquad\qquad\qquad\qquad\qquad\qquad\qquad\qquad\qquad\qquad\quad | \\ CH_3-(CH_2)_4-CH{=}CH-CH_2-CH{=}CH-(CH_2)_7-\overset{\displaystyle O}{\overset{\|}{C}}-O-CH_2 \end{array}$$

## ANSWERS

**True–False**

1. True
2. True
3. True. Fats and oils have the general formula

$$\begin{array}{l} R-\overset{\displaystyle O}{\overset{\|}{C}}-O-CH_2 \\ \qquad\qquad\quad | \\ R-\overset{\displaystyle O}{\overset{\|}{C}}-O-CH \\ \qquad\qquad\quad | \\ R-\overset{\displaystyle O}{\overset{\|}{C}}-O-CH_2 \end{array}$$

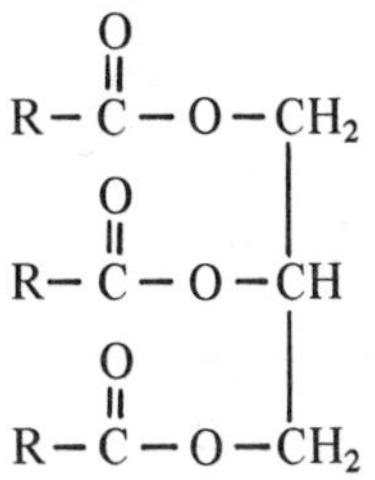

   Waxes have the general formula of a simple ester,

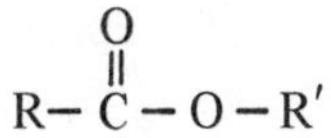

$$R-\overset{\displaystyle O}{\overset{\|}{C}}-O-R'$$

4. True. The alkyl portions of the acid components of the glyceride molecule (the R's in the general formula in answer 3 above) are different in mixed glycerides.

5. True
6. True
7. True
8. True. They are oleic, linoleic, and linolenic acid.
9. False. Oleic acid is an unsaturated acid. The iodine value is a measure of the extent of unsaturation.
10. True. The sodium and potassium salts of the long-chain fatty acids, usually called soaps, are soluble in water, but the salts of the heavier metals are insoluble.
11. True
12. True
13. True
14. False. Soaps are insoluble in hard water, but synthetic detergents are soluble. This is a major advantage that synthetic detergents have over soaps.
15. True
16. True

**Multiple Choice**

17. (b) Soaps are salts of long-chain fatty acids.
18. (b) Waxes are esters of long-chain, unbranched fatty acids and long-chain alcohols.
19. (a) The acid components of a simple glyceride are identical.
20. (c) An oil is a glyceride in which the acid component has various degrees of unsaturation.
21. (d) The anabolism of fats in the animal system appears to involve a sequence of reactions in which the chain length is always increased by a 2-carbon unit. Thus a fatty acid which contains an odd number of carbon atoms is unlikely.
22. (b)
23. (c)
24. (c)
25. (b)
26. (c)

**Completion**

27. waxes
28. glycerides
29. fats
30. oils
31. iodine value
32. soaps
33. syndets
34. rancidity
35. hard
36. lipases
37. catabolism, anabolism
38. *-ase*

39. dehydrogenase
40. oxidase
41. prostaglandins

**Structural Formulas**

42. $$\begin{array}{ccc} CH_2 - & CH - & CH_2 \\ | & | & | \\ OH & OH & OH \end{array}$$

43. $$CH_3-(CH_2)_{14}-\overset{\overset{\large O}{\|}}{C}-OH$$

44. $$\begin{array}{l} CH_3-(CH_2)_7-CH=CH-(CH_2)_7-\overset{\overset{\large O}{\|}}{C}-O-CH_2 \\ \qquad\qquad\qquad\qquad\qquad\qquad\qquad\qquad\qquad\quad | \\ CH_3-(CH_2)_7-CH=CH-(CH_2)_7-\overset{\overset{\large O}{\|}}{C}-O-CH \\ \qquad\qquad\qquad\qquad\qquad\qquad\qquad\qquad\qquad\quad | \\ CH_3-(CH_2)_7-CH=CH-(CH_2)_7-\overset{\overset{\large O}{\|}}{C}-O-CH_2 \end{array}$$

45. $$CH_3-(CH_2)_{16}-\overset{\overset{\large O}{\|}}{C}-O^- \; Na^+$$

46. $$\begin{array}{r} CH_3-(CH_2)_{10}-\overset{\overset{\large O}{\|}}{C}-O-CH_2 \\ | \\ CH_3-(CH_2)_{14}-\overset{\overset{\large O}{\|}}{C}-O-CH \\ | \\ CH_3-(CH_2)_7-CH=CH-(CH_2)_7-\overset{\overset{\large O}{\|}}{C}-O-CH_2 \end{array}$$

47. $$CH_3-(CH_2)_{11}-O-\underset{\underset{\large O}{\downarrow}}{\overset{\overset{\large O}{\uparrow}}{S}}-O^- \; Na^+$$

48. $$\begin{array}{c} CH_2-CH_3 \\ | \\ \text{(benzene ring, para-substituted)} \\ | \\ O \leftarrow S \rightarrow O \\ | \\ O^- \; Na^+ \end{array}$$

**Reactions**

49.

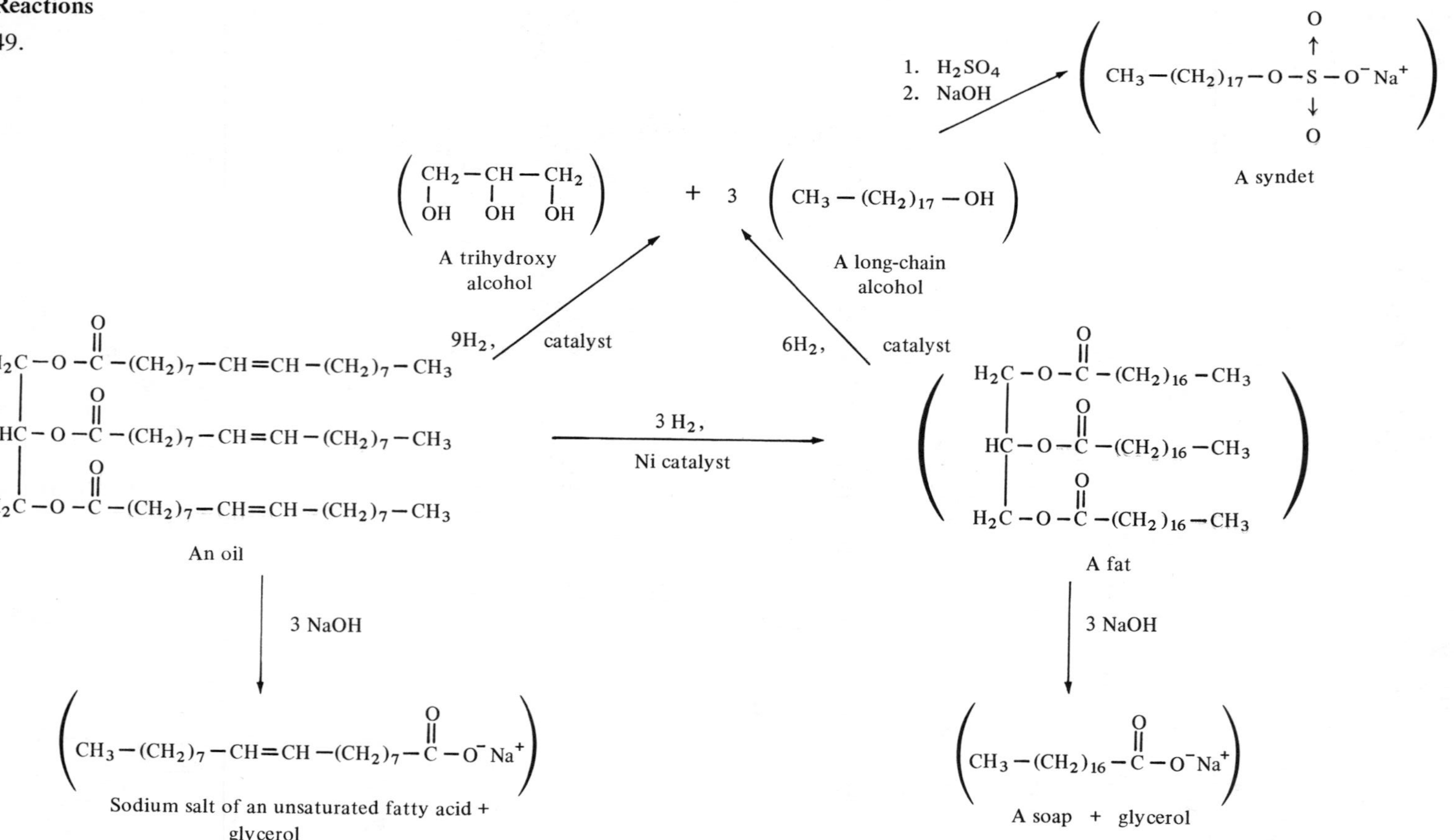

**Miscellaneous**

50. The saponification number of a fat or an oil is the number of milligrams of potassium hydroxide required to saponify one gram of the fat or oil.

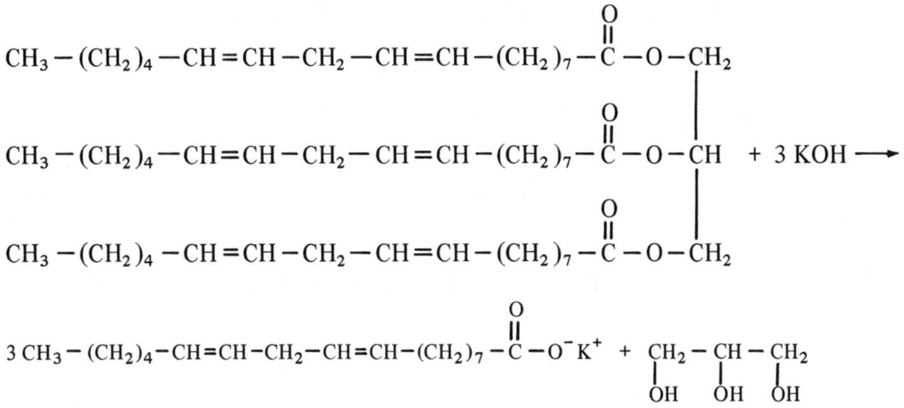

$$\begin{array}{l} CH_3-(CH_2)_4-CH=CH-CH_2-CH=CH-(CH_2)_7-\overset{\overset{\large O}{\|}}{C}-O-CH_2 \\ CH_3-(CH_2)_4-CH=CH-CH_2-CH=CH-(CH_2)_7-\overset{\overset{\large O}{\|}}{C}-O-CH \quad + 3\,KOH \longrightarrow \\ CH_3-(CH_2)_4-CH=CH-CH_2-CH=CH-(CH_2)_7-\overset{\overset{\large O}{\|}}{C}-O-CH_2 \end{array}$$

$$3\,CH_3-(CH_2)_4-CH=CH-CH_2-CH=CH-(CH_2)_7-\overset{\overset{\large O}{\|}}{C}-O^-K^+ + \underset{OH}{CH_2}-\underset{OH}{CH}-\underset{OH}{CH_2}$$

Molecular formula of trilinolein $= C_{57}H_{98}O_6$

$$\begin{aligned} \text{Molecular weight} &= 12\,(57) + 1\,(98) + 16\,(6) \\ &= 684 + 98 + 96 \\ &= 878 \end{aligned}$$

Three gram formula weights of potassium hydroxide are required to saponify 878 g of trilinolein.

$$\begin{aligned} \text{Three gram formula weights of KOH} &= 3\,(56) \\ &= 168 \text{ g} \\ &= 168{,}000 \text{ mg} \end{aligned}$$

If 168,000 mg of potassium hydroxide are needed to saponify 878 g of trilinolein, how many mg of potassium hydroxide are required to saponify 1 g of trilinolein?

$$\frac{168{,}000 \text{ mg}}{878 \text{ g}} = \frac{x \text{ mg}}{1 \text{ g}}$$

$$x \text{ mg} = \frac{168{,}000 \text{ mg } (1 \text{ g})}{878 \text{ g}}$$

$$= 191.3 \text{ mg}$$

Thus, 191.3 is the saponification value of trilinolein.

The iodine value is the number of grams of iodine which will add to 100 g of fat or oil.

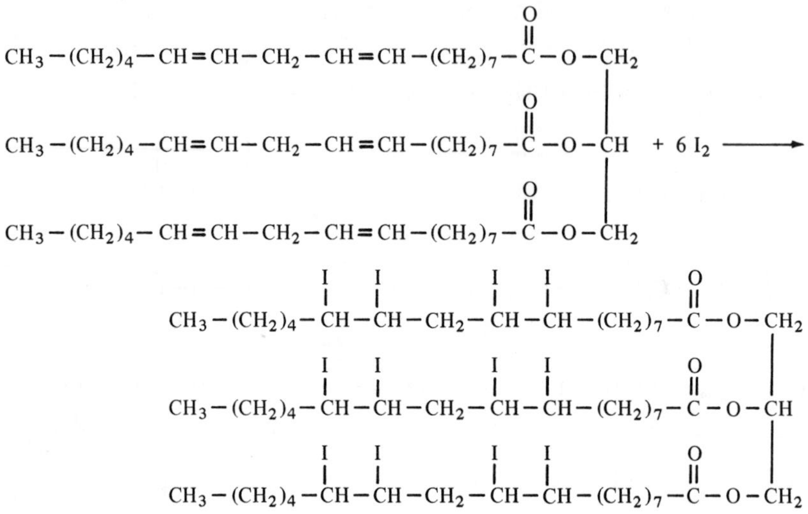

$$
\begin{array}{l}
CH_3-(CH_2)_4-CH{=}CH-CH_2-CH{=}CH-(CH_2)_7-\overset{\overset{\displaystyle O}{\|}}{C}-O-CH_2 \\
CH_3-(CH_2)_4-CH{=}CH-CH_2-CH{=}CH-(CH_2)_7-\overset{\overset{\displaystyle O}{\|}}{C}-O-CH \quad + 6\,I_2 \longrightarrow \\
CH_3-(CH_2)_4-CH{=}CH-CH_2-CH{=}CH-(CH_2)_7-\overset{\overset{\displaystyle O}{\|}}{C}-O-CH_2
\end{array}
$$

$$
\begin{array}{l}
CH_3-(CH_2)_4-\underset{}{\overset{I}{\overset{|}{C}}}H-\overset{I}{\overset{|}{C}}H-CH_2-\overset{I}{\overset{|}{C}}H-\overset{I}{\overset{|}{C}}H-(CH_2)_7-\overset{\overset{\displaystyle O}{\|}}{C}-O-CH_2 \\
CH_3-(CH_2)_4-\overset{I}{\overset{|}{C}}H-\overset{I}{\overset{|}{C}}H-CH_2-\overset{I}{\overset{|}{C}}H-\overset{I}{\overset{|}{C}}H-(CH_2)_7-\overset{\overset{\displaystyle O}{\|}}{C}-O-CH \\
CH_3-(CH_2)_4-\overset{I}{\overset{|}{C}}H-\overset{I}{\overset{|}{C}}H-CH_2-\overset{I}{\overset{|}{C}}H-\overset{I}{\overset{|}{C}}H-(CH_2)_7-\overset{\overset{\displaystyle O}{\|}}{C}-O-CH_2
\end{array}
$$

Six moles of iodine will add to 878 g of trilinolein.

$$
\begin{aligned}
\text{Six moles of } I_2 &= 12\,(126.9)\text{ g} \\
&= 1522.8\text{ g}
\end{aligned}
$$

1522.8 g of $I_2$ will add to 8.78 (100) g of trilinolein.

If 1522.8 g of $I_2$ will add to 8.78 (100) g of trilinolein, how many grams of iodine will add to 1 (100) g of trilinolein?

$$\frac{1522.8\text{ g}}{8.78\,(100)\text{ g}} = \frac{x\text{ g}}{1\,(100)\text{ g}}$$

$$x\text{ g} = \frac{1\,(100)\text{ g}\,(1522.8)\text{ g}}{8.78\,(100)\text{ g}}$$

$$= 173.4\text{ g}$$

Thus, the iodine value of trilinolein is 173.4

# 12

# *Bifunctional Acids*

## INTRODUCTION

In this chapter we shall study carboxylic acids that contain in addition to the carboxyl another reactive group such as a second carboxyl group, a carbonyl group, a halogen atom, a hydroxyl group, or a carbon — carbon double bond.

Actually, we shall not have a great deal of new chemistry to consider here because we shall find that a second functional group may be incorporated into the acid molecule by some of the same general methods that we used for the preparation of the monocarboxylic acids, the alkyl halides, the alcohols, and the olefins. One of the reactions that we shall want to pay particular attention to is the **malonic ester synthesis.** This reaction is used for the synthesis of substituted acetic acids, substituted malonic acids, and substituted malonic esters. The latter are used for the preparation of an important class of drugs known as barbiturates, which are commonly used in medicine as hypnotics. The use of a similar reaction, the **acetoacetic ester synthesis**, to prepare $\beta$-keto acids and substituted acetone derivatives also will be illustrated.

Finally, we shall review briefly some of the reactions of the bifunctional acids. We shall be agreeably surprised to learn that the chemical behavior of the bifunctional acids, for the most part, is the behavior characteristic of each functional group separately and thus will be largely a review of some chemistry we have previously covered.

We shall defer until Chapter 15 the study of another class of important bifunctional acids, the $\alpha$-amino acids, because it will serve our purpose better to consider these building blocks of proteins in the chapter dealing with this important nutrient.

## PROBLEMS

**True–False**

1. T F The first two members of the dicarboxylic acid family, oxalic acid and malonic acid, are much stronger than acetic acid because of the electron withdrawing effect of the second carboxyl group.
2. T F The first ionization constant of a decarboxylic acid is larger than the second ionization constant.
3. T F Maleic and fumaric acid are examples of optical isomers.
4. T F Adipic acid is used in the preparation of nylon.
5. T F Lactic acid is an $\alpha$-amino acid.
6. T F Cyclic anhydrides are produced when succinic and glutaric acids are heated.
7. T F The Hell-Volhard-Zelinsky reaction is an important starting point for many $\alpha$-substituted acids.
8. T F $\alpha$-Hydroxy acids undergo an intermolecular diesterification to form a lactide when heated.
9. T F $\beta$-Hydroxy acids, when heated strongly, undergo a intramolecular esterification to form a lactone.
10. T F Maleic acid forms an anhydride when heated to a high temperature, but fumaric acid does not because of its spacial configuration.

**Matching** (Match both a correct common name and an IUPAC name to corresponding structural formulas.)

| Common Names | IUPAC Names |
|---|---|
| (a) Adipic acid | (i) Propanedioic acid |
| (b) Oxalic acid | (j) *cis*-Butenedioic acid |
| (c) Succinic acid | (k) Hexanedioic acid |
| (d) Phthalic acid | (l) Ethanedioic acid |
| (e) Glutaric acid | (m) 1,2-Benzenedicarboxylic acid |
| (f) Maleic acid | (n) Pentanedioic acid |
| (g) Malonic acid | (o) *trans*-Butenedioic acid |
| (h) Fumaric acid | (p) Butanedioic acid |

| | Common | IUPAC |
|---|---|---|
| 11. HOOC—COOH | ____ | ____ |
| 12. HOOC—$CH_2$—COOH | ____ | ____ |
| 13. HOOC—$(CH_2)_2$—COOH | ____ | ____ |

14. $HOOC-(CH_2)_3-COOH$ _____ _____

15. $HOOC-(CH_2)_4-COOH$ _____ _____

16. (benzene ring with two adjacent COOH groups) _____ _____

COOH
COOH

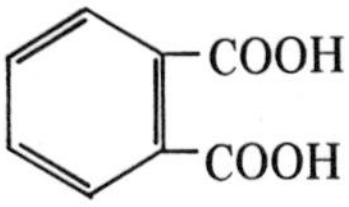

17. _____ _____

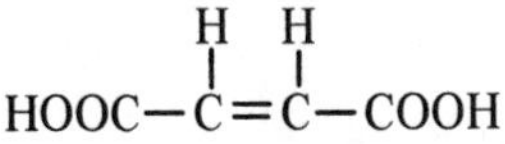

$$HOOC-\overset{H}{\overset{|}{C}}=\overset{H}{\overset{|}{C}}-COOH$$

18. _____ _____

$$HOOC-\overset{H}{\overset{|}{C}}=\underset{H}{\underset{|}{C}}-COOH$$

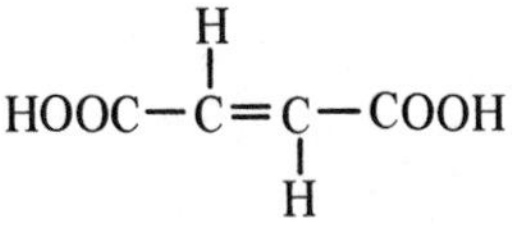

(Match each compound with an appropriate name or description.)

(a) (benzene ring with adjacent COOH and OH groups)

COOH
OH

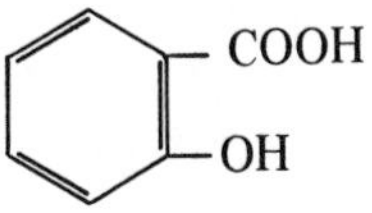

(b)

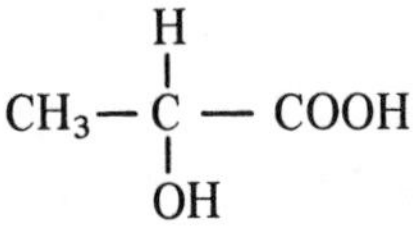

$$CH_3-\overset{H}{\overset{|}{\underset{OH}{\underset{|}{C}}}}-COOH$$

(c) (six-membered ring: $H_3C$, H on C; O; C=O; H, $CH_3$ on C; O; O=C)

$H_3C$ O
C C=O
H H
O=C C
O $CH_3$

(d) (benzene ring with COOH groups at para positions)

COOH
COOH

(e) (six-membered ring)

O
‖
C
$(C_2H_5)_2C$ N—H
C C
O N O
H

(f)

$$CH_3-\overset{H}{\overset{|}{C}}=\overset{H}{\overset{|}{C}}-COOH$$

(g) (benzene ring with adjacent $O-\overset{O}{\overset{\|}{C}}-CH_3$ and COOH groups)

$O-\overset{O}{\overset{\|}{C}}-CH_3$
COOH

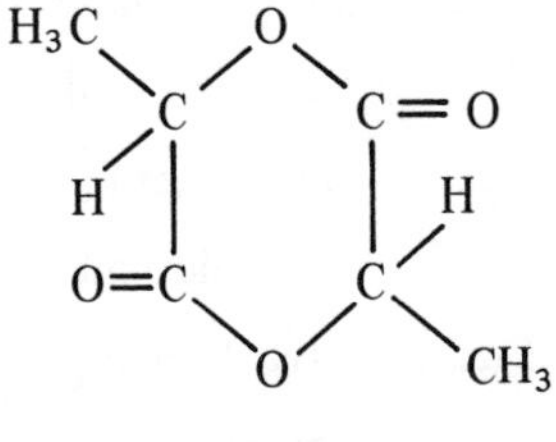

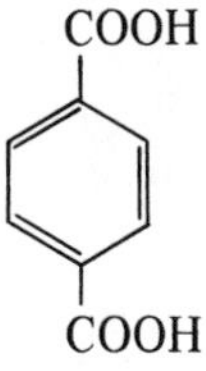

(h)

$CH_2-CH_2$
$CH_2$ C=O
$CH_2-O$

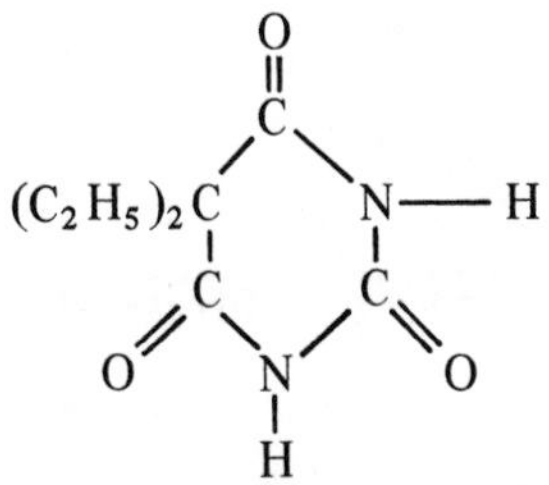

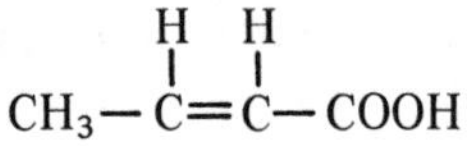

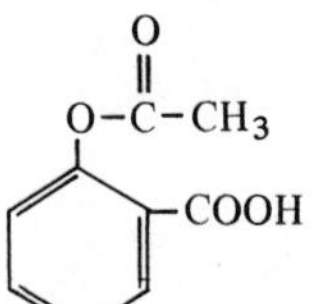

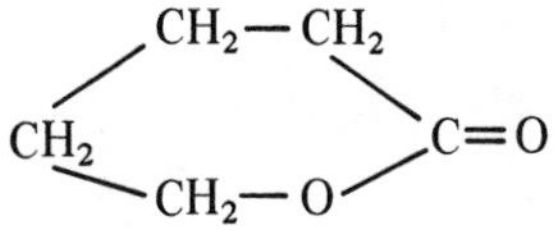

(i)

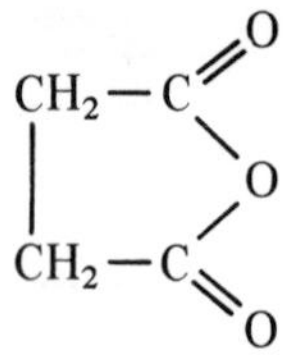

$$\begin{array}{l} CH_2-C(=O) \\ | \quad\quad\quad\ \ \backslash O \\ CH_2-C(=O) / \end{array}$$

(j) $HO-CH_2-COOH$

(k)

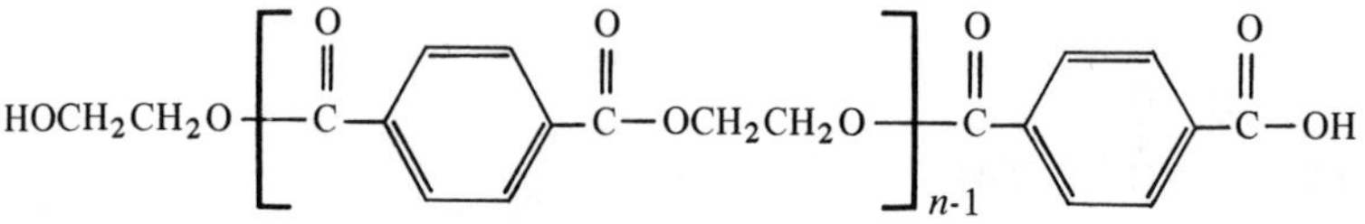

$$HOCH_2CH_2O\left[-\overset{O}{\overset{\|}{C}}-C_6H_4-\overset{O}{\overset{\|}{C}}-OCH_2CH_2O-\right]_{n-1}\overset{O}{\overset{\|}{C}}-C_6H_4-\overset{O}{\overset{\|}{C}}-OH$$

(l) $o\text{-}HO-C_6H_4-\overset{O}{\overset{\|}{C}}-OCH_3$

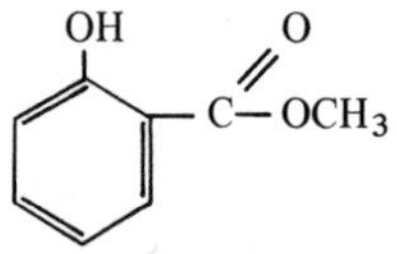

(m)

$$CH_2=\overset{H}{\overset{|}{C}}-COOH$$

(n)

$$CH_3-\overset{O}{\overset{\|}{C}}-COOH$$

19. _____ A cyclic anhydride
20. _____ A lactide
21. _____ A δ-lactone
22. _____ A barbiturate
23. _____ Lactic acid
24. _____ Dacron
25. _____ Pyruvic acid
26. _____ Aspirin
27. _____ Oil of wintergreen
28. _____ Salicyclic acid
29. _____ Acrylic acid
30. _____ Crotonic acid
31. _____ Glycolic acid
32. _____ Terephthalic acid

**Reactions** (Complete the following:)

33.

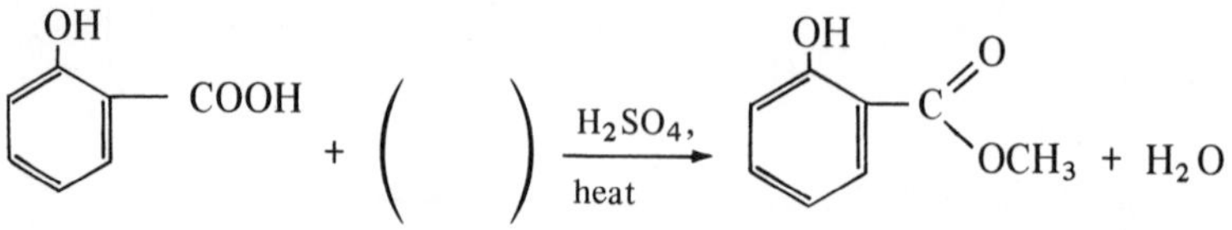

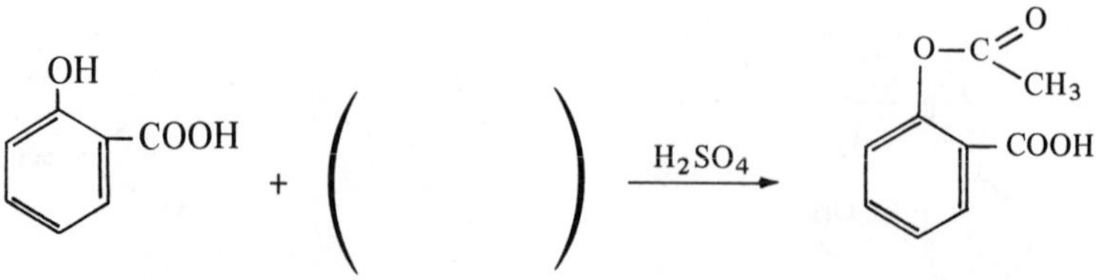

$$o\text{-}HO-C_6H_4-COOH + (\quad) \xrightarrow[\text{heat}]{H_2SO_4,} o\text{-}HO-C_6H_4-\overset{O}{\overset{\|}{C}}-OCH_3 + H_2O$$

$$o\text{-}HO-C_6H_4-COOH + (\quad\quad) \xrightarrow{H_2SO_4} o\text{-}(CH_3\overset{O}{\overset{\|}{C}}-O)-C_6H_4-COOH$$

34.

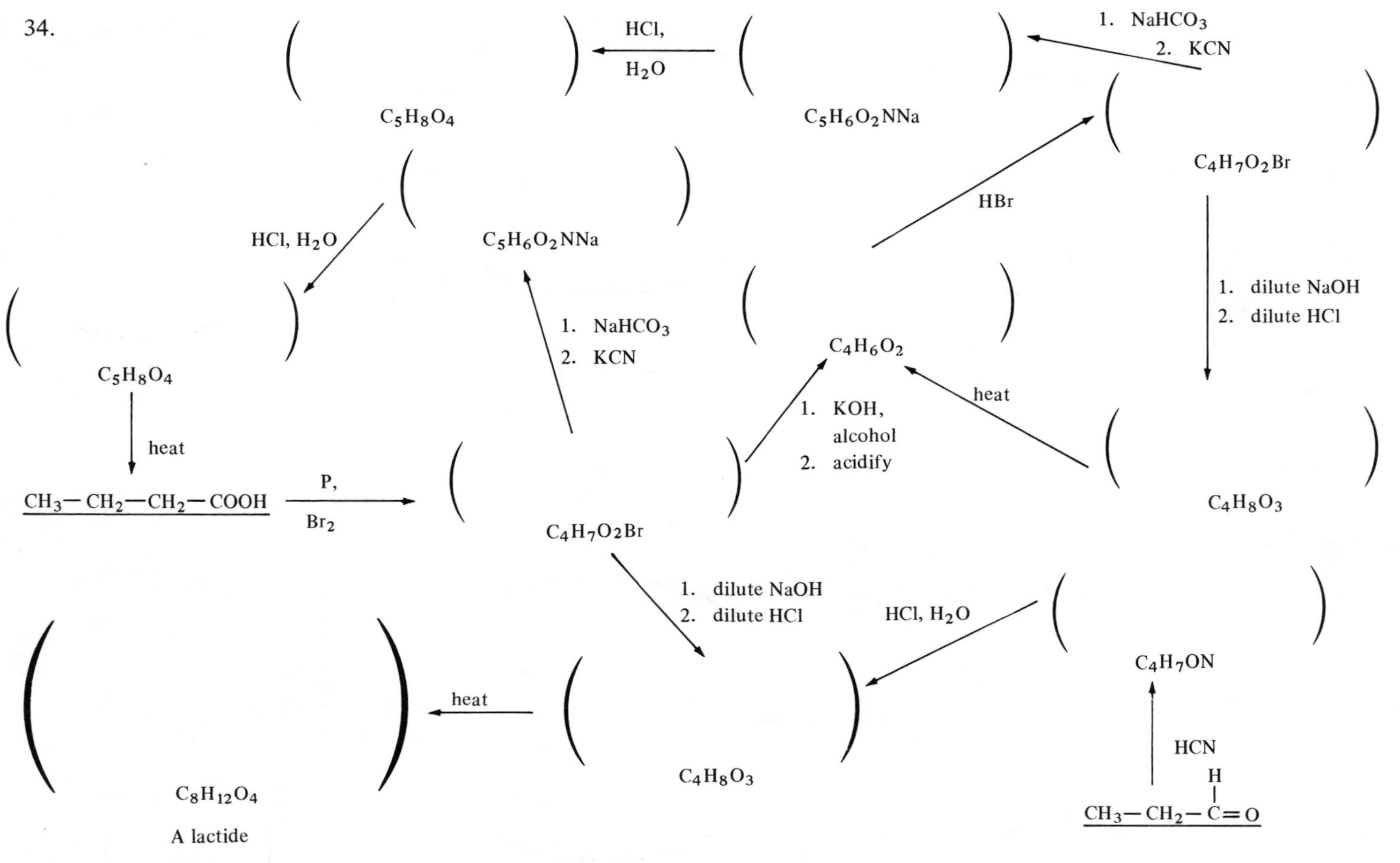

HCl,
$H_2O$
1. $NaHCO_3$
2. KCN
$C_5H_8O_4$
$C_5H_6O_2NNa$
$C_4H_7O_2Br$
HCl, $H_2O$
$C_5H_6O_2NNa$
HBr
1. dilute NaOH
2. dilute HCl
1. $NaHCO_3$
2. KCN
$C_4H_6O_2$
$C_5H_8O_4$
heat
heat
1. KOH, alcohol
2. acidify
$CH_3-CH_2-CH_2-COOH$
P,
$Br_2$
$C_4H_8O_3$
$C_4H_7O_2Br$
1. dilute NaOH
2. dilute HCl
HCl, $H_2O$
$C_4H_7ON$
heat
HCN
$CH_3-CH_2-CH=O$
$C_8H_{12}O_4$
A lactide
$C_4H_8O_3$

35.

$CH_3-CH_2-CH(COOH)-COOH$ $\xrightarrow{2\ SOCl_2}$ ( ) $C_5H_6O_2Cl_2$

$C_5H_6O_2Cl_2$ $\xrightarrow{CH_3-CH_2-OH}$ ( ) $C_9H_{16}O_4$

$C_9H_{16}O_4$ $\xrightarrow{CH_3-CH_2-O^-Na^+}$ ( ) $C_9H_{15}O_4Na$

$C_9H_{15}O_4Na$ $\xrightarrow{CH_3-CH(CH_3)-CH_2-CH_2-I}$ ( ) $C_{14}H_{26}O_4$

$C_{14}H_{26}O_4$ $\xrightarrow{H_2N-C(=O)-NH_2}$ ( ) $C_{11}H_{18}N_2O_3$

$C_{14}H_{26}O_4$ $\xrightarrow{1.\ NaOH\quad 2.\ HCl}$ ( ) $C_{10}H_{18}O_4$

$C_{10}H_{18}O_4$ $\xrightarrow{heat}$ ( ) $C_9H_{18}O_2$

## Miscellaneous

36. Compound (A) contained only carbon, hydrogen, and oxygen, and was insoluble in water but soluble in sodium bicarbonate solution. Titration with a standardized base gave a neutralization equivalent of 114 ± 1. Compound (A) also decolorized bromine in carbon tetrachloride. Oxidation produced propionic and malonic acid. What was the structure of (A)?

37. An acid (A) has a molecular formula $C_4H_8O_3$. What is its structure if it is optically active but, on heating in the presence of acid, loses one mole of water to give a new acid (B), $C_4H_6O_2$? Compound (B) is one of a pair of geometric isomers.

38. A hydrocarbon (A), $C_{10}H_{14}$, when oxidized strongly gives an acid (B) with a neutralization equivalent of 83 ± 1. Strong heating causes (B) to lose water to form the anhydride. Nitration of either (A) or (B) may result in only two mononitrated products. What is compound (A)?

39. Write an acceptable mechanism for each of the following.

(a) Malonic ester synthesis

$$C_2H_5O{-}C({=}O){-}CH_2{-}C({=}O){-}OC_2H_5 + CH_3CH_2CH_2Cl \xrightarrow{Na^+\ {}^-OC_2H_5} C_2H_5O{-}C({=}O){-}CH(CH_2CH_2CH_3){-}C({=}O){-}OC_2H_5$$

(b) Acetoacetic ester synthesis

$$C_2H_5O{-}C({=}O){-}CH_2{-}C({=}O){-}CH_3 + CH_3CH_2CH_2Cl \xrightarrow{Na^+\ {}^-OC_2H_5} C_2H_5O{-}C({=}O){-}CH(CH_2CH_2CH_3){-}C({=}O){-}CH_3$$

40. Write an acceptable mechanism for the addition of HBr to each of the following.

(a)

$$R{-}CH{=}CH{-}\overset{O}{\overset{\|}{C}}{-}R$$

(b)

$$R-CH=CH-\overset{\overset{O}{\|}}{C}-OH$$

## ANSWERS

**True-False**

1. True
2. True
3. False.

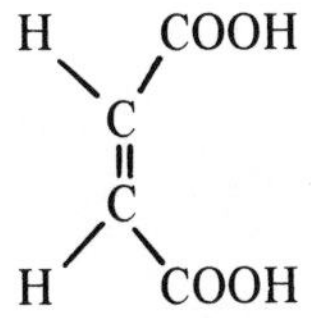

$$\begin{array}{c} H \diagdown \quad \diagup COOH \\ C \\ \| \\ C \\ H \diagup \quad \diagdown COOH \end{array} \qquad \begin{array}{c} H \diagdown \quad \diagup COOH \\ C \\ \| \\ C \\ HOOC \diagup \quad \diagdown H \end{array}$$

Maleic acid 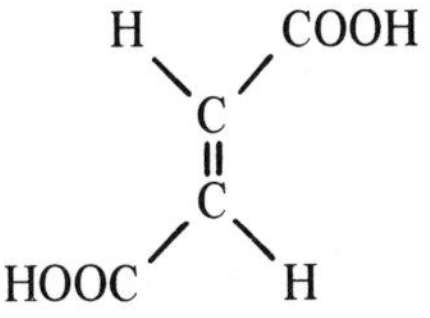 — Fumaric acid

These are examples of *cis,trans*-isomers, but not optical isomers.

4. True. Adipic acid, $HOOC-(CH_2)_4-COOH$, is polymerized with hexamethylenediamine, $H_2N-(CH_2)_6-NH_2$, to produce the polyamide, Nylon-66.
5. False. Lactic acid is an $\alpha$-hydroxy acid.

$$CH_3-\underset{\underset{OH}{|}}{CH}-COOH$$

6. True.

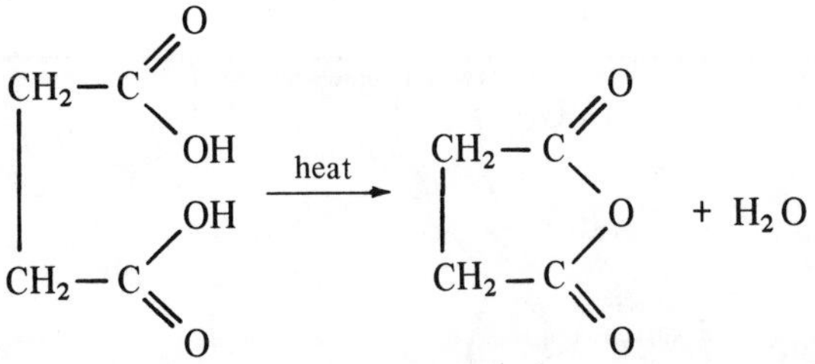

$$\begin{array}{l} CH_2-C(=O)OH \\ | \\ CH_2-C(=O)OH \end{array} \xrightarrow{\text{heat}} \begin{array}{l} CH_2-C(=O) \\ | \qquad\quad O \\ CH_2-C(=O) \end{array} + H_2O$$

7. True. $\alpha$-Halogen acids are prepared via the Hell-Volhard-Zelinsky reaction. $\alpha$-Halogen acids are often converted to other acids such as $\alpha,\beta$-unsaturated acids, $\alpha$-hydroxy acids, and $\alpha$-amino acids.

8. True.

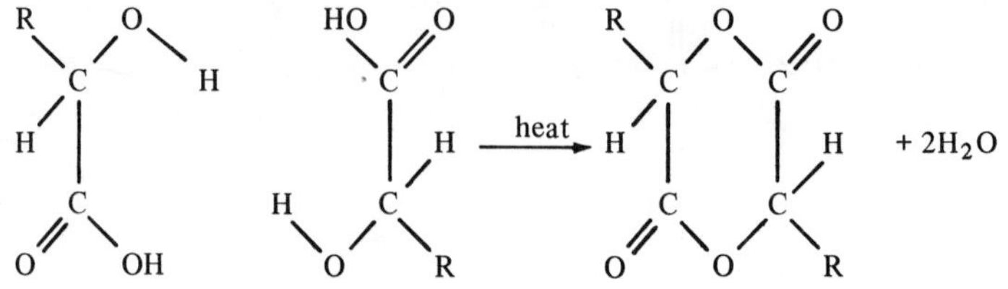

$$2\ R{-}CH(OH){-}COOH \xrightarrow{heat} \text{cyclic dimer (R, H substituted ring: C, O, C(=O), C, O, C(=O))} + 2H_2O$$

A lactide

9. False. $\beta$-Hydroxy acids, when heated strongly, form $\alpha,\beta$-unsaturated acids. The $\gamma$- and $\delta$-hydroxy acids can react intramolecularly to produce lactones.

10. True. The carboxyl groups in the structure of fumaric acid are *trans* and thus incapable of intramolecular anhydride formation. At high temperatures, however, fumaric acid rearranges to maleic acid which then yields the anhydride.

## Matching

| | Common | IUPAC | | Common | IUPAC |
|---|---|---|---|---|---|
| 11. | (b) | (l) | 15. | (a) | (k) |
| 12. | (g) | (i) | 16. | (d) | (m) |
| 13. | (c) | (p) | 17. | (f) | (j) |
| 14. | (e) | (n) | 18. | (h) | (o) |

| | | | |
|---|---|---|---|
| 19. | (i) | 26. | (g) |
| 20. | (c) | 27. | (l) |
| 21. | (h) | 28. | (a) |
| 22. | (e) | 29. | (m) |
| 23. | (b) | 30. | (f) |
| 24. | (k) | 31. | (j) |
| 25. | (n) | 32. | (d) |

## Reactions

33.

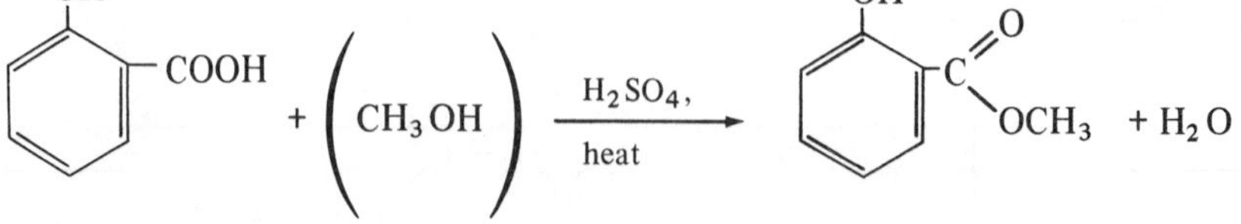

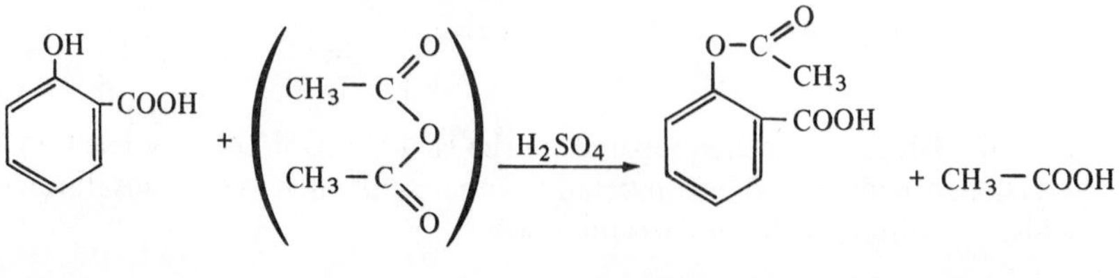

$$C_6H_4(OH)COOH + (CH_3OH) \xrightarrow[heat]{H_2SO_4,} C_6H_4(OH){-}C({=}O)OCH_3 + H_2O$$

$$C_6H_4(OH)COOH + \left(\begin{matrix} CH_3{-}C{=}O \\ O \\ CH_3{-}C{=}O \end{matrix}\right) \xrightarrow{H_2SO_4} C_6H_4(O{-}C({=}O)CH_3)COOH + CH_3{-}COOH$$

34.

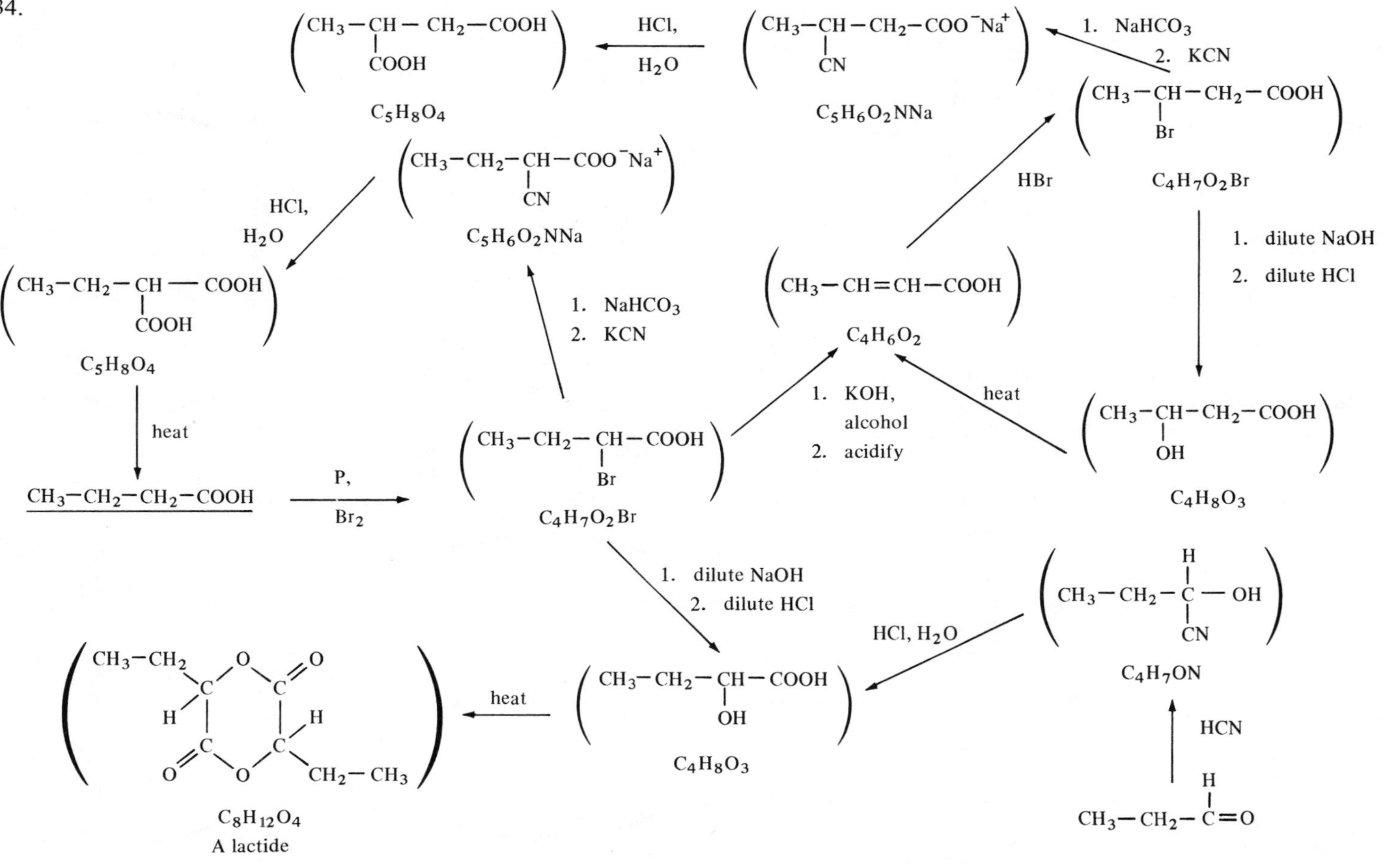
CH3−CH−CH2−COOH (COOH)
C5H8O4
HCl, H2O
CH3−CH−CH2−COO⁻Na⁺ (CN)
C5H6O2NNa
1. NaHCO3
2. KCN
CH3−CH−CH2−COOH (Br)
C4H7O2Br
HBr
CH3−CH2−CH−COO⁻Na⁺ (CN)
C5H6O2NNa
HCl, H2O
CH3−CH2−CH−COOH (COOH)
C5H8O4
heat
CH3−CH2−CH2−COOH
P, Br2
CH3−CH2−CH−COOH (Br)
C4H7O2Br
1. NaHCO3
2. KCN
CH3−CH=CH−COOH
C4H6O2
1. KOH, alcohol
2. acidify
heat
1. dilute NaOH
2. dilute HCl
CH3−CH−CH2−COOH (OH)
C4H8O3
1. dilute NaOH
2. dilute HCl
CH3−CH2−CH−COOH (OH)
C4H8O3
HCl, H2O
CH3−CH2−C(H)(CN)−OH
C4H7ON
HCN
CH3−CH2−C(H)=O
heat
C8H12O4
A lactide

35.

$$CH_3-CH_2-CH(COOH)-COOH \xrightarrow{2\ SOCl_2} CH_3-CH_2-CH(COCl)-COCl$$

$C_5H_6O_2Cl_2$

$$\xrightarrow{CH_3-CH_2-OH} CH_3-CH_2-CH(COOCH_2CH_3)-COOCH_2CH_3$$

$C_9H_{16}O_4$

$$\xrightarrow{CH_3-CH_2-O^-\,Na^+} CH_3-CH_2-\overset{-}{C}(Na^+)(COOCH_2CH_3)-COOCH_2CH_3$$

$C_9H_{15}O_4Na$

$$\xrightarrow{(CH_3)_2CH-CH_2-CH_2-I} (CH_3)_2CH-CH_2-CH_2-C(CH_2CH_3)(COOCH_2CH_3)_2$$

$C_{14}H_{26}O_4$

$$\xrightarrow{H_2N-\overset{O}{\overset{\|}{C}}-NH_2} \text{5-ethyl-5-(3-methylbutyl)barbituric acid ring: } (CH_3)_2CH-CH_2-CH_2 \text{ and } CH_3-CH_2 \text{ on C, ring } C(=O)-NH-C(=O)-NH-C(=O)$$

$C_{11}H_{18}N_2O_3$

From $C_{14}H_{26}O_4$:

$$\xrightarrow{1.\ NaOH\quad 2.\ HCl} (CH_3)_2CH-CH_2-CH_2-C(CH_2CH_3)(COOH)-COOH$$

$C_{10}H_{18}O_4$

$$\xrightarrow{heat} (CH_3)_2CH-CH_2-CH_2-CH(CH_2CH_3)-COOH$$

$C_9H_{18}O_2$

**Miscellaneous**

36.

$$HOOC{-}CH_2{-}\overset{H}{\overset{|}{C}}{=}\overset{H}{\overset{|}{C}}{-}CH_2{-}CH_3 \xrightarrow{KMnO_4} HOOC{-}CH_2{-}COOH + HOOC{-}CH_2{-}CH_3$$

Malonic acid  Propionic acid

37.

$$CH_3CH(OH)CH_2COOH \xrightarrow{\text{acid, heat}} CH_3{-}\underset{\underset{H}{|}}{C}{=}\overset{H}{\overset{|}{C}}{-}COOH$$

3-Hydroxybutanoic acid
($\beta$-Hydroxybutyric acid)
(A)

*trans*-2-Butenoic acid
(Crotonic acid)
(B)

2-Hydroxybutanoic acid is isomeric with 3-hydroxybutanoic and also optically active, but would yield a lactide when heated in the presence of acid.

38.

$$C_6H_4(C_2H_5)_2 \xrightarrow{[O]} C_6H_4(COOH)_2 \xrightarrow{\text{heat}} C_6H_4(CO)_2O$$

(A)
*o*-Diethylbenzene

(B)
Phthalic acid

Phthalic anhydride

39. (a)

$$C_2H_5O{-}C(=O){-}CH_2{-}C(=O){-}OC_2H_5 \xrightarrow{Na^+\ {}^-OC_2H_5} C_2H_5O{-}C(=O){-}\overset{-}{C}H(Na^+){-}C(=O){-}OC_2H_5 + C_2H_5OH$$

$$\downarrow CH_3CH_2CH_2Cl$$

$$C_2H_5O{-}C(=O){-}CH(CH_2CH_2CH_3){-}C(=O){-}OC_2H_5 + NaCl$$

(b)

$$CH_3COCH_2COOC_2H_5 \xrightarrow{Na^+\ ^-OC_2H_5} CH_3CO\overset{-}{C}H(Na^+)COOC_2H_5 + C_2H_5OH$$

$$CH_3CO\overset{-}{C}H(Na^+)COOC_2H_5 \xrightarrow{CH_3CH_2CH_2Cl} CH_3COCH(CH_2CH_2CH_3)COOC_2H_5 + NaCl$$

40. (a)

$$R{-}CH{=}CH{-}\overset{O}{\overset{\|}{C}}{-}R + H^+ \longrightarrow \left[ R{-}CH{=}CH{-}\overset{^+OH}{\overset{\|}{C}}{-}R \longleftrightarrow R{-}\overset{+}{C}H{-}CH{=}\overset{OH}{\overset{|}{C}}{-}R \right]$$

$$\xrightarrow{Br^-} R{-}\underset{Br}{\underset{|}{C}H}{-}CH{=}\overset{OH}{\overset{|}{C}}{-}R \rightleftharpoons R{-}\underset{Br}{\underset{|}{C}H}{-}CH_2{-}\overset{O}{\overset{\|}{C}}{-}R$$

(b)

$$R{-}CH{=}CH{-}\overset{O}{\overset{\|}{C}}{-}OH + H^+ \longrightarrow \left[ R{-}CH{=}CH{-}\overset{^+OH}{\overset{\|}{C}}{-}OH \longleftrightarrow R{-}\overset{+}{C}H{-}CH{=}\overset{OH}{\overset{|}{C}}{-}OH \right]$$

$$\xrightarrow{Br^-} R{-}\underset{Br}{\underset{|}{C}H}{-}CH{=}\overset{OH}{\overset{|}{C}}{-}OH \longrightarrow R{-}\underset{Br}{\underset{|}{C}H}{-}CH_2{-}\overset{O}{\overset{\|}{C}}{-}OH$$

# 13

# Amines and Other Nitrogen Compounds; Sulfur Compounds

## INTRODUCTION

Up to this point our study of organic compounds has been devoted largely to compounds containing the three elements: carbon, hydrogen, and oxygen. In this chapter we shall deal mainly with the preparation and reactions of nitrogen-containing compounds known as **amines**. Amines are classified as **primary**, **secondary**, or **tertiary** and have the following general formulas:

$$R-\ddot{N}\begin{matrix}H\\H\end{matrix} \qquad R-\ddot{N}\begin{matrix}R\\H\end{matrix} \qquad R-\ddot{N}\begin{matrix}R\\R\end{matrix}$$

A primary amine — A secondary amine — A tertiary amine

In order to synthesize any nitrogen compound in the laboratory we must find some means of attaching nitrogen to carbon. We usually do this by obtaining nitrogen from ammonia ($NH_3$), nitric acid ($HNO_3$), or a metal cyanide such as sodium cyanide (NaCN). We have already studied a number of reactions involving the use of these inorganic reagents in Chapters 4, 7, and 10 where we respectively nitrated benzene, prepared alkyl cyanides, and formed amides. We shall see now how such intermediates may be converted into amines and how the amines, through special reactions, may be used to prepare other useful compounds.

We also will study those organic sulfur compounds which are of greatest importance in the chemistry of life processes—the **thiols**, **sulfides**, and **disulfides**.

| R—S—H | R—S—R | R—S—S—R |
|---|---|---|
| Thiol | Sulfide | Disulfide |

## PROBLEMS

### True–False

1. T F Amines are weakly basic compounds.
2. T F Amine salts are usually water-soluble solids.
3. T F Electron-releasing groups bonded to an amine nitrogen increase its basicity.
4. T F Most amines have pleasant odors.
5. T F Aniline is the most important aromatic amine.
6. T F An N-alkyl sulfonamide is insoluble in sodium hydroxide because it has no acidic hydrogen.
7. T F Nitrous acid is unstable, but can be prepared in solution by the addition of a mineral acid to sodium nitrite.
8. T F Primary and secondary amines react with nitrous acid to form diazonium salts.
9. T F Diazonium salts are often used as intermediates to prepare compounds that are not easily prepared by other methods.
10. T F An amino group is a powerful *ortho-para* director.
11. T F Diazonium salts readily couple with phenols and aromatic amines to form azo compounds.
12. T F Thiols are acidic enough to form salts with aqueous sodium hydroxide.

### Multiple Choice

13. Dimethylethylamine is classified as a

    (a) primary amine
    (b) secondary amine
    (c) tertiary amine
    (d) a quarternary amine

14. The reaction between aniline and nitrous acid at low temperature yields

    (a) a N-nitroso amine
    (b) a diazonium salt
    (c) a nitrile
    (d) an amine nitrite salt

15. Which of the following produces the weakest conjugate acid?

    (a) $NH_3$ (b) $CH_3—NH_2$

(c) $(CH_3)_2NH$

(d) $(CH_3)_3N$

(e) $C_6H_5{-}NH_2$

16. Which reaction does not yield an amine?

(a) $R{-}X + NH_3 \longrightarrow$

(b) $R{-}\overset{\displaystyle H}{\overset{|}{C}}{=}N{-}OH + [H] \xrightarrow[C_2H_5OH]{Na,}$

(c) $RCN + H_2O \xrightarrow{H^+}$

(d) $R{-}\overset{\displaystyle O}{\overset{\|}{C}}{-}NH_2 + 4[H] \xrightarrow{LiAlH_4}$

(e) $R{-}CH_2{-}NO_2 + 6[H] \xrightarrow[HCl]{Fe,}$

17. Which reaction will not yield an amide?

(a) $CH_3{-}\overset{\displaystyle O}{\overset{\|}{C}}{-}Cl + NH_3 \longrightarrow$

(b) $CH_3{-}\overset{\displaystyle O}{\overset{\|}{C}}{-}O{-}\overset{\displaystyle O}{\overset{\|}{C}}{-}CH_3 + CH_3{-}NH_2 \longrightarrow$

(c) $CH_3{-}\overset{\displaystyle O}{\overset{\|}{C}}{-}Cl + (CH_3)_3N \longrightarrow$

(d) $CH_3{-}\overset{\displaystyle O}{\overset{\|}{C}}{-}O{-}\overset{\displaystyle O}{\overset{\|}{C}}{-}CH_3 + CH_3{-}\overset{\displaystyle H}{\overset{|}{N}}{-}CH_2{-}CH_3 \longrightarrow$

18. Which compound is benzenesulfonyl chloride?

(a) $C_6H_5{-}\underset{\displaystyle O}{\underset{\downarrow}{\overset{\displaystyle O}{\overset{\uparrow}{S}}}}{-}Cl$

(b) $C_6H_5{-}S{-}Cl$

(c) $C_6H_5{-}\underset{\displaystyle O}{\underset{\downarrow}{\overset{\displaystyle O}{\overset{\uparrow}{S}}}}{-}C_6H_4{-}Cl$

(d) $C_6H_5{-}\overset{\displaystyle O}{\overset{\uparrow}{S}}{-}C_6H_4{-}Cl$

*(cont'd)*

(e) Cl–$C_6H_4$–S(→O)(→O)–O–H

19. Which amine forms a sulfonamide that is insoluble in an alkaline solution?

(a) $CH_3-N(CH_3)-CH_3$

(b) $C_6H_5-NH_2$

(c) $CH_3-CH_2-NH-CH_2-CH_3$

(d) $CH_3-CH(CH_3)-NH_2$

20. Which amine forms a diazonium salt that is reasonably stable at 0–5°.

(a) m-$CH_3-C_6H_4-NH_2$

(b) $C_6H_5-N(CH_3)-H$

(c) $C_6H_5-N(CH_3)-CH_3$

(d) $CH_3-NH_2$

(e) $CH_3-NH-CH_2-CH_3$

**Matching** (Match each structure with its class of compound.)

(a) R–S–H

(b) R–C≡N

(c) R–C(=O)–N(H)(Br)

(d) H(R)C=H–OH

(e) 1-(N=N–$C_6H_5$)-2-naphthol (naphthalene ring with N=N–phenyl and OH)

(f) R–N=C=O

(g) $C_6H_5-SO_2-NH-R$

(h) $C_6H_5-N_2^+\,Cl^-$

(i) $R_2N-N=O$

(j) $R-C(=O)-NH-C_6H_5$

(k) 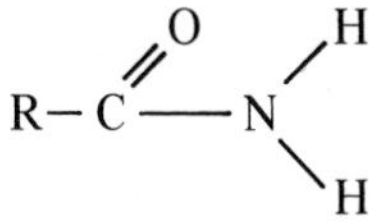

(l) 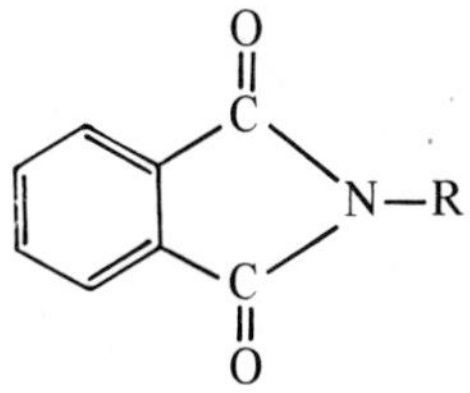

(m) R—S—R

(n) R—S—S—R

21. ____ A disulfide
22. ____ An aldoxime
23. ____ An amide
24. ____ A nitrile
25. ____ A N-bromoamide
26. ____ An anilide
27. ____ An isocyanate
28. ____ A N-alkyl sulfonamide
29. ____ A N-alkyl phthalimide
30. ____ A diazonium salt
31. ____ A N-nitroso amine
32. ____ A sulfide or thioether
33. ____ An azo compound
34. ____ A thiol or mercaptan

**Structural Formulas** (Write the correct structural formula for each of the following compounds.)

35. Aniline

36. *sec*-Butylamine

37. Dimethylammonium chloride

38. Succinimide

39. Tetraethylammonium hydroxide

40. N-Ethylformamide

41. Acetanilide

42. Sulfanilamide

43. Ethylenediamine

44. Phenylhydrazine

45. Lithium diethylamide

46. 2-Propanethiol

47. Diisopropyl sulfide

48. Diisopropyl disulfide

**Reactions** (Complete the following.)

49. $$CH_3-\overset{\displaystyle CH_3 \atop |}{NH} + HCl \longrightarrow (\qquad) \xrightarrow{NaOH} (\qquad) + NaCl + H_2O$$

50.

$$CH_3-CH_2-CH_2-C(=O)OH \xrightarrow{SOCl_2} (\qquad) \xrightarrow{NH_3} (\qquad) \xrightarrow[(NaOBr)]{NaOH,\ Br_2} (\qquad)$$

51.

$$\text{Phthalimide (C}_6\text{H}_4(CO)_2N-H) + KOH \longrightarrow (\qquad)$$

$$\downarrow \; Cl-CH(CH_3)-CH_2-CH_3$$

$$(\qquad) \xrightarrow{KOH} (\qquad)$$

An N-alkyl phthalimide

A primary amine

$$\downarrow \; CH_3CH_2CH_2CH_2{}^- \, Li^+$$

$$(\qquad)$$

52.

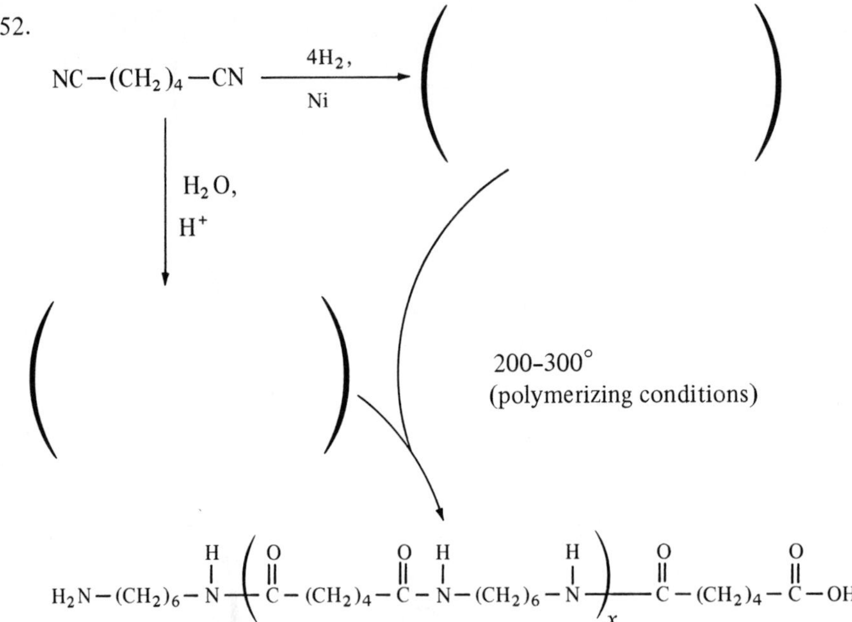

NC—(CH2)4—CN
4H2,
Ni
H2O,
H+
200–300°
(polymerizing conditions)
H2N—(CH2)6—NH—(C(=O)—(CH2)4—C(=O)—NH—(CH2)6—NH)x—C(=O)—(CH2)4—C(=O)—OH


53.

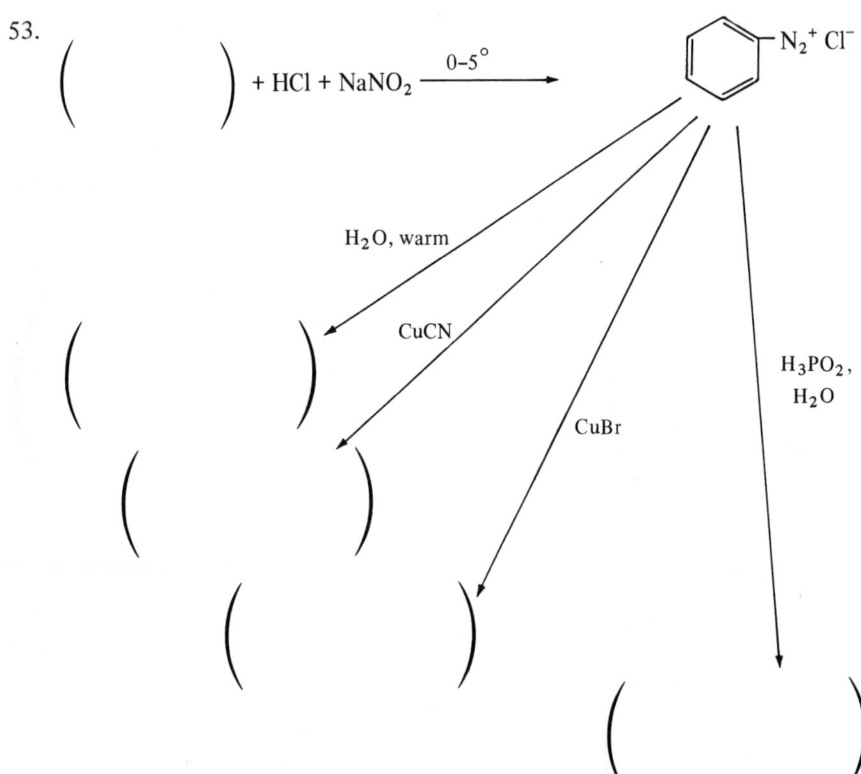

+ HCl + NaNO2
0–5°
N2+ Cl−
H2O, warm
CuCN
CuBr
H3PO2,
H2O

54. $CH_3-CH(CH_3)-CH_2-Br \xrightarrow{Mg} (\quad) \xrightarrow[\text{ether}]{S} (\quad)$

$\downarrow$ $(H_2N)_2C{=}S$ from the starting bromide to ( ); $\xrightarrow{H_2O, H^+}$ from the second product to ( )

$(\quad) \xrightarrow[2.\ H^+]{1.\ H_2O,\ NaOH} (\quad)$

$(\quad) \underset{LiAlH_4}{\overset{I_2}{\rightleftarrows}} (\quad)$

$(\quad) \xrightarrow{NaOH} (\quad) \xrightarrow{CH_3CH_2Br} (\quad)$

**Syntheses** (Outline syntheses leading to the preparation of the following compounds from the suggested starting materials. Use any other reagents that are necessary.)

55. Prepare *p*-nitroacetanilide from benzene.

56. Prepare *n*-propylamine from ethyl chloride.

57. Prepare *n*-butylamine from butyric acid.

**Miscellaneous**

58. Explain how the Hinsberg test enables one to distinguish between the following amines.

$CH_3NH_2$ $(CH_3)_2NH$ $(CH_3)_3N$

59. Azo compounds usually are brightly colored. What series of reactions could easily distinguish between a primary aliphatic amine and a primary aromatic amine?

60. Which compound in each of the following pairs is the stronger base?

(a) I: p-aminoacetophenone ($NH_2$ and $C(=O)CH_3$ on a benzene ring: $NH_2$, $H_3C$, C, O) or II: p-toluidine ($NH_2$, $CH_3$)

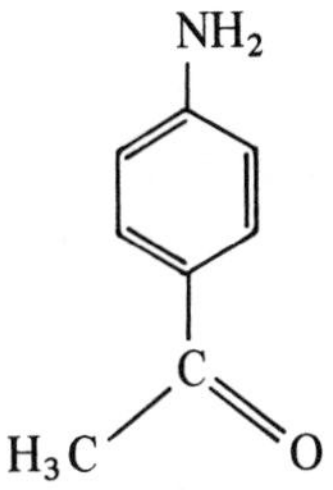

I II

(b) I: $C_6H_5-CH_2-NH_2$ or II: $CH_3-C_6H_4-NH_2$ (m-toluidine)

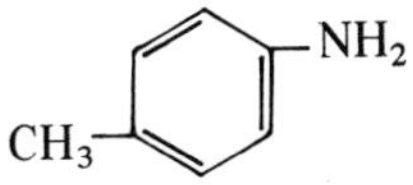

I II

(c) I: cyclohexylamine ($NH_2$, H) or II: aniline ($C_6H_5-NH_2$)

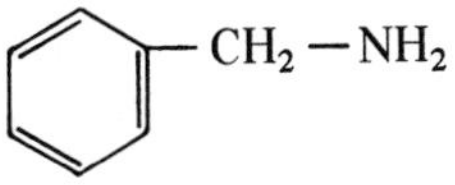

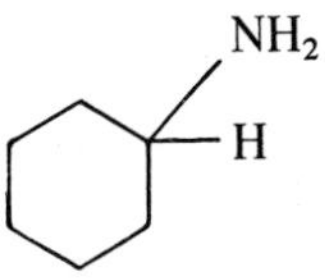

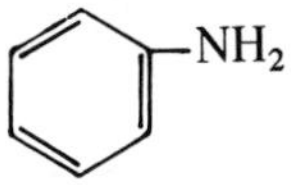

I II

61. Draw a structure for a compound, $C_8H_{11}N$, consistent with each of the following sets of facts:

(a) It is soluble in HCl. Treatment of a cold, acidified solution with aqueous $NaNO_2$, followed by the addition of phenol produces a colored compound. If, the diazotization step is followed by treatment with $H_3PO_2$ a new compound, $C_8H_{10}$, is formed. Oxidation of this hydrocarbon gives an acid (N.E. = 83) which, on heating loses one mole of water to form an anhydride.

(b) It is an optically active, acid-soluble compound. Treatment of an acidified solution with sodium nitrite liberates nitrogen gas and forms a new compound, $C_8H_{10}O$, which upon oxidation yields a white solid, m.p., 121, N.E. 122 ± 1.

62. (a) Give the structures of the conjugate acids of $NH_3$ and $CH_3NH_2$.

(b) The $pK_b$ of $NH_3$ is 4.76 and of $CH_3NH_2$ 3.38. Which is the strongest base?

__________

(c) What is the $pK_a$ of the conjugate acid of $NH_3$? __________ of $CH_3NH_2$? __________

(d) Which conjugate acid is the strongest? __________

63. Give a structure consistent with each of the following sets of nmr data.

(a) $C_7H_9N$

*a* $\delta$ 1.8, singlet, 2H

*b* $\delta$ 3.7, singlet, 2H

*c* $\delta$ 7.2, singlet, 5H

(b) $C_7H_9N$

*a* $\delta$ 2.7, singlet 3H

*b* $\delta$ 3.4, singlet, 1H

*c* $\delta$ 6.8, multiplet, 5H

(c) $C_7H_9N$

*a* $\delta$ 2.3, singlet, 3H

*b* $\delta$ 3.4, singlet, 2H

*c* $\delta$ 6.8, multiplet, 4H

## ANSWERS

### True–False

1. True. An amine nitrogen has two unshared electrons. This allows it to function as an electron-pair donor.
2. True
3. True
4. False. Most amines have unpleasant fishlike odors, but their salts are odorless.
5. True

6. False

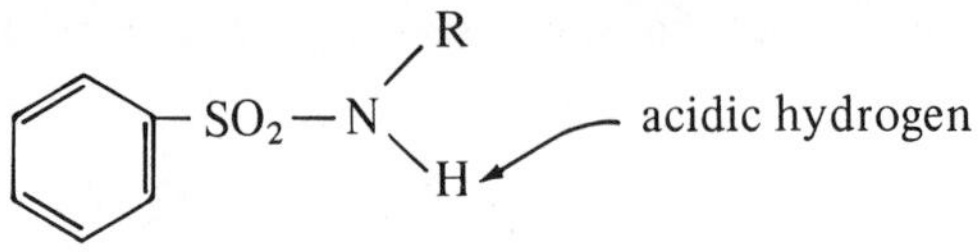

$C_6H_5-SO_2-N(R)(H)$ ← acidic hydrogen

This is the basis for the Hinsberg test. Primary amines form N-alkyl sulfonamides which have a replaceable hydrogen and are soluble in base. Secondary amines form N,N-dialkyl sulfonamides which have no replaceable hydrogens and are insoluble in base. Tertiary amines do not form sulfonamides.

7. True
8. False. Primary amines form diazonium salts, but secondary amines form N-nitroso amines. Diazonium salts from **primary aromatic amines** are moderately stable intermediates **if kept cold**, but diazonium salts from primary aliphatic amines decompose immediately when formed.
9. True
10. True
11. True
12. True

**Multiple Choice**

13. (c)
14. (b)

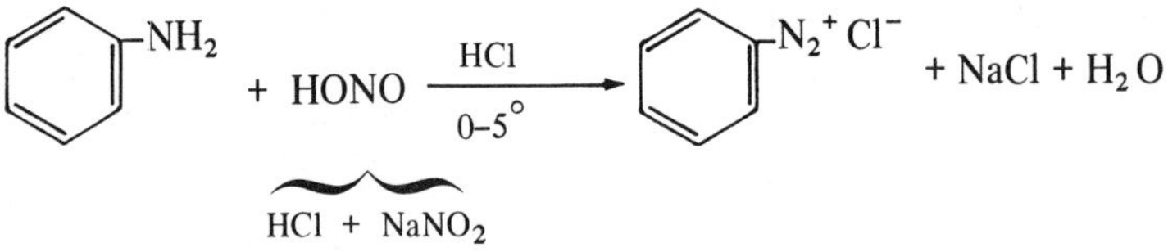

$$C_6H_5-NH_2 + \underbrace{HONO}_{HCl\ +\ NaNO_2} \xrightarrow[0\text{–}5^{\circ}]{HCl} C_6H_5-N_2^+\,Cl^- + NaCl + H_2O$$

15. (c)

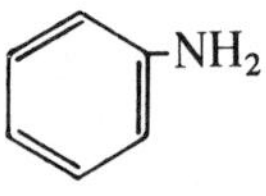

$$C_6H_5-NH_2 < NH_3 < (CH_3)_3N < CH_3NH_2 < (CH_3)_2NH_2$$

increasing basicity →

← increasing acidity of conjugate acid

The methyl group is inductively electron-releasing (+I effect). The unshared electron-pair is more readily available for bonding and thus the basic strength of the amine is increased. Resonance stabilization of aniline relative to the anilinium ion makes aniline a weaker base than ammonia. The stronger base forms the weakest conjugate acid.

16. (c) $R-C\equiv N + 2\ H_2O \xrightarrow{H^+} R-COOH + NH_3$

17. (c) The tertiary amine, $(CH_3)_3N$, has no replaceable hydrogens.
18. (a)
19. (c)

$$C_6H_5-SO_2-N(CH_2-CH_3)_2$$

There is no replaceable hydrogen on the nitrogen atom.

20. (a) Only primary aromatic amines form diazonium salts that are reasonably stable at 0–5°.

**Matching**

21. (n)
22. (d)
23. (k)
24. (b)
25. (c)
26. (j)
27. (f)
28. (g)
29. (l)
30. (h)
31. (i)
32. (m)
33. (e)
34. (a)

**Structural Formulas**

35. $C_6H_5-NH_2$

36. $CH_3-CH_2-CH(CH_3)-NH_2$

37. $[CH_3-N(CH_3)(H)-H]^+ \; Cl^-$

37. $H_2C-C(=O)-N(H)-C(=O)-CH_2$ (ring)

39. $(CH_3-CH_2)_4N^+ \; OH^-$

40. $H-C(=O)-N(H)-CH_2-CH_3$

41. $C_6H_5-N(H)-C(=O)-CH_3$

42. $NH_2-C_6H_4-SO_2-NH_2$ (para)

43. $H_2N-CH_2-CH_2-NH_2$

44. $C_6H_5-NH-NH_2$

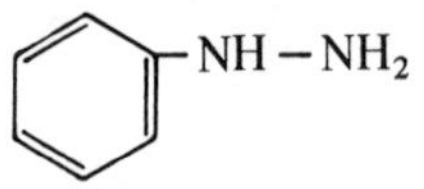

45. $(CH_3CH_2)_2\ddot{N}:^- Li^+$

46. $CH_3-CH(CH_3)-SH$

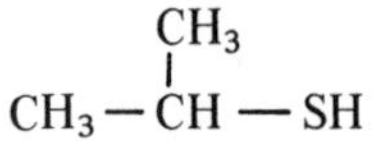

47. $CH_3-CH(CH_3)-S-CH(CH_3)-CH_3$

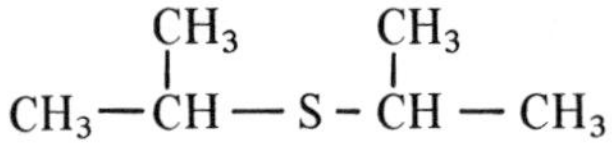

48. $CH_3-CH(CH_3)-S-S-CH(CH_3)-CH_3$

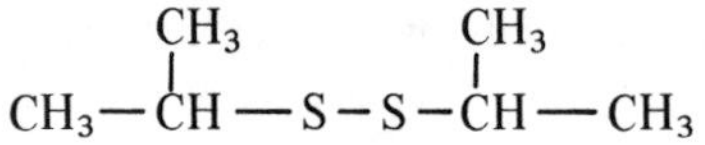

49.

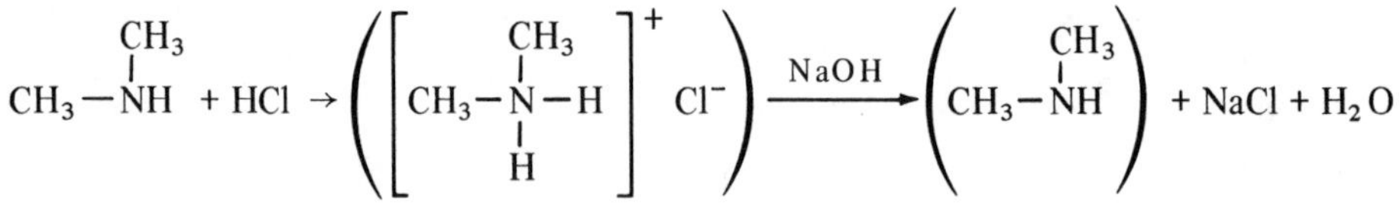

$$CH_3-NH(CH_3) + HCl \rightarrow \left(\left[CH_3-N(CH_3)(H)-H\right]^+ Cl^-\right) \xrightarrow{NaOH} \left(CH_3-NH(CH_3)\right) + NaCl + H_2O$$

50.

$$CH_3-CH_2-CH_2-C(=O)OH \xrightarrow{SOCl_2} \left(CH_3-CH_2-CH_2-C(=O)Cl\right) \xrightarrow{NH_3} \left(CH_3-CH_2-CH_2-C(=O)NH_2\right) \xrightarrow[(NaOBr)]{NaOH,\ Br_2} \left(CH_3-CH_2-CH_2-NH_2\right)$$

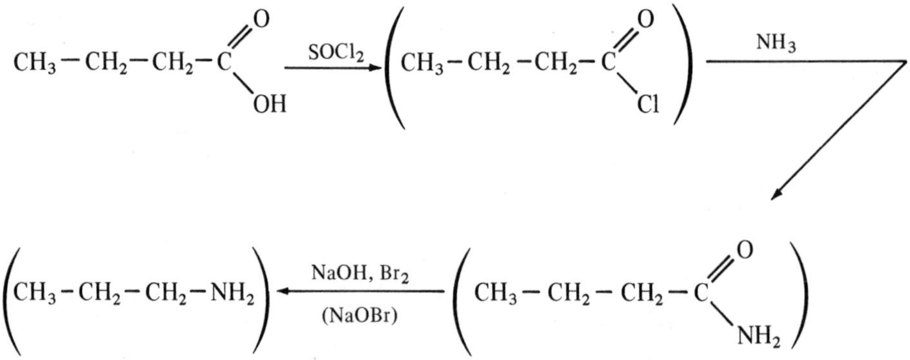

The last step is a Hofmann amide hypohalite degradation.

51.

Phthalimide N–H + KOH ⟶ (phthalimide anion $N^{-}$ $K^{+}$)

$$\xrightarrow{Cl-CH(CH_3)-CH_2-CH_3}$$

(N-alkyl phthalimide: phthalimide N–CH($CH_3$)–$CH_2$–$CH_3$)

An N-alkyl phthalimide

$$\xrightarrow{KOH}$$

($H_2N-CH(CH_3)-CH_2-CH_3$)

A primary amine

$$\xrightarrow{CH_3CH_2CH_2CH_2^- \; Li^+}$$

($Li^+ \; {}^-NH-CH(CH_3)-CH_2-CH_3$)

The synthesis of the primary amine, *sec*-butylamine, is an example of Gabriel's phthalimide synthesis.

52.

$$NC-(CH_2)_4-CN \xrightarrow[Ni]{4H_2,} (H_2N-(CH_2)_6-NH_2)$$

$$NC-(CH_2)_4-CN \xrightarrow{H_2O,\ H^+} (HOOC-(CH_2)_4-COOH)$$

200-300° (polymerizing conditions)

$$H_2N-(CH_2)_6-\overset{H}{N}\left(\overset{O}{\overset{\|}{C}}-(CH_2)_4-\overset{O}{\overset{\|}{C}}-\overset{H}{N}-(CH_2)_6-\overset{H}{N}\right)_x\overset{O}{\overset{\|}{C}}-(CH_2)_4-\overset{O}{\overset{\|}{C}}-OH$$

Nylon 66

53.

$$\left(C_6H_5\text{-}NH_2\right) + HCl + NaNO_2 \xrightarrow{0\text{–}5^\circ} C_6H_5\text{-}N_2^+\,Cl^-$$

$C_6H_5\text{-}N_2^+\,Cl^-$ with $H_2O$, warm $\rightarrow$ $\left(C_6H_5\text{-}OH\right)$

$C_6H_5\text{-}N_2^+\,Cl^-$ with CuCN $\rightarrow$ $\left(C_6H_5\text{-}C{\equiv}N\right)$

$C_6H_5\text{-}N_2^+\,Cl^-$ with CuBr $\rightarrow$ $\left(C_6H_5\text{-}Br\right)$

$C_6H_5\text{-}N_2^+\,Cl^-$ with $H_3PO_2$, $H_2O$ $\rightarrow$ $\left(C_6H_6\right)$

54.

$$CH_3-CH(CH_3)-CH_2-Br \xrightarrow{Mg} \left(CH_3-CH(CH_3)-CH_2-MgBr\right) \xrightarrow[\text{ether}]{S} \left(CH_3-CH(CH_3)-CH_2-S-MgBr\right)$$

$$\left(CH_3-CH(CH_3)-CH_2-S-MgBr\right) \xrightarrow{H_2O,\ H^+} \left(CH_3-CH(CH_3)-CH_2-SH\right)$$

$$CH_3-CH(CH_3)-CH_2-Br \xrightarrow{(H_2N)_2C{=}S} \left((H_2N^+{=})(H_2N)C-S-CH_2-CH(CH_3)-CH_3\quad Br^-\right)$$

$$\left((H_2N^+{=})(H_2N)C-S-CH_2-CH(CH_3)-CH_3\quad Br^-\right) \xrightarrow[2.\ H^+]{1.\ H_2O,\ NaOH} \left(CH_3-CH(CH_3)-CH_2-SH\right)$$

$$\left(CH_3-CH(CH_3)-CH_2-SH\right) \underset{LiAlH_4}{\overset{I_2}{\rightleftharpoons}} \left(CH_3-CH(CH_3)-CH_2-S-S-CH_2-CH(CH_3)-CH_3\right)$$

$$\left(CH_3-CH(CH_3)-CH_2-SH\right) \xrightarrow{NaOH} \left(CH_3-CH(CH_3)-CH_2-S^-Na^+\right) \xrightarrow{CH_3CH_2Br} \left(CH_3-CH(CH_3)-CH_2-S-CH_2-CH_3\right)$$

## Syntheses

55.

$$C_6H_6 + HNO_3 + H_2SO_4 \rightarrow C_6H_5{-}NO_2 \xrightarrow[\text{catalyst}]{H_2,} C_6H_5{-}NH_2$$

$$C_6H_5{-}NH_2 \xrightarrow{CH_3{-}C(=O){-}Cl} C_6H_5{-}N(H){-}C(=O){-}CH_3$$

$$C_6H_5{-}N(H){-}C(=O){-}CH_3 \xrightarrow[H_2SO_4]{HNO_3,} O_2N{-}C_6H_4{-}N(H){-}C(=O){-}CH_3$$

56. $CH_3-CH_2-Cl + KCN \rightarrow CH_3-CH_2-C{\equiv}N \xrightarrow[\text{catalyst, heat, pressure}]{H_2,} CH_3-CH_2-CH_2-NH_2$

57.

$$CH_3-CH_2-CH_2-C(=O)OH \xrightarrow{SOCl_2} CH_3-CH_2-CH_2-C(=O)Cl \xrightarrow{NH_3} CH_3-CH_2-CH_2-C(=O)NH_2$$

$$CH_3-CH_2-CH_2-C(=O)NH_2 \xrightarrow{LiAlH_4} CH_3-CH_2-CH_2-CH_2-NH_2$$

## Miscellaneous

58.

$$C_6H_5{-}SO_2{-}Cl + H_2N{-}CH_3 \rightarrow C_6H_5{-}SO_2{-}N(H){-}CH_3 + HCl$$

Benzenesulfonyl chloride

Replaceable hydrogen (on N)

(Soluble in base)

$$C_6H_5{-}SO_2{-}Cl + H{-}N(CH_3){-}CH_3 \rightarrow C_6H_5{-}SO_2{-}N(CH_3)_2 + HCl$$

No replaceable hydrogen on amide

(Insoluble in base)

$C_6H_5{-}SO_2{-}Cl + (CH_3)_2N{-}CH_3 \longrightarrow$ No Reaction

59.

$C_6H_5{-}NH_2 \xrightarrow[NaNO_2,\ 0\text{-}5^\circ]{HCl} C_6H_5{-}N_2^+\ Cl^- \xrightarrow{\beta\text{-naphthol (OH)}}$ 1-(phenylazo)-2-naphthol ($N{=}N{-}C_6H_5$, OH)

An orange-red azo dye

Primary aliphatic amines do not form a stable diazonium salt and therefore do not couple with phenols such as $\beta$-naphthol to give the colored azo dye.

60. (a) II. The electron-releasing methyl group exerts a base-strengthening effect. The electron-withdrawing carbonyl group exerts a base-weakening effect.

(b) I. The electron pair of nitrogen can not delocalize and enter into the resonance of the ring as it can in compound II. The unshared electron pair is more available in I.

(c) I. The electron pair of nitrogen can not delocalize as it can in II. The unshared electron pair is more available in I.

61. (a)

2,3-dimethylaniline ($NH_2$, $CH_3$, $CH_3$) or 3,4-dimethylaniline ($H_2N$, $CH_3$, $CH_3$)

Example:

$NH_2$-C6H3(CH3)2 (2,3-dimethylaniline) $\xrightarrow{HCl,\ NaNO_2}$ $N_2^+Cl^-$-C6H3(CH3)2 $\xrightarrow{C_6H_5OH}$ azo compound: $H_3C$, $CH_3$-C6H3-N=N-C6H4-OH

An azo compound

Diazonium salt $\xrightarrow{H_3PO_2}$ C6H4($CH_3$)($CH_3$) $\xrightarrow{\text{oxidation}}$ C6H4(COOH)(COOH)

$C_8H_{10}$ (N.E. = 83)

$\xrightarrow{\text{heat}}$ phthalic anhydride (C6H4 with C(=O)–O–C(=O))

(b) C6H5-CH($NH_2$)-$CH_3$ $\xrightarrow{HCl,\ NaNO_2}$ C6H5-CH(OH)-$CH_3$ + $N_2$

$C_8H_{10}O$

$\xrightarrow{\text{oxidation}}$ C6H5-COOH

(N.E. = 122)

62. (a) $NH_3 + H^+ \longrightarrow NH_4^+$ (conjugate acid)

$CH_3NH_2 + H^+ \longrightarrow CH_3NH_3^+$ (conjugate acid)

(b) $CH_3NH_2$ The stronger the base, the larger the $K_b$, but the smaller the p$K_b$.

(c) 9.24, 10.62

$pK_a = 14 - pK_b$

(d) $NH_4^+$

increasing basicity of the base →

increasing value of $K_b$
decreasing value of p$K_b$
decreasing acidity of conjugate acid
decreasing value of $K_a$ of conjugate acid
increasing value of p$K_a$ of conjugate acid

63. (a)

$C_6H_5$—$CH_2$—$NH_2$

*b* *a*

*c*
(aromatic hydrogens)

(b)

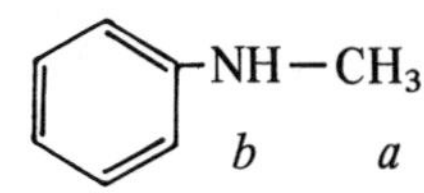


*b* *a*

*c*
(aromatic hydrogens)

(c)

$NH_2$

$CH_3$ *b*

*a* *c*
(aromatic hydrogens)

# 14

# *Carbohydrates*

## INTRODUCTION

The carbohydrates constitute a major class of naturally occurring organic compounds synthesized by green plants through a process known as **photosynthesis.** Carbohydrates include sugars, starches, and cellulose and may be described as polyhydroxy aldehydes or polyhydroxy ketones, or substances which, when hydrolyzed, give polyhydroxy aldehydes or polyhydroxy ketones. We shall find that the carbonyl group in a carbohydrate does not normally occur as such but instead combines intramolecularly with a hydroxyl group to form a **hemiacetal** structure which can exist in either an $\alpha$ or $\beta$ form. These two forms are called **anomers** and the hemiacetal carbon is called the anomeric carbon. We shall study in detail the structure and chemical properties of **glucose**, the most important **monosaccharide.** Monosaccharides may be bonded together by an **acetal** linkage (a **glycosidic** linkage) to form **oligosaccharides** or **polysaccharides.** The structures and properties of some of the important oligosaccharides (the **disaccharides–sucrose, maltose, lactose,** and **cellobiose**) will be discussed as will the important polysaccharides (**starch, glycogen,** and **cellulose**). A review of stereochemistry (Chapter 5) may be necessary for a better understanding of the properties of the carbohydrates.

## PROBLEMS

### True–False

1. T F Most common monosaccharides contain five or six carbon atoms in their structure.

2. T F Oligosaccharides contain more monosaccharide units joined together through glycosidic linkages than are to be found in polysaccharides.
3. T F Dextrose is another name for glucose.
4. T F Living animal organisms depend upon glucose as a main source of energy.
5. T F The D-series of aldohexoses is related, from a configurational standpoint, to D-glyceraldehyde.
6. T F The linkage between two monosaccharides to form a disaccharide is called a glycosidic linkage.
7. T F Methyl glucosides are capable of reducing Fehling's solution or Tollens' reagent.
8. T F $\alpha$-D-Glucose and $\beta$-D-glucose have different optical properties.
9. T F The enzyme zymase acts upon fructose to yield ethyl alcohol and carbon dioxide.
10. T F The monosaccharide fructose is ordinary table sugar.
11. T F Fructose is sweeter than glucose and is largely responsible for the sweetness of honey.
12. T F The difference between cellobiose and maltose is that the glucose units are joined by a $\beta$-linkage in cellobiose and by an $\alpha$-linkage in maltose.
13. T F Aldoses that differ only in the configuration about the No. 2 carbon atom are called epimers.
14. T F Both starch and glycogen yield D-glucose upon complete hydrolysis.
15. T F Man is capable of using cellulose for food.
16. T F Carbohydrates, in meeting the energy requirements of the body, are eventually oxidized to carbon dioxide, water, and energy.
17. T F Glycogen is synthesized in the body and can be stored until energy is needed.

**Multiple Choice**

18. Glucose cannot be classified as

    (a) a hexose
    (b) a carbohydrate
    (c) an oligosaccharide
    (d) an aldose
    (e) a monosaccharide

19. Which statement about glucose is false?

    (a) Glucose is dextrorotatory.
    (b) Glucose forms the same osazone as galactose.
    (c) Glucose reacts with acetic anhydride to produce a pentaacetate.
    (d) Glucose is a reducing sugar.
    (e) Glucose can exist in a hemiacetal form to give a pyranose ring.

20. A carbohydrate not naturally occurring is
    (a) gulose
    (b) galactose
    (c) mannose
    (d) glucose
    (e) fructose

21. All of the following monosaccharides give the same osazone except
    (a) fructose
    (b) galactose
    (c) mannose
    (d) glucose

22. Which structure has the architectural form of an acetal?

(a)

```
       H
       |
       C=O
       |
   H - C - OH
       |
  HO - C - H
       |
   H - C - OH
       |
   H - C - OH
       |
       CH2OH
```

(b)

```
       ┌──────────┐
   H - C - OH     │
       |          │
   H - C - OH     │
       |          │
  HO - C - H      │
       |          │
   H - C - OH     │
       |          │
   H - C - O ─────┘
       |
       CH2OH
```

(c)

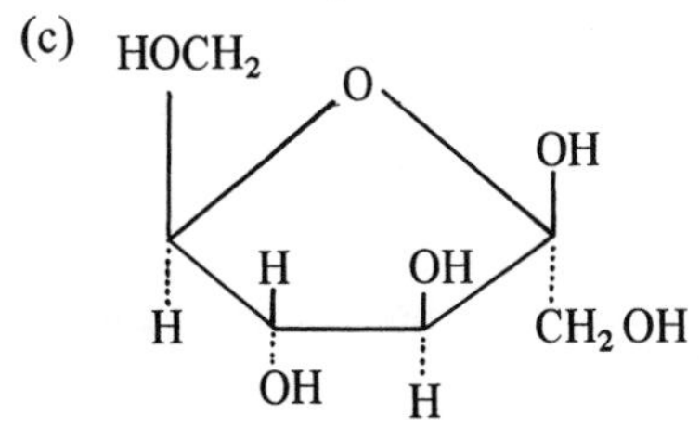


(d)

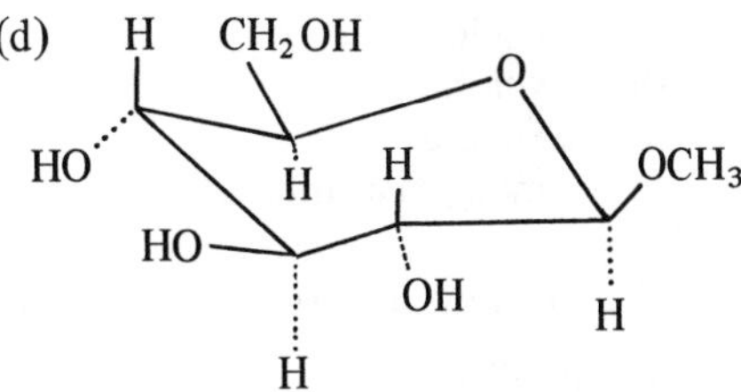


23. Which structure has the form of a hemiacetal?

(a)

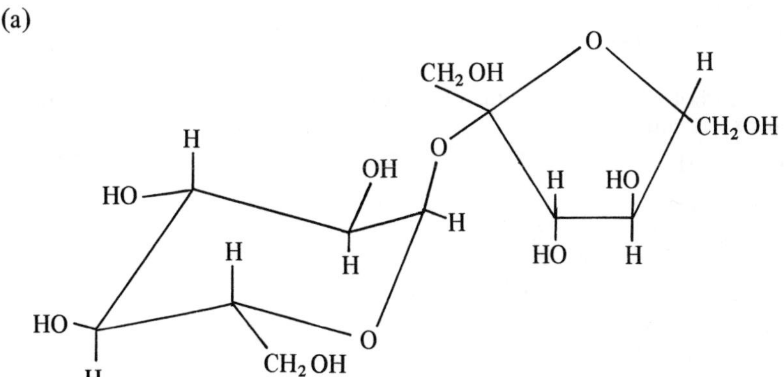

(b)

```
       H
       |
       C=O
       |
   H—C—OH
       |
  HO—C—H
       |
   H—C—OH
       |
   H—C—OH
       |
       CH2OH
```

(c)

```
       H        H
       |        |
       C=N—N—C6H5
       |
       |        H
       |        |
       C=N—N—C6H5
       |
  HO—C—H
       |
   H—C—OH
       |
   H—C—OH
       |
       CH2OH
```

(d)

```
  H   CH2OH          O
HO                H        H
        H
HO               OH
                          OH
      H
```

(e)

```
   H—C—OCH3 ┐
     |       |
   H—C—OH    |
     |       |
  HO—C—H     |
     |       |
   H—C—OH    |
     |       |
   H—C—O ────┘
     |
     CH2OH
```

24. Fructose can be named or classified as any one of the following except
   (a) an aldose
   (b) levulose
   (c) a hexose
   (d) a monosaccharide

25. The 3-carbon acid which is a product of the enzymatic breakdown of glucose is
   (a) pyruvic acid
   (b) citric acid
   (c) succinic acid
   (d) acetic acid
   (e) oxalacetic acid

26. Which carbohydrate is nonreducing, that is, which will not reduce Fehling's or Tollens' reagent?

(a)

```
       H
       |
       C=O
       |
   H—C—OH
       |
  HO—C—H
       |
   H—C—OH
       |
   H—C—OH
       |
       CH2OH
```

(b)

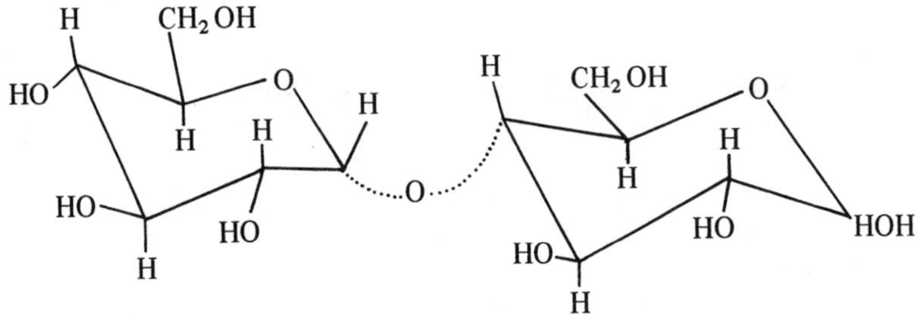

(c)

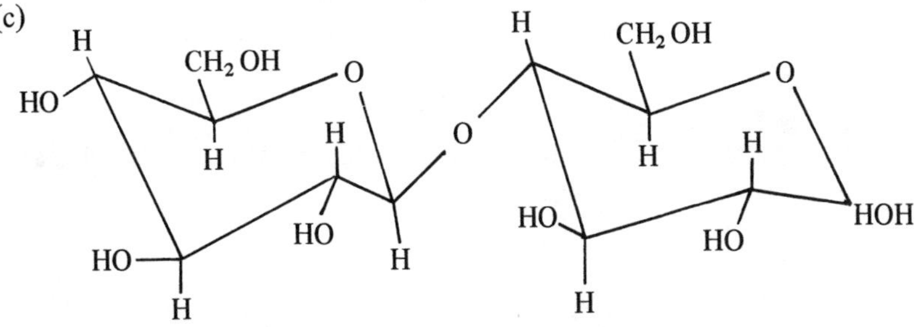

(d)

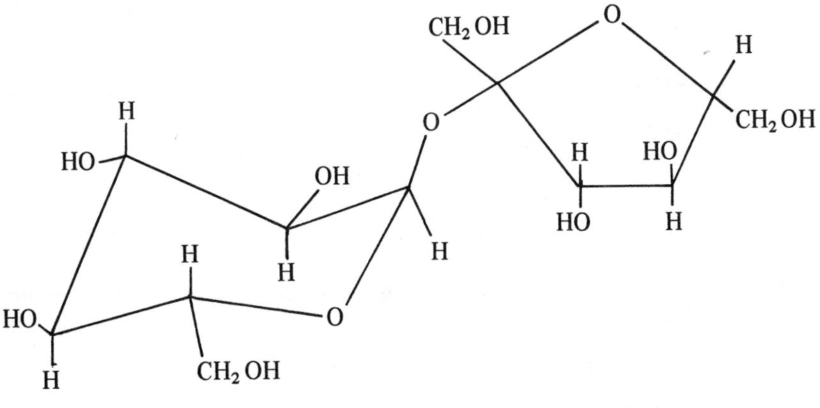

27. Which of the following is not a product derived from cellulose?
    (a) rayon
    (b) guncotton
    (c) paper
    (d) insulin
    (e) celluloid

**Completion**

28. Polyhydroxy aldehydes or ketones are called ____________________ .
29. Monosaccharides containing five carbon atoms are called ____________________ and those containing six carbon atoms are called ____________________ .

30. Carbohydrates that can be hydrolyzed to give two monosaccharides are called ____________________.

31. A glycoside produced from glucose is specifically called a ____________________.

32.

```
   ┌──────┐
H—C—OR    │
  |       │
H—C—OH    │
  |       │
HO—C—H    │
  |       │
H—C—OH    │
  |       │
H—C—O─────┘
  |
 CH2OH
```

The nonsugar portion (R) of the above formula for a glucoside is called an ____________________.

33.

```
   ┌──────┐                                 ┌──────┐
H—C—OH    │           H               HO—C—H     │
  |       │           |                  |       │
H—C—OH    │          C=O               H—C—OH    │
  |       │           |                  |       │
HO—C—H    │   ⇌     H—C—OH      ⇌     HO—C—H     │
  |       │           |                  |       │
H—C—OH    │        HO—C—H              H—C—OH    │
  |       │           |                  |       │
H—C—O─────┘         H—C—OH             H—C—O─────┘
  |                   |                  |
 CH2OH              H—C—OH              CH2OH
                      |
                     CH2OH
```

α-D-Glucose　　　　　D-Glucose　　　　　β-D-Glucose

A freshly prepared aqueous solution of α-D-glucose has a specific rotation of +113°. The rotation changes to the equilibrium value of +52° upon standing. Such a change in rotation with time is called ____________________.

34. Glucose exists to a large extent in its hemiacetal form. This six-membered cyclic ether is called a ____________________ ring.

35. The five-membered cyclic ether of the hemiacetal form of fructose is referred to as a ____________________ ring.

36. Sugars react with an excess of phenylhydrazine to form a derivative called an ____________________.

37. The hydrolysis mixture of sucrose is called ____________________ sugar because of the inversion or change in sign of the specific rotation during hydrolysis.

38. The disaccharide found in mammalian milk is ____________________.

39. The reserve carbohydrate of most plants is ____________________ and that of animals is ____________________.

40. Starch consists of two structurally different polysaccharides. One of the polysaccharides is soluble in water and consists of glucose units joined through alpha 1-4 linkages. It is called ____________________. The other polysaccharide is insoluble in water and in addition to the alpha 1-4 linkages has branches joined between carbon 1 and carbon 6. It is called ____________________.

41. The polysaccharide which comprises the skeletal material of plants is ____________________.

42. Stereoisomers which differ in the configuration of the hemiacetal carbon of the carbohydrate are called ____________________.

**Matching** (Match each structure to the appropriate name or description.)

(a)

```
       H
       |
       C=O
       |
  HO—C—H
       |
   H—C—OH
       |
   H—C—OH
       |
       CH2OH
```

(b)

(c)

(d)

(e)

(f)

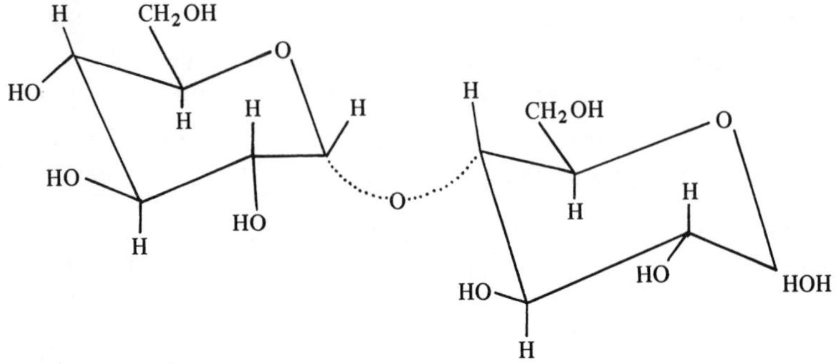

(g)

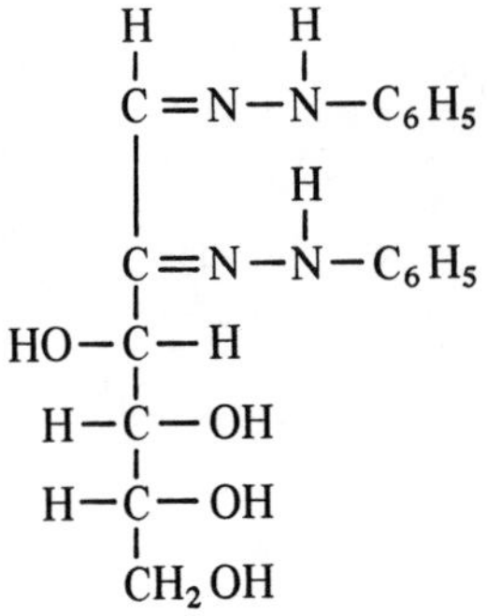

$H-C=N-NH-C_6H_5$

$C=N-NH-C_6H_5$

$HO-C-H$

$H-C-OH$

$H-C-OH$

$CH_2OH$

(h)

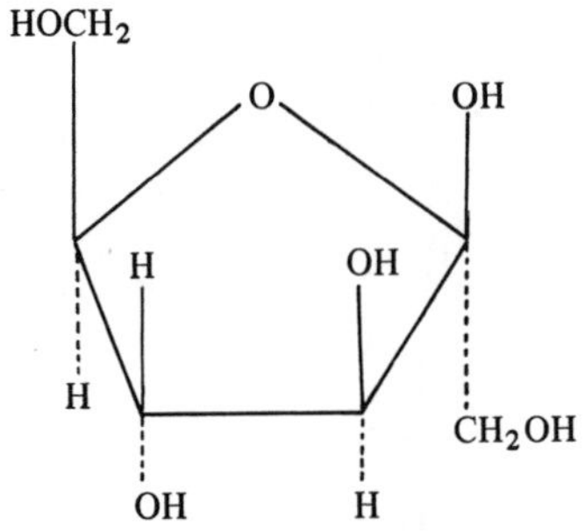

(i)

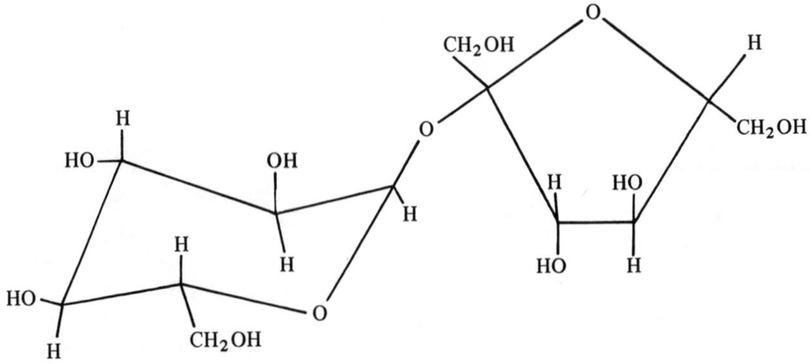

(j)

H, HO, $CH_2OH$, O, H, H, $OCH_3$, HO, H, H, OH

(k)

H H
| |
$C{=}N{-}N{-}C_6H_5$
| H
| |
$C{=}N{-}N{-}C_6H_5$
|
$HO{-}C{-}H$
|
$HO{-}C{-}H$
|
$H{-}C{-}OH$
|
$CH_2OH$

(l)

HO, $CH_2OH$, H, O, H, H, O, HO, HO, H, H, H, $CH_2OH$, O, H, H, HO, HO, HOH, H

(m) $CO_2H$
|
$(CHOH)_4$
|
$CH_2OH$

43. ______ An aldonic acid
44. ______ $\beta$-Methylglucoside
45. ______ $\beta$-D-Glucopyranose
46. ______ $\alpha$-D-Glucose
47. ______ $\beta$-D-Fructopyranose
48. ______ $\beta$-D-Fructofuranose
49. ______ A pentose
50. ______ Osazone of D-galactose
51. ______ Osazone of D-glucose, D-mannose, and D-fructose
52. ______ A disaccharide consisting of a glucose and fructose unit
53. ______ A disaccharide consisting of a glucose and galactose unit
54. ______ A disaccharide consisting of two glucose units joined through a $\beta$-linkage
55. ______ A disaccharide consisting of two glucose units joined through an $\alpha$-linkage

(Match the word(s) or phrases that name or describe the same compound.)

(n) D-Glucose
(o) D-Fructose
(p) Sucrose
(q) Maltose
(r) Lactose
(s) Lactase
(t) Invert sugar
(u) Cellobiose
(v) Maltase

56. _____ Levulose
57. _____ Disaccharide obtained from partial hydrolysis of cellulose
58. _____ Malt sugar
59. _____ Mixture obtained from the hydrolysis of sucrose
60. _____ Dextrose
61. _____ Milk sugar
62. _____ Enzyme which breaks maltose down into two D-glucose units
63. _____ Enzyme which breaks lactose down into D-glucose and D-galactose
64. _____ Table sugar

**Reactions** (Complete the following.)

65.

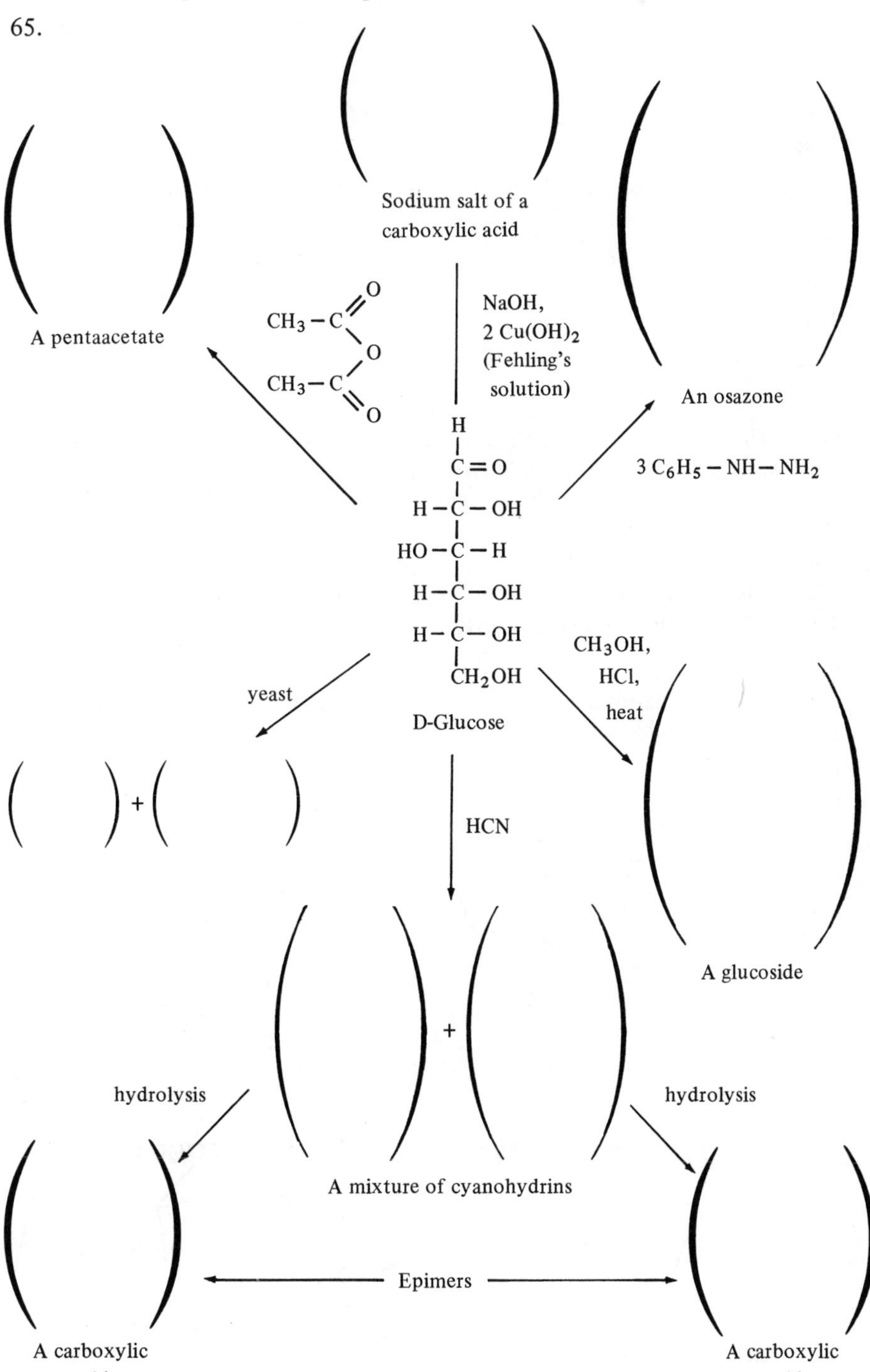

## ANSWERS

**True–False**

1. True
2. False. Oligosaccharides consist of two or more (but a relatively small number) monosaccharide units joined by glycosidic linkages, but polysaccharides consist of hundreds of monosaccharide units joined together.
3. True
4. True
5. True
6. True
7. False. The aldehyde group of glucose is easily oxidized by Fehling's solution or Tollens' reagent, but the aldehyde group no longer exists in a glucoside. A glucoside, therefore, is incapable of reducing Fehling's solution or Tollens' reagent.
8. True. The configuration of groups attached to carbon (1), the anomeric carbon, of α-D-glucose is different from that of carbon (1) of β-D-glucose. Therefore they exhibit different optical properties.
9. True
10. False. The disaccharide sucrose, composed of one glucose unit and one fructose unit, is common table sugar.
11. True
12. True.

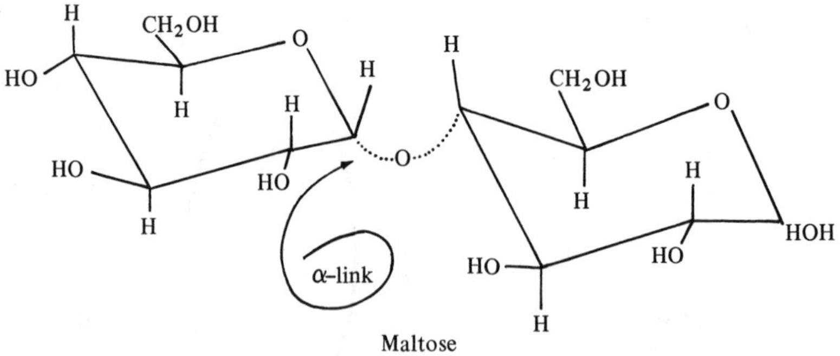

Maltose

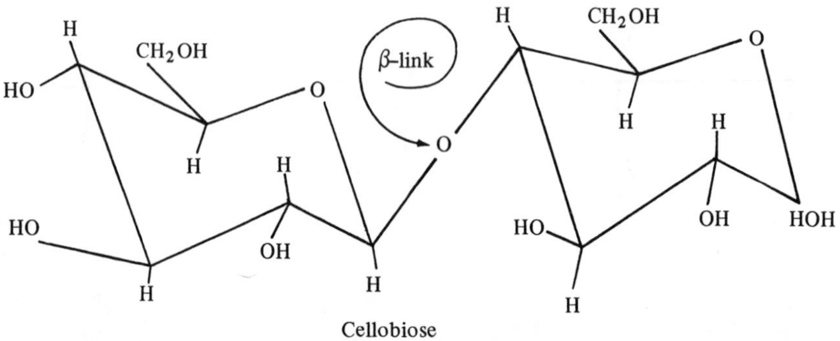

Cellobiose

13. True
14. True
15. False. Man lacks the enzymes capable of hydrolyzing the $\beta$-glycosidic linkage. Cellulose is composed of glucose units joined by $\beta$-links. Starch is composed of glucose units joined by $\alpha$-links and can be utilized by man.
16. True
17. True

**Multiple Choice**

18. (c) Oligosaccharides consist of two or more (but a relatively small number) monosaccharide units joined by glycosidic linkages.
19. (b) The same osazone is given by carbohydrates which differ only in the configuration of carbons (1) and (2). Glucose and galactose differ in the configuration of carbon (4).
20. (a)
21. (b) The configuration of carbons (3), (4), (5), and (6) of glucose, mannose, and fructose are the same and thus they give the same osazone.
22. (d) An acetal has the general structure

$$\begin{array}{c} \text{H} \\ | \\ \text{R}'-\text{C}-\text{O}-\text{R}'' \\ | \\ \text{O}-\text{R}''' \end{array}$$

23. (d) A hemiacetal has the general structure

$$\begin{array}{c} \text{H} \\ | \\ \text{R}'-\text{C}-\text{O}-\text{R}'' \\ | \\ \text{OH} \end{array}$$

24. (a) Fructose is a ketose.
25. (a)
26. (d) Sucrose is a nonreducing sugar because there is no potential aldehyde group to be oxidized. There is no hemiacetal group in sucrose.
27. (d) Insulin is a protein.

**Completion**

28. carbohydrates
29. pentoses, hexoses
30. disaccharides or oligosaccharides
31. glucoside
32. aglycone
33. mutarotation
34. pyranose
35. furanose
36. osazone
37. invert
38. lactose
39. starch, glycogen
40. amylose, amylopectin
41. cellulose
42. anomers

**Matching**

43. (m)
44. (j)
45. (b)
46. (e)
47. (d)
48. (h)
49. (a)
50. (k)
51. (g)
52. (i)
53. (l)
54. (c)
55. (f)
56. (o)
57. (u)
58. (q)
59. (t)
60. (n)
61. (r)
62. (v)
63. (s)
64. (p)

## Reactions

65.

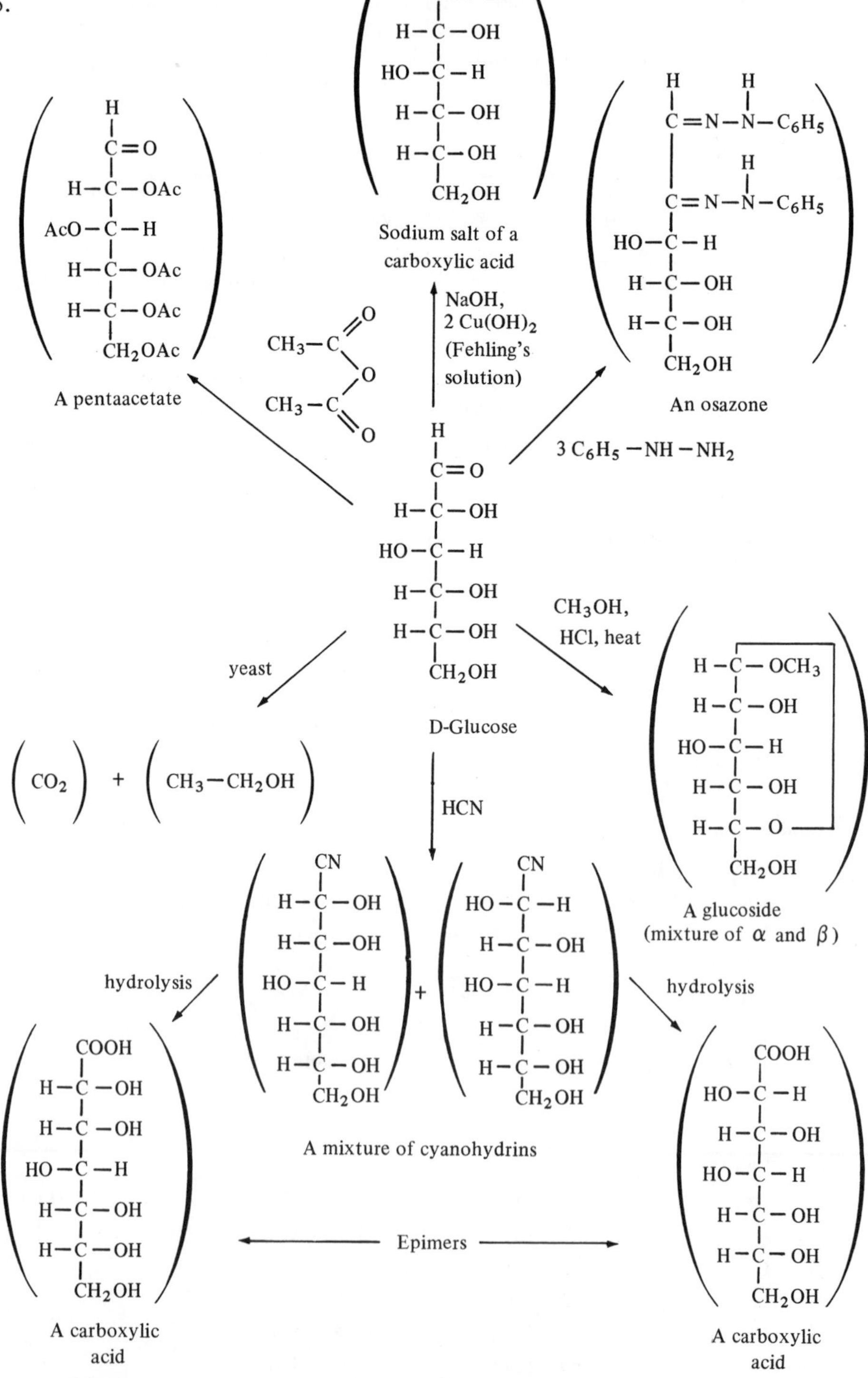

# 15

# *Amino Acids, Peptides, and Proteins*

## INTRODUCTION

This chapter probably would appear in a biochemistry text book near the beginning because "protein" comes from the Greek word meaning "first". The chapter on the proteins is a very important one because we are dealing with complex chemistry that takes place within the living cell.

To learn something about the macromolecules which comprise protein matter, we will find it necessary to begin our study with the properties of the small molecules of which these large polymeric substances are made—namely, the **α-amino acids**. After we have studied the physical and chemical properties of the α-amino acids, we shall see how they can be assembled into **peptides** and **proteins**.

One of the most interesting parts of this chapter is the portion describing how the sequence of amino acids in a peptide is determined.

## PROBLEMS

### True–False

1. T F Proteins make up the bulk of the body's nonbony structure.
2. T F β-Amino acids are as common in living tissue as α-amino acids.
3. T F Amino acids of protein origin possess the L-configuration.
4. T F All animals synthesize the same amino acids.

5. T F Many amino acids are colorless liquids.
6. T F Amino acids are joined together in a very specific sequence to form proteins.
7. T F The molecular weights of proteins are very high.
8. T F A positive biuret test is given by proteins, but not by α-amino acids.
9. T F Proteins, like fats and carbohydrates, are used primarily to supply heat and energy to the body.
10. T F Protein material eaten by animals provides the amino acids which are used by the cells to synthesize other proteins.
11. T F Amino acids not needed by the body for the synthesis of proteins can be ultimately oxidized to provide energy.
12. T F Only certain plants are capable of "fixing" atmospheric nitrogen–that is, converting atmospheric nitrogen into organic nitrogen compounds.
13. T F The $pK_a$ of the carboxylic acid group of an amino acid is always less than the $pK_a$ of the ammonium ion.

**Multiple Choice**

14. Which of the following compounds is an amino acid?

(a) $CH_3-CH_2-\overset{\overset{O}{\|}}{C}-O^-NH_4^+$

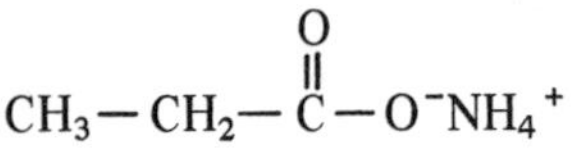

(b) $CH_3-CH_2-\overset{\overset{O}{\|}}{C}-NH_2$

(c) $CH_3-\underset{\underset{NH_2}{|}}{CH}-\overset{\overset{O}{\|}}{C}-OH$

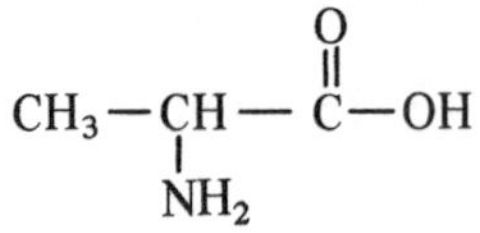

(d) $CH_3-\underset{\underset{NH_2}{|}}{CH}-\overset{\overset{O}{\|}}{C}-Cl$

15. The material which contains the least amount of protein is
   (a) muscle
   (b) enzyme
   (c) bone
   (d) hair
   (e) antibodies

16. Which of the following amino acids is not optically active?

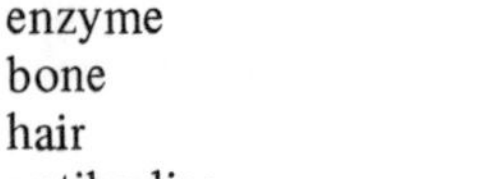

(a) $CH_3-\overset{\overset{H}{|}}{\underset{\underset{NH_2}{|}}{C}}-COOH$

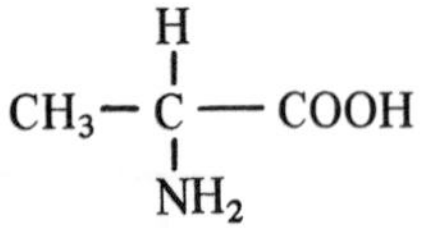

(b) $HO-\overset{\overset{H}{|}}{\underset{\underset{H}{|}}{C}}-\overset{\overset{H}{|}}{\underset{\underset{NH_2}{|}}{C}}-COOH$

(c) $H_2C$——$CH_2$ / $H_2C$ $CHCOOH$ joined through N–H (five-membered ring: $H_2C-CH_2$, $H_2C$, $CHCOOH$, $\underset{H}{N}$)

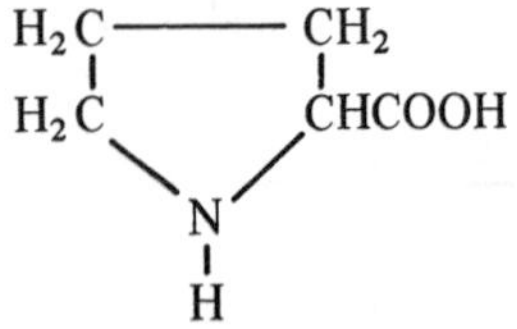

(d) indole ring (with N–H) $-\overset{\overset{H}{|}}{\underset{\underset{H}{|}}{C}}-\overset{\overset{H}{|}}{\underset{\underset{NH_2}{|}}{C}}-COOH$

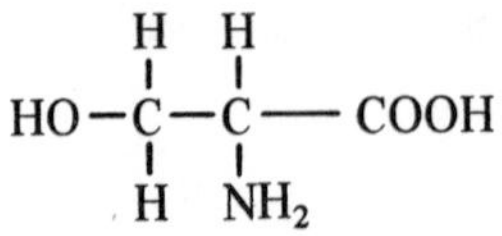

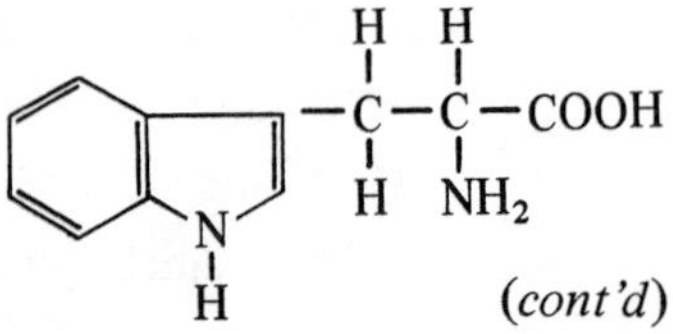

*(cont'd)*

(e) $H-\underset{NH_2}{\overset{H}{C}}-COOH$

17. Which of the following amino acids is not considered to be a neutral amino acid?

(a) $\underset{NH_2}{CH_2}-CH_2-CH_2-CH_2-\underset{NH_2}{\overset{H}{C}}-COOH$

(b) $H-\underset{NH_2}{\overset{H}{C}}-COOH$

(c) $H_2C-CH_2$ / $H_2C$ $CHCOOH$ / N–H (pyrrolidine ring: $H_2C-CH_2-CHCOOH$, with $H_2C$ and $CHCOOH$ joined through N–H)

(d) $CH_3-\underset{H}{\overset{CH_3}{C}}-\underset{H}{\overset{H}{C}}-\underset{NH_2}{\overset{H}{C}}-COOH$

(e) $CH_3-\underset{OH}{\overset{H}{C}}-\underset{NH_2}{\overset{H}{C}}-COOH$

18. Which amino acid is classified as a basic amino acid?

(a) Imidazole ring (H–C, N, NH, H–C=C) attached: $-CH_2-\underset{NH_2}{\overset{H}{C}}-COOH$

(b) $HS-CH_2-\underset{NH_2}{\overset{H}{C}}-COOH$

(c) $CH_3-\underset{H}{\overset{CH_3}{C}}-\underset{NH_2}{\overset{H}{C}}-COOH$

(d) $HOOC-CH_2-\underset{NH_2}{\overset{H}{C}}-COOH$

(e) $H_2C-CH_2$ / $H_2C$ $CHCOOH$ / N–H (pyrrolidine ring: $H_2C-CH_2-CHCOOH$, with $H_2C$ and $CHCOOH$ joined through N–H)

19. The reaction sequence that does not yield an $\alpha$-amino acid is

(a) $R-CH{=}CH-COOH + NH_3 \xrightarrow{\text{heat}}$

(b) $R-\underset{Br}{CH}-COOH + 2\ NH_3 \longrightarrow$

(c)

$$\text{C}_6\text{H}_4(\text{C}{=}\text{O})_2\text{N:}^-\ \text{K}^+ + \text{X}-\underset{\text{H}}{\overset{\text{R}}{\text{C}}}-\overset{\text{O}}{\overset{\|}{\text{C}}}-\text{OC}_2\text{H}_5 \rightarrow \text{C}_6\text{H}_4(\text{C}{=}\text{O})_2\text{N}-\underset{\text{H}}{\overset{\text{R}}{\text{C}}}-\overset{\text{O}}{\overset{\|}{\text{C}}}-\text{OC}_2\text{H}_5 \xrightarrow{\text{hydrolysis}}$$

$$+ \text{K}^+\text{X}^-$$

(d)

$$\text{R}-\overset{\text{H}}{\text{C}}{=}\text{O} + \text{HCN} + \text{NH}_3 \longrightarrow \text{R}-\underset{\text{NH}_2}{\text{CH}}-\text{CN} \xrightarrow{2\ \text{H}_2\text{O}}$$

20. The reaction sequence in question 18 which is known as the Strecker synthesis is____________.

21. Which amino acid will not liberate nitrogen gas when treated with nitrous acid?

(a) $\text{CH}_3-\underset{\text{NH}_2}{\overset{\text{H}}{\text{C}}}-\text{COOH}$

(b) $\text{H}_2\text{C}-\text{CH}_2$ / $\text{H}_2\text{C}$, $\text{CHCOOH}$ ring closed through $\text{N}-\text{H}$

(c) $\underset{\text{NH}_2}{\text{CH}_2}-\text{CH}_2-\text{CH}_2-\text{CH}_2-\underset{\text{NH}_2}{\overset{\text{H}}{\text{C}}}-\text{COOH}$

(d) $\text{HO}\overset{\text{O}}{\overset{\|}{\text{C}}}-\text{CH}_2-\underset{\text{NH}_2}{\overset{\text{H}}{\text{C}}}-\text{COOH}$

(e) $\text{CH}_3-\text{S}-\text{CH}_2-\text{CH}_2-\underset{\text{NH}_2}{\overset{\text{H}}{\text{C}}}-\text{COOH}$

22. Six different amino acids can be combined in different sequences to form how many different hexapeptides?

(a) 6
(b) 12
(c) 36
(d) 72
(e) 720

23.

$$\text{H}_2\text{N}-\text{CH}_2-\overset{\text{O}}{\overset{\|}{\text{C}}}-\overset{\text{H}}{\text{N}}-\underset{\text{CH}_2\text{C}_6\text{H}_5}{\overset{\text{H}}{\text{C}}}-\overset{\text{O}}{\overset{\|}{\text{C}}}-\overset{\text{H}}{\text{N}}-\underset{\text{CH}_3}{\text{CH}}-\overset{\text{O}}{\overset{\|}{\text{C}}}-\text{OH}$$

The name of the tripeptide above is

(a) glycyl-phenylalanyl-alanine

(b) glycyl-alanyl-phenylalanine
(c) alanyl-phenylglycyl-valine
(d) alanyl-phenylalanyl-glycine

24. Precipitation or coagulation of proteins may be caused by

(a) heat
(b) changes in pH
(c) heavy metal salts
(d) all of these

25. Which amino acid will not give a positive xanthoproteic test?

(a)
$$\begin{array}{c} \quad\ \ H\quad\ H \\ \quad\ \ |\quad\ \ | \\ C_6H_5-C-C-COOH \\ \quad\ \ |\quad\ \ | \\ \quad\ \ H\quad NH_2 \end{array}$$

(b)
$$\begin{array}{l} \quad\ \ H \\ \quad\ \ | \\ \quad\ \ C \\ N \quad\quad NH \quad\quad\ H \\ | \quad\quad\quad | \quad\quad\quad\ | \\ H-C=C-CH_2-C-COOH \\ \quad\quad\quad\quad\quad\quad\quad\ | \\ \quad\quad\quad\quad\quad\quad\ NH_2 \end{array}$$

(c)
$$\begin{array}{c} \quad\quad\quad H\quad\ H \\ \quad\quad\quad |\quad\ \ | \\ \text{(indol-3-yl)}-C-C-COOH \\ \quad\quad\quad |\quad\ \ | \\ \quad\quad\quad H\quad NH_2 \\ N \\ | \\ H \end{array}$$

(d)
$$\begin{array}{c} \quad\quad\quad\ \ H\quad\ H \\ \quad\quad\quad\ \ |\quad\ \ | \\ HO-C_6H_4-C-C-COOH \\ \quad\quad\quad\ \ |\quad\ \ | \\ \quad\quad\quad\ \ H\quad NH_2 \end{array}$$

26. Which amino acid gives a positive Millon test?

(a)
$$\begin{array}{c} \quad\ \ H\quad\ H \\ \quad\ \ |\quad\ \ | \\ HO-C-C-COOH \\ \quad\ \ |\quad\ \ | \\ \quad\ \ H\quad NH_2 \end{array}$$

(b)
$$\begin{array}{c} \quad\quad\quad\ \ H \\ \quad\quad\quad\ \ | \\ HS-CH_2-C-COOH \\ \quad\quad\quad\ \ | \\ \quad\quad\quad NH_2 \end{array}$$

(c)
$$\begin{array}{c} \quad\quad\quad\ \ H\quad\ H \\ \quad\quad\quad\ \ |\quad\ \ | \\ HO-C_6H_4-C-C-COOH \\ \quad\quad\quad\ \ |\quad\ \ | \\ \quad\quad\quad\ \ H\quad NH_2 \end{array}$$

(d)
$$\begin{array}{c} \quad\ \ H\quad\ H \\ \quad\ \ |\quad\ \ | \\ C_6H_5-C-C-COOH \\ \quad\ \ |\quad\ \ | \\ \quad\ \ H\quad NH_2 \end{array}$$

(e)
$$\begin{array}{c} H_2C-CH_2 \\ |\quad\quad\quad | \\ H_2C\quad CHCOOH \\ N \\ | \\ H \end{array}$$

27. Amino nitrogen removed during the deamination process is eventually eliminated from most mammals by way of the urine as

(a) ammonia
(b) sodium nitrate
(c) hippuric acid
(d) urea
(e) alanine

**Completion**

28. Amino acids that are required in the diet because the animal organism is incapable of synthesizing them are classified as ____________ ____________ ____________.

29. The hydrogen ion concentration at which a neutral amino acid exists as a dipolar ion $\left( R{-}\underset{\substack{| \\ ^{+}NH_3}}{CH}{-}CO_2^- \right)$ is called the ____________ ____________.

30. High molecular weight natural polymers composed of various combinations of α-amino acids are called ____________.
Partial hydrolysis of these polymers yield ____________.

31. The amide linkages by which amino acids are joined together are called ____________ links.

32. Conjugated proteins, upon hydrolysis, yield nonprotein material called ____________ ____________.

33. Irreversible precipitation or coagulation of proteins is called ____________.

34. $$R{-}\underset{\substack{| \\ NH_2}}{CH}{-}COOH + R'{-}\underset{\substack{\| \\ O}}{C}{-}COOH \rightleftharpoons R{-}\underset{\substack{\| \\ O}}{C}{-}COOH + R'{-}\underset{\substack{| \\ NH_2}}{CH}{-}COOH$$

The above reaction is referred to in biochemistry as ____________.

35. $$R{-}\underset{\substack{| \\ NH_2}}{CH}{-}COOH + \tfrac{1}{2}\,O_2 \longrightarrow R{-}\underset{\substack{\| \\ O}}{C}{-}COOH + NH_3$$

The above process is referred to in biochemistry as ____________.

36. The $pK_1$ and $pK_2$ of alanine are 2.29 and 9.74 respectively. The isoelectric pH is approximately ____________.

37. The secondary structure of the β-keratins has been designated as the ______-conformation or ____________ structure.

38. The secondary structure of an α-keratin is that of an ____________.

**Matching** (Match each compound with an appropriate description and an appropriate name or abbreviation.)

(a) $C_6H_5{-}CH_2{-}O{-}\overset{\substack{O \\ \|}}{C}{-}Cl$

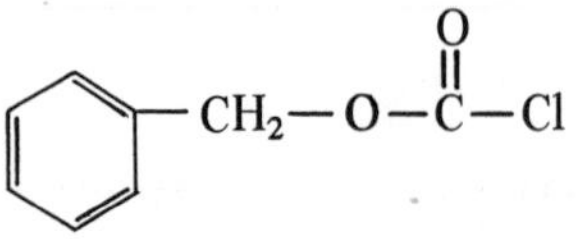

(b) $C_6H_5{-}N{=}C{=}S$

(c) benzene ring bearing F, $NO_2$ (ortho to F), and $O_2N$ (para to F): $O_2N{-}C_6H_3(F){-}NO_2$

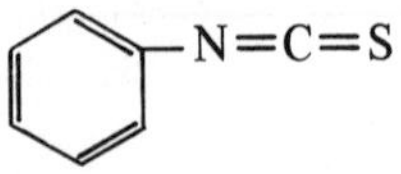

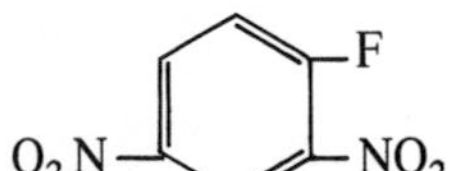

*(cont'd)*

(d) 1-dimethylaminonaphthalene-5-sulfonyl chloride: $SO_2Cl$ on naphthalene, $N(CH_3)_2$

(e) $C_6H_5$–$CH_2OH$

(f) $C_6H_5$–N=C=N–$C_6H_{11}$

39. _____ a carboxyl group protecting agent
40. _____ a reagent which can be used to establish the sequence of amino acids from the N-terminus in a polypeptide chain, forms a phenylthiohydantoin
41. _____ an amino group protecting agent
42. _____ a coupling reagent used in peptide bond formation
43. _____ a reagent which yields an easily detected fluorescent product when used to tag the N-terminal group of a peptide
44. _____ a reagent used to tag the N-terminal group of a peptide, also called Sanger's reagent
45. _____ dansyl chloride
46. _____ DCC
47. _____ FDNB
48. _____ benzyl alcohol
49. _____ Cbz
50. _____ phenylisothiocyanate

**Miscellaneous** (Place a check in the square under the class of compound for which each of the following statements is true.)

| Proteins | Carbohydrates | Fats | |
|---|---|---|---|
| | | | 51. Composed principally of carbon, hydrogen, and oxygen. |
| | | | 52. An important class of foodstuffs. |
| | | | 53. Primarily a supply of heat and energy. |
| | | | 54. Used mainly to repair and replace worn-out tissue. |
| | | | 55. May be stored in a body depot. |
| | | | 56. Esters of glycerol. |
| | | | 57. High molecular weight polymers containing amide linkages. |
| | | | 58. Polyhydroxy aldehydes or ketones, or substances which, when hydrolyzed, yield polyhydroxy aldehydes or ketones. |

59. A peptide yields upon complete hydrolysis one residue, each, of proline, phenylalanine, valine, and leucine and three residues each of glycine and serine. A partial hydrolysis yields Gly-Phe-Ser-Val, Ser-Gly-Phe, Val-Leu-Ser-Gly, and Gly-Pro-Ser. Give the abbreviated formula for the decapeptide.

60. The titration curve for a hypothetical amino acid, $R{-}\underset{\displaystyle NH_2}{\underset{|}{C}H}{-}COOH$, is given below.

What is the $pK_1$, $pK_2$, and the isoelectric pH? What are the predominant ionic species at *A, B, C, D,* and *E*?

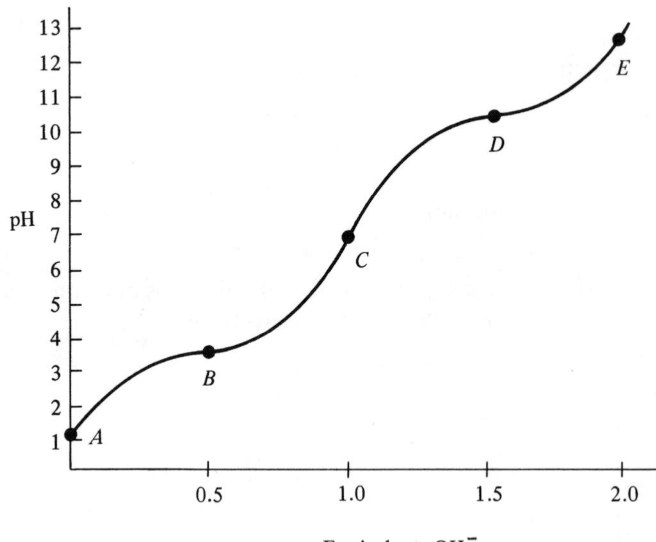

## ANSWERS

### True–False

1. True
2. False
3. True
4. False. Different amino acids are required in the diet of different animals because the animals are incapable of synthesizing every α–amino acid they require. Such required amino acids are classified as essential amino acids.
5. False. Amino acids are colorless crystalline solids which have high melting points.
6. True
7. True. The molecular weights of proteins range from 10,000 to 10,000,000.
8. True. The biuret test is specific for multiple peptide links and therefore is not given by α–amino acids.
9. False. Fats and carbohydrates are used primarily to supply heat and energy, but proteins are used mainly to repair and replace worn-out tissue.
10. True
11. True
12. True. Animals obtain their protein by eating plants or by eating other animals that have eaten plants.
13. True. The carboxylic acid group is a stronger acid than the ammonium ion.

### Multiple Choice

14. (c)
15. (c)
16. (e) Optical isomerism is possible whenever an asymmetric carbon atom is part of a molecular structure. Glycine does not have an asymmetric carbon atom.
17. (a) The extra amino group in lysine makes it a basic amino acid.
18. (a) Histidine has a basic amino group other than the α–amino group and thus is classified as a basic amino acid.
19. (a)
20. (d)
21. (b) The amino group in proline is a secondary amine and thus does not liberate nitrogen gas when treated with nitrous acid. Recall that secondary amines react with nitrous acid to form N-nitroso derivatives.
22. (e) Six different amino acids can combine to form 6 factorial, (6!), hexapeptides.
23. (a)
24. (d)
25. (b) The xanthoproteic test is a nitration of the aromatic ring of certain amino acids (tyrosine, phenylalanine, tryptophan).

26. (c) Any phenolic amino acid will give a positive Millon test.

27. (d) The structure of urea is

$$H_2N-\overset{\overset{\displaystyle O}{\|}}{C}-NH_2$$

**Completion**

28. essential amino acids
29. isoelectric point
30. proteins, peptides
31. peptide
32. prosthetic groups
33. denaturation
34. transamination
35. deamination
36. 6, $\dfrac{pK_1 + pK_2}{2}$ = isoelectric pH
37. $\beta$, pleated sheet
38. $\alpha$-helix

**Matching**

39. (e)
40. (b)
41. (a)
42. (f)
43. (d)
44. (c)
45. (d)
46. (f)
47. (c)
48. (e)
49. (a)
50. (b)

**Miscellaneous**

| Proteins | Carbohydrates | Fats | |
|---|---|---|---|
| | ✓ | ✓ | 51. |
| ✓ | ✓ | ✓ | 52. |
| | ✓ | ✓ | 53. |
| ✓ | | | 54. |
| | ✓ | ✓ | 55. |
| | | ✓ | 56. |
| ✓ | | | 57. |
| | ✓ | | 58. |

59. Gly-Pro-Ser-Gly-Phe-Ser-Val-Leu-Ser-Gly

Since only one residue each of Pro, Phe, Val, and Leu are isolated, part of the peptide sequence can be established as follows:

Gly-Phe-Ser-Val
Ser-Gly-Phe
Val-Leu-Ser-Gly

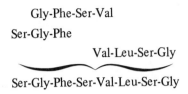

Ser-Gly-Phe-Ser-Val-Leu-Ser-Gly

Two more amino acids are needed to complete the structure of the decapeptide. They must be Pro and Gly. The only hydrolysis residue containing Pro was Gly-Pro-Ser. This residue must be added to the left side of the octapeptide to give Gly-Pro-Ser-Gly-Phe-Ser-Val-Leu-Ser-Gly.

60. $pK_1$ is 3.5, $pK_2$ is 10.5, isoelectric point is 7.0

Predominant ionic species:

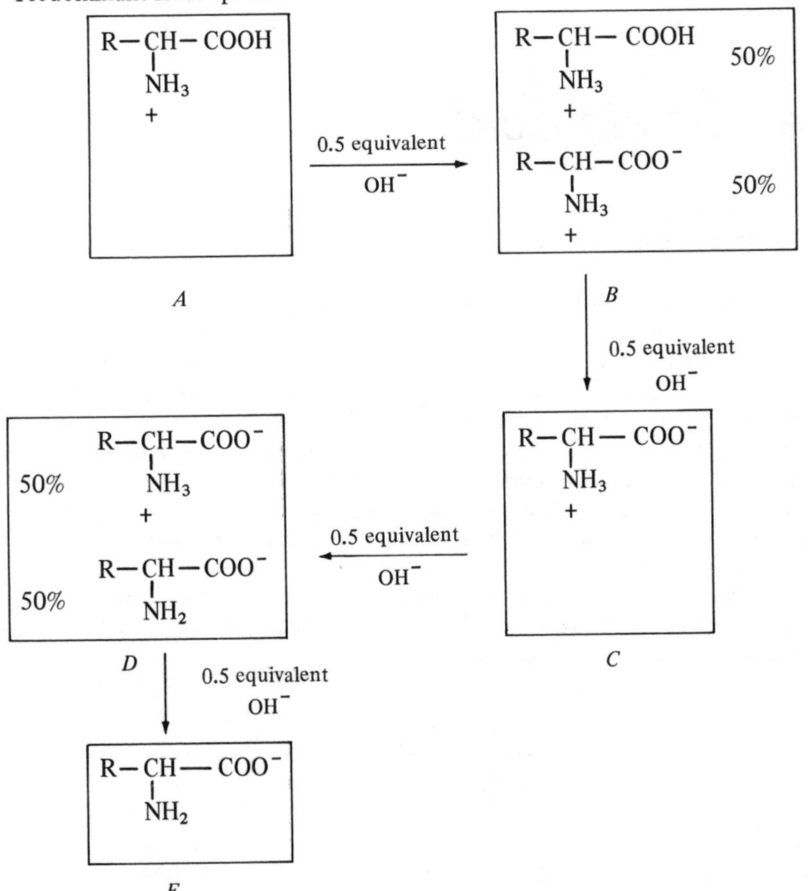

# 16

# *The Nucleic Acids*

## INTRODUCTION

Some of the most intriguing problems that perplexed biochemists for many years were those concerning the vitally important chemistry involved in the life processes. It had been known for some time that protein material was synthesized from α-amino acids, but how were these protein building blocks directed into hair, or into skin, or into any one of the many other cellular structures comprising the complex living organism? Equally as perplexing was the question of how individual characteristics and traits were transferred from parent to progeny.

The answers to these fascinating questions were found in the discovery and study of the nucleic acids. The present chapter outlines briefly the developments in this exciting area of chemistry.

The nucleic acids are high molecular weight, natural polymers composed of **orthophosphoric acid**, a **pentose**, and a **purine** or a **pyrimidine base**. The fundamental nucleic acid structure is

$$\sim\!\sim\; \underset{}{\overset{\text{Base}}{\overset{|}{\text{Sugar}}}} - O - \underset{\underset{OH}{|}}{\overset{\overset{O}{\|}}{P}} - O - \overset{\text{Base}}{\overset{|}{\text{Sugar}}} - O - \underset{\underset{OH}{|}}{\overset{\overset{O}{\|}}{P}} - O \;\sim\!\sim$$

We will find there are two families of nucleic acids, **RNA** and **DNA**. Their structures and functions differ. It is the function of DNA to transmit the **genetic code** from parent to daughter cell and to transfer this information to **messenger RNA**.

**Messenger RNA, ribosomal RNA**, and **transfer RNA**, on the other hand, are directly involved in protein synthesis.

Other new terms with which we must become familiar are **codons, anticodons, nucleosides, nucleotides, replication**, and **mutations.**

## PROBLEMS

### True-False

1. T F Nucleic acids direct the synthesis of specialized cellular protein according to a predetermined code and also control the genetic continuity of the species.
2. T F The sugar portion of RNA is 2-deoxy-D-ribose.
3. T F A nucleotide contains all three components of a nucleic acid—base, sugar, and phosphoric acid unit.
4. T F The ratio of adenine to thymine and guanine to cytosine in DNA is almost 1:1.
5. T F Adenine and guanine are purine bases and are found in both RNA and DNA.
6. T F The concept of complementary base pairs offers a mechanism whereby genetic information can be transmitted from one generation to the next.
7. T F RNA and DNA are responsible for transmission of genetic information from parent to daughter cells.
8. T F Codons are combinations of three base units in a *m*-RNA strand.
9. T F The secondary structure of RNA is the same as that of DNA and thus the pyrimidine-purine ratio in RNA is also 1:1.
10. T F Messenger RNA carries the transcribed genetic code from the DNA in the nucleus of the cell to the protein-building site in the ribosomes.
11. T F Mutations may prove beneficial as well as detrimental.
12. T F Transfer RNAs characteristically contain a number of minor bases as well as the major bases, A, G, U, and C.
13. T F Transfer RNAs have a cloverleaf structure, sometimes with an extra arm.

### Multiple Choice

14. Which type of chemical bond or linkage joins the sugar and phosphoric acid units in a polynucleotide?
    (a) peptide
    (b) hemiacetal
    (c) glycosidic
    (d) ester
    (e) hydrogen bond

15. The double helical structure of DNA is held together by
    (a) sulfur-sulfur bridging
    (b) hydrogen bonding
    (c) peptide bonding
    (d) ester linkages
    (e) glycosidic bonds

16. All of the following are pyrimidine bases except
    (a) guanine
    (b) thymine
    (c) cytosine
    (d) uracil
    (e) all are pyrimidine bases

17. Which type of nucleic acid contains an anticodon for a specific amino acid and selects and carries that amino acid to the correct site for protein synthesis?
    (a) transfer RNA
    (b) DNA
    (c) messenger RNA
    (d) ribosomal RNA

18. Which sequence of bases in the complementary strand of DNA will pair with adenine-guanine-thymine?
    (a) guanine-adenine-cytosine
    (b) adenine-guanine-thymine
    (c) uracil-cytosine-adenine
    (d) thymine-guanine-adenine
    (e) thymine-cytosine-adenine

19. Hydrolysis of DNA would not be expected to yield
    (a) cytosine
    (b) phosphoric acid
    (c) ribose
    (d) adenine
    (e) deoxyribose

20. What will the codon be in *m*-RNA if the sequence of bases in DNA is ATC?
    (a) TAG
    (b) GCT
    (c) UAG
    (d) UUG
    (e) TGA

21. What will the anticodon be in *t*-RNA if the codon in *m*-RNA is UAG? (Assume there are no minor bases in the coding triplet.)
    (a) ATC
    (b) TUG
    (c) CUT
    (d) GUT
    (e) AUC

22. Hydrolysis of RNA could yield all of the following except
    (a) uracil
    (b) ribose
    (c) thymine
    (d) phosphoric acid
    (e) adenine

**Matching** (Match each structure to the appropriate name or description)

(a)

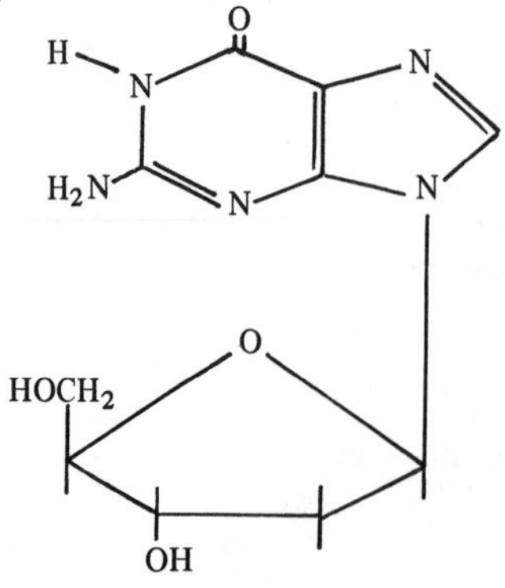

(b)

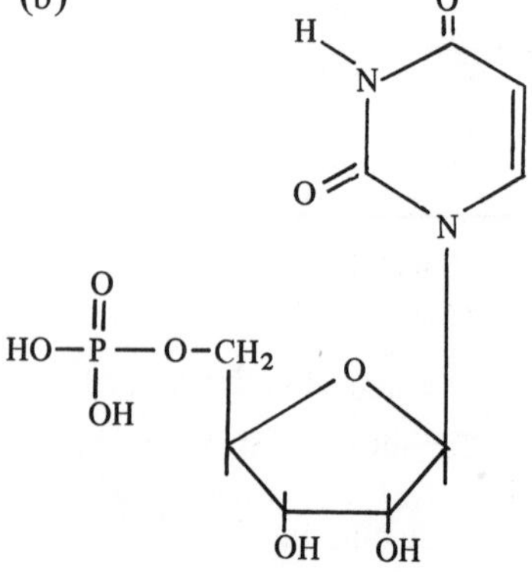

(c)

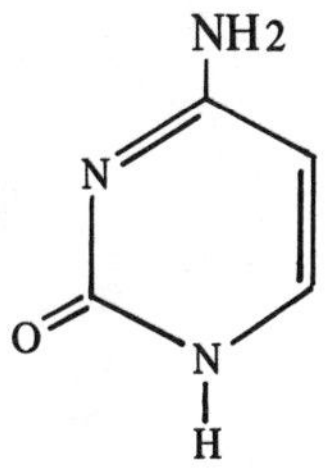

NH2
N
O
N
H

(d)

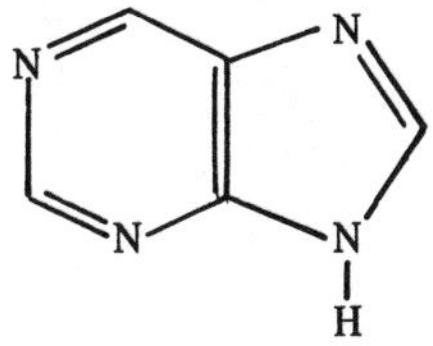

N
N
N
N
H

(e)

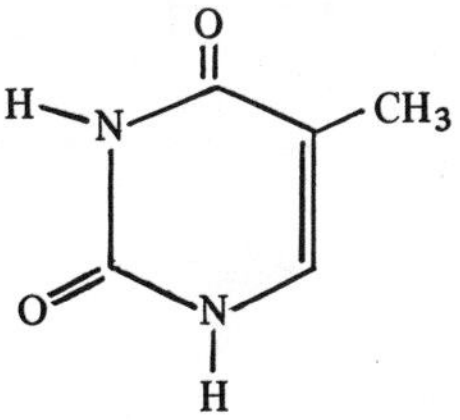

O
H
N
CH3
O
N
H

(f)

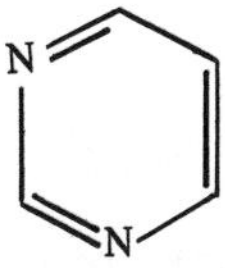

N
N

(g)

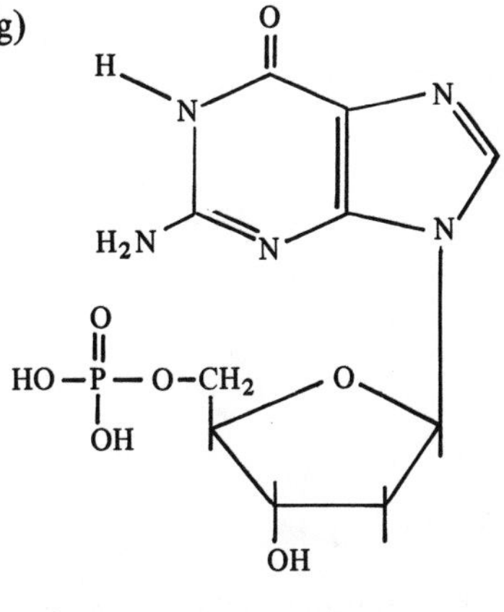

O
H
N
N
$H_2N$
N
N
O
HO — P — O — $CH_2$
OH
O
OH

(h)

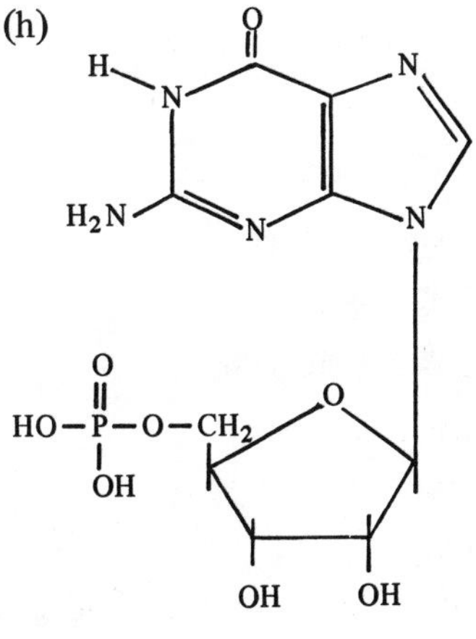

O
H
N
N
$H_2N$
N
N
O
HO — P — O — $CH_2$
OH
O
OH OH

(i)

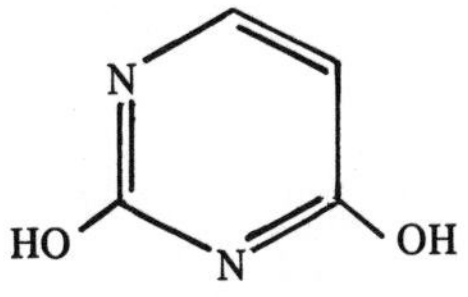

N
HO
N
OH

(j)

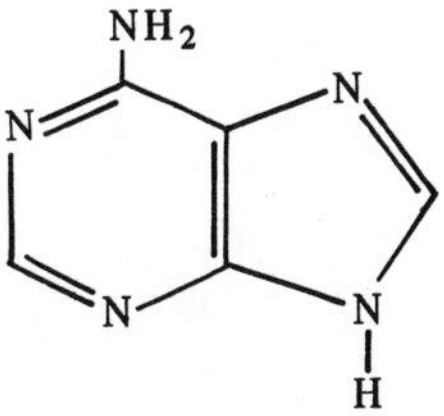

$NH_2$
N
N
N
N
H

(k)

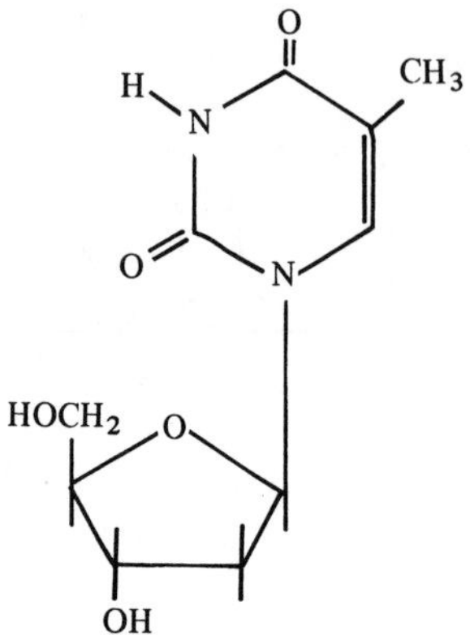

(l)

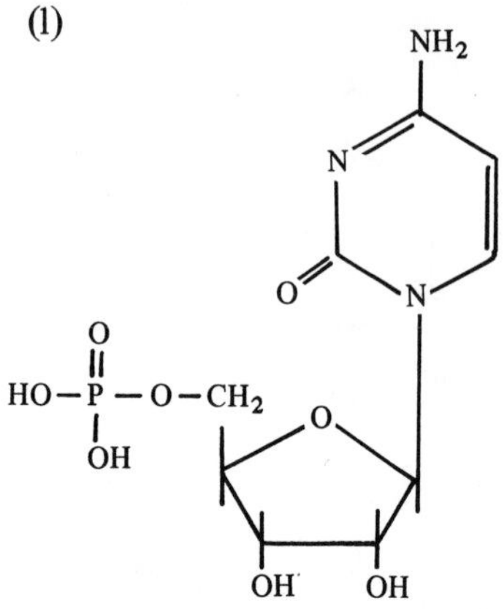

23. ______ Uracil (enol form)
24. ______ Pyrimidine
25. ______ Deoxyguanosine
26. ______ Adenine
27. ______ Cytosine
28. ______ Deoxyguanylic acid
29. ______ Cytidylic acid
30. ______ Uridylic acid
31. ______ Guanylic acid
32. ______ Thymine
33. ______ Deoxythymidine
34. ______ Purine

**Miscellaneous**

35. Draw the structure of ribose and clearly indicate the position where each phosphoric acid molecule is attached to form the polynucleotide "backbone".

36. Draw the structure of deoxyribose and clearly indicate the position where the nitrogen base is attached to form a nucleoside.

37. Draw structures for cytosine and guanine and show how they are hydrogen-bonded to each other.

38. Number each element in the following ring systems and indicate the position at which a pentose unit can attach to each system.

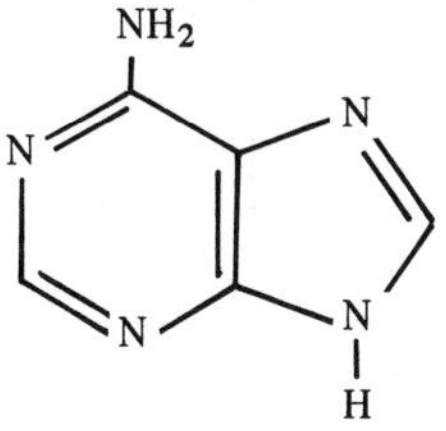

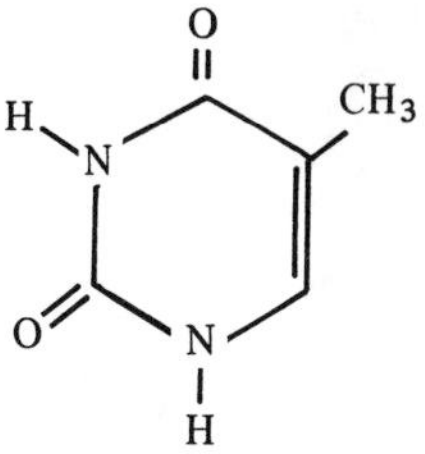

39. The code of the DNA strand is read from the 5′ to the 3′ direction. Indicate in which direction the *m*-RNA, *t*-RNA, and peptide strands would line up in respect to DNA in protein synthesis. (Circle the correct answer.)

| | | |
|---|---|---|
| 5′ end | DNA | 3′-end |
| (3′-end, 5′-end) | *m*-RNA | (3′-end, 5′-end) |
| (3′-end, 5′-end) | *t*-RNA | (3′-end, 5′-end) |
| (carboxyl end, amino end) | peptide | (carboxyl end, amino end) |

## ANSWERS

### True–False

1. True
2. False. The sugar portion of ribonucleic acid (RNA) is D-ribose and of deoxyribonucleic acid (DNA), 2-deoxy-D-ribose.
3. True. The nucleoside consists of only the base and sugar units.

4. True. This is because of the hydrogen-bonding between these complementary base pairs and the double helix structure of DNA.
5. True
6. True
7. False. DNA is responsible for transmission of the genetic code, but the various RNAs are responsible for transcribing this code into specific protein material.
8. True
9. False. The secondary structure of DNA is that of a double helix. RNA molecules exist mainly as single strands.
10. True
11. True. Beneficial mutations could result in a stronger species while detrimental mutation could result in a weaker species which will eventually cease to exist.
12. True. The transfer RNAs contain as much as 10% minor bases and sometimes they are part of the coding triplets.
13. True

**Multiple Choice**

14. (d)
15. (b)
16. (a)
17. (a)
18. (e)
19. (c)
20. (c)
21. (e)
22. (c)

**Matching**

23. (i)
24. (f)
25. (a)
26. (j)
27. (c)
28. (g)
29. (l)
30. (b)
31. (h)
32. (e)
33. (k)
34. (d)

**Miscellaneous**

35. Phosphoric acid molecules are attached by ester linkages at these positions

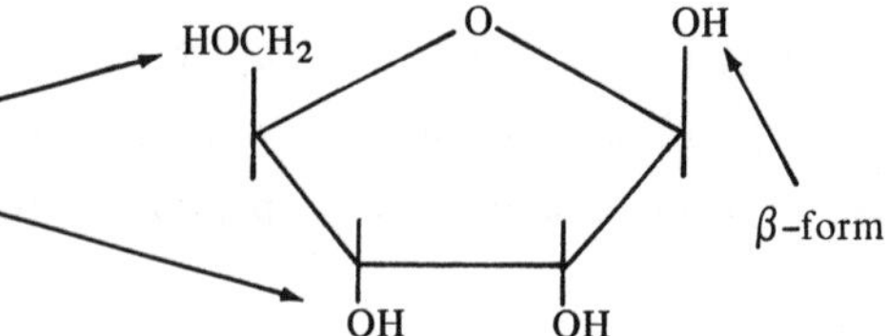


36. HOCH$_2$ O OH

OH

The nitrogen base is attached at this position

37.

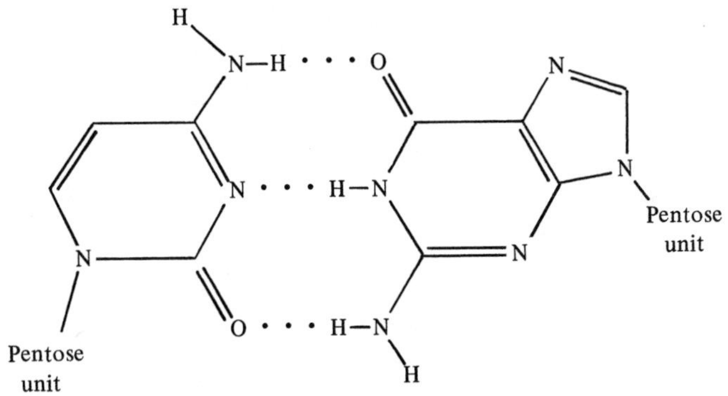

3 hydrogen bonds (dotted lines)

38.

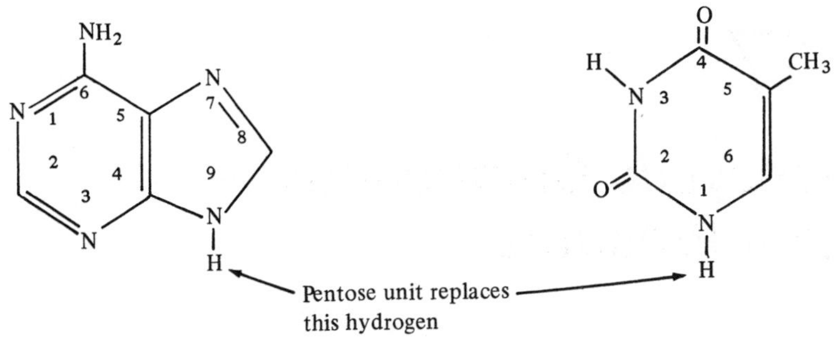

Pentose unit replaces this hydrogen

39.

| 5′-end | DNA | 3′-end |
|---|---|---|
| 3′-end | *m*-RNA | 5′-end |
| 5′-end | *t*-RNA | 3′-end |
| carboxyl end | peptide | amino end |

# 17

# *Heterocyclic Compounds — Natural Products*

## INTRODUCTION

Heterocyclic compounds are ring compounds containing atoms other than carbon in the ring. The most common hetero atoms are nitrogen, oxygen, and sulfur. Heterocyclic compounds are very abundant in nature, and in this chapter we shall study those heterocycles most often encountered in natural products. However, all natural products do not contain heterocyclic rings.

The natural products discussed in this chapter are the **alkaloids**, **vitamins**, **terpenes**, **steroids**, and **antibiotics**. We shall learn to recognize some of the important compounds in each of these groups and also learn about their properties and uses.

## PROBLEMS

### True–False

1. T F Heterocyclic compounds are seldom found as natural products.
2. T F The numbering of a heterocyclic compound begins with the hetero atom.
3. T F Furan, pyrrole, and thiophene undergo the Friedel-Crafts reaction in a manner similar to that of benzene.
4. T F Many valuable drugs are obtained from alkaloids.
5. T F Insufficient amounts of certain vitamins in man's diet result in diseases such as pellagra and beriberi.

6. T F Heterocycles are most commonly five- or six-membered rings.
7. T F Four piperidine nuclei are found in the porphyrin structure.
8. T F Heme and chlorophyll-a are similar in that they both contain a prophyrin structure.
9. T F The six-membered heterocycle most widely occurring in nature is pyridine.
10. T F Methadone is a powerful analgesic which does not lead to addiction.
11. T F The local anesthetic procaine is a synthetic substitute for cocaine.
12. T F Codeine is a derivative of morphine isolated from the opium poppy.
13. T F Sunlight does not supply vitamin D, but merely supplies the radiation energy necessary for its formation.
14. T F The body is incapable of synthesizing most vitamins and therefore must have them supplied in the daily diet.
15. T F Natural rubber is a polyterpene.
16. T F Cholesterol does not occur naturally in animal tissue.
17. T F A number of penicillins have been prepared which have the same skeletal structures, but with different side chains.

**Multiple Choice**

18. Which of the following compounds is not a heterocycle?

(a) $H_2C$ — $CH_2$; C; $CH_2$; H; N; N; $CH_3$

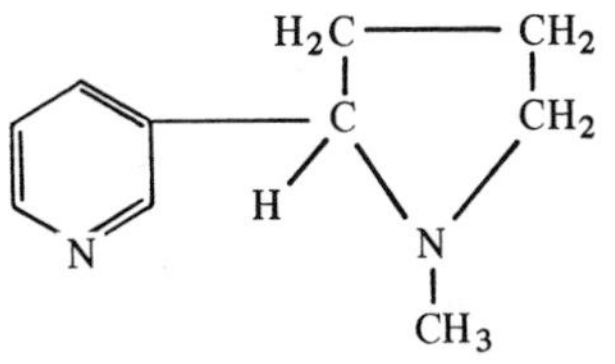

(b) HO—C═C—OH; O═C; CH— CHOH— $CH_2OH$; O

(c) $HOCH_2$; O; OH; H; OH; H; $CH_2OH$; OH; H

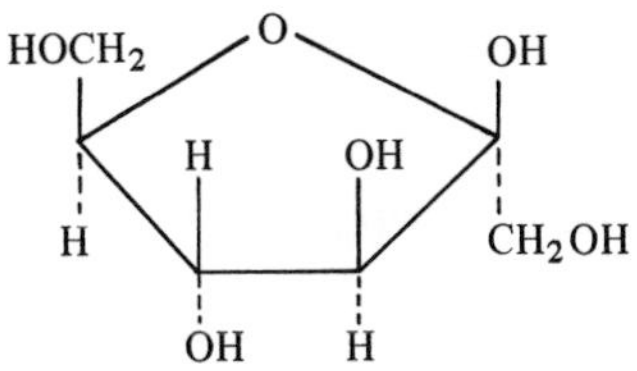

(d) O H H H S; $CH_3$; R—C—N—C—C; C; C—N; $CH_3$; O; C; H; COOH

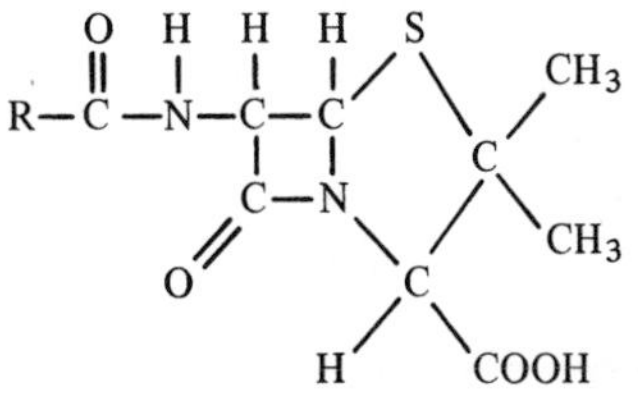

(e) $H_3C$ OH; $N(CH_3)_2$; OH; $CONH_2$; OH O OH O; OH

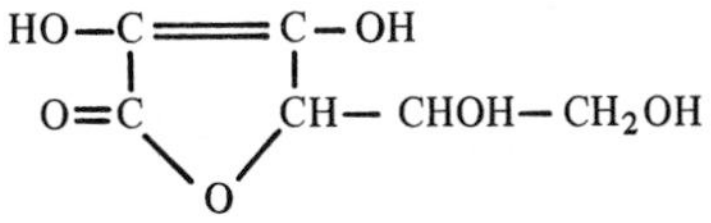

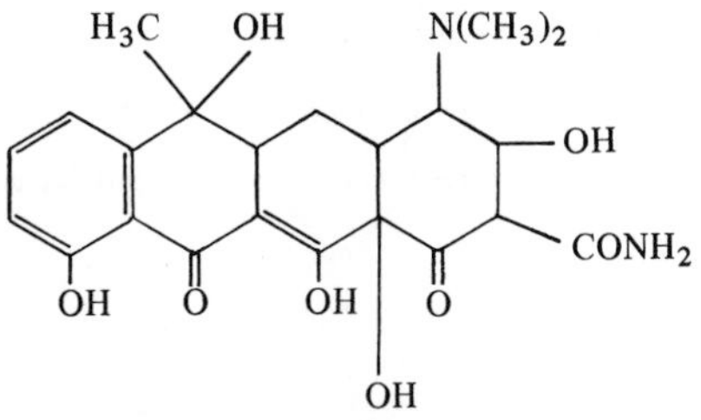

19. Which of the following heterocycles is pyrrole?

(a)

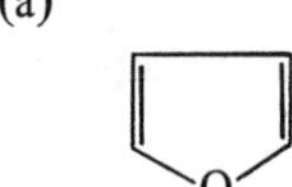


(b)

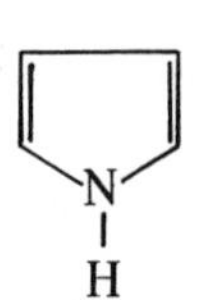


(c) S

(d)

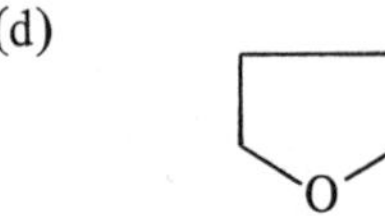


(e) O O

20. Which of the following would not be classified as a steroid?
   (a) female sex hormones
   (b) vitamin C
   (c) male sex hormones
   (d) cholesterol
   (e) bile acids

21. Which of the following compounds is a monoterpene?

(a) $CH_2{=}C(CH_3){-}CH{=}CH_2$

(b) $CH_3{-}C(CH_3){=}CH{-}CH_2{-}CH_2{-}C(CH_3){=}CH{-}CH_3$

(c) $CH_3{-}(C(CH_3){=}CH{-}CH_2{-}CH_2)_2{-}C(CH_3){=}CH{-}CH_3$

(d) $CH_3{-}(C(CH_3){=}CH{-}CH_2{-}CH_2)_3{-}C(CH_3){=}CH{-}CH_3$

(e) $CH_3{-}(C(CH_3){=}CH{-}CH_2{-}CH_2)_4{-}C(CH_3){=}CH{-}CH_3$

22. Which compound would not be classified as an alkaloid?

(a)

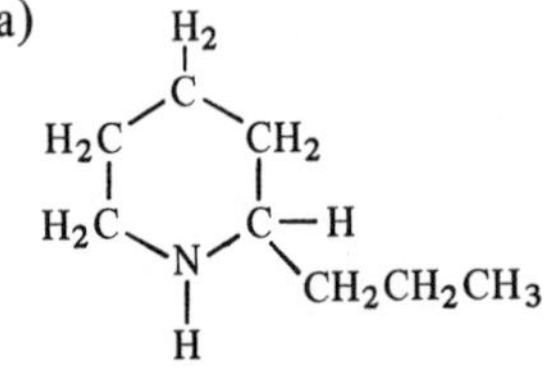


(b) HO—C═C—OH
O═C CH—CHOH—$CH_2OH$
O

(c)

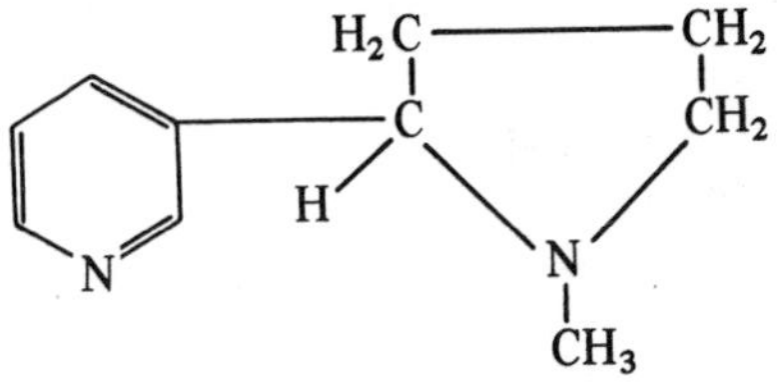

$H_2C$—$CH_2$, C, $CH_2$, H, N, $CH_3$, N

(d)

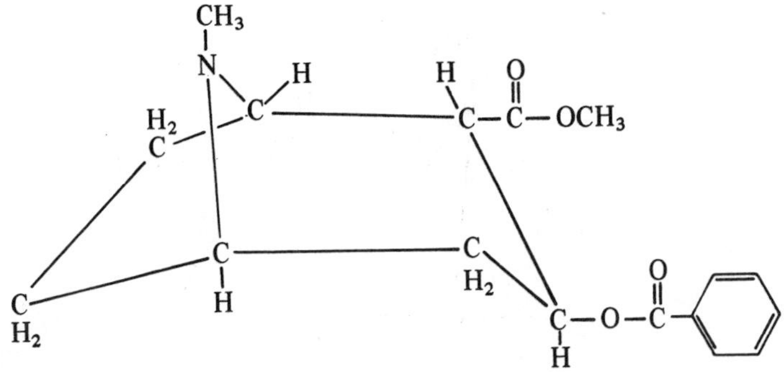

$CH_3$, N, H, H, O, C—C—$OCH_3$, $H_2$ C, C, C, H, C $H_2$, C $H_2$, O, C—O—C, H

(e) HO

N—$CH_3$, O, HO

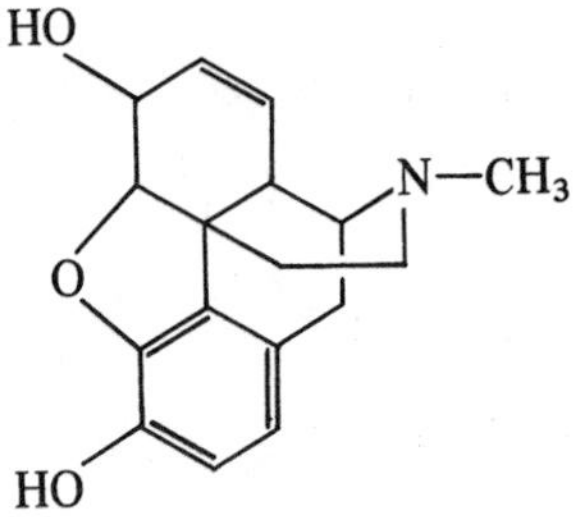

23. Which heterocycle does not contain a piperidine ring in its structure?

(a)

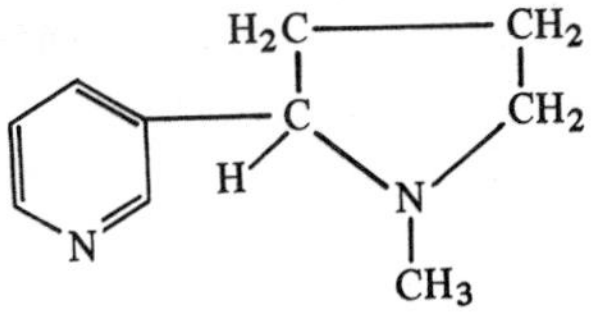

$H_2C$—$CH_2$, C, $CH_2$, H, N, $CH_3$, N

(b)

$H_2$ C, $H_2C$, $CH_2$, $H_2C$, C—H, N, H, $CH_2CH_2CH_3$

(c)

$CH_3$, N, H, $H_2$ C, C—$CH_2$, C $H_2$, C, H, C $H_2$, C, H, O, O, C, H C, $CH_2OH$

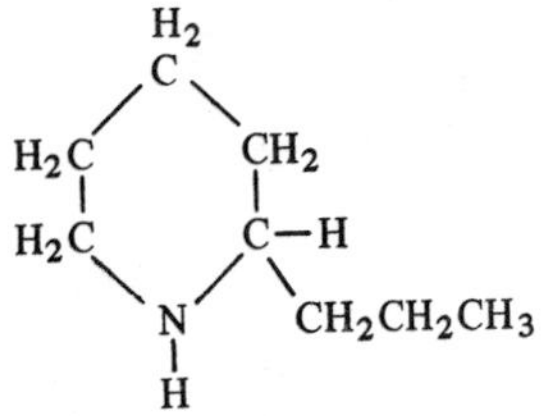

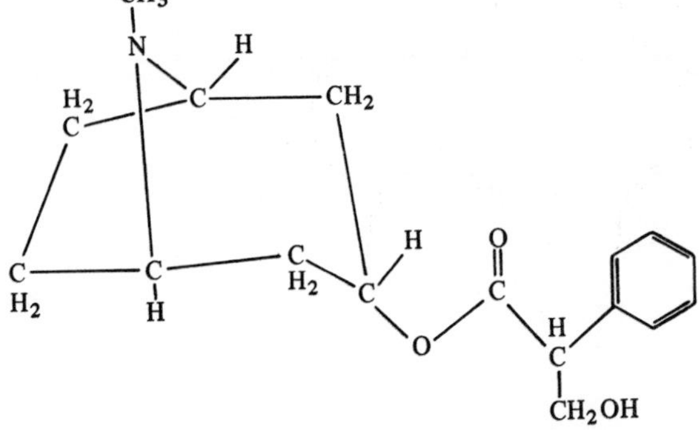

(d)

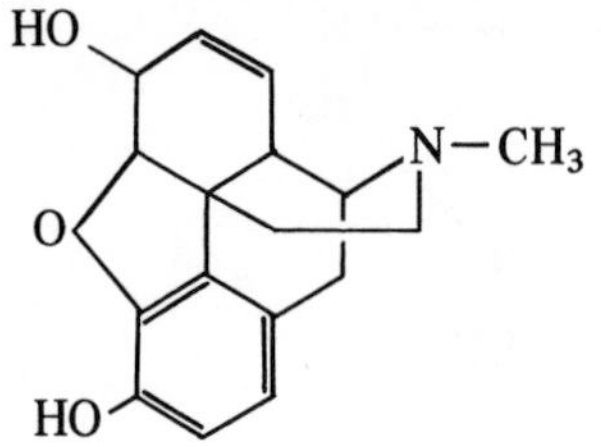

(e)

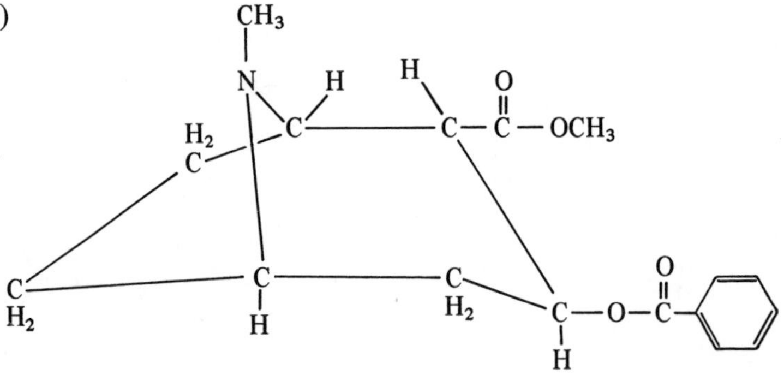

24. Which of the following heterocycles would not be classified as a vitamin?

(a)

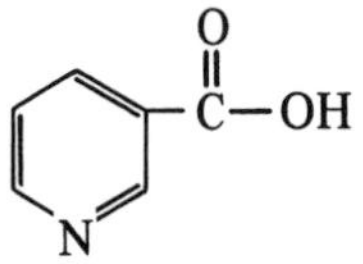

(b)

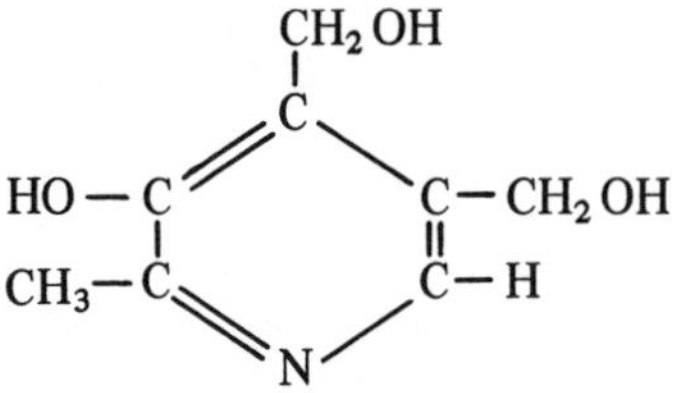

(c)

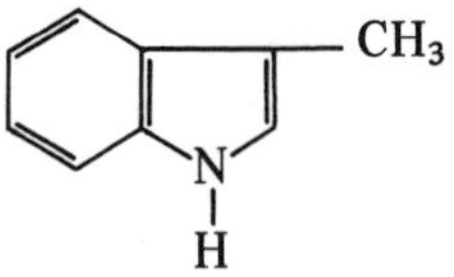

(d)

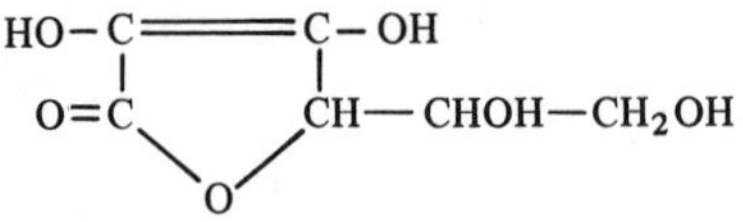

(e)

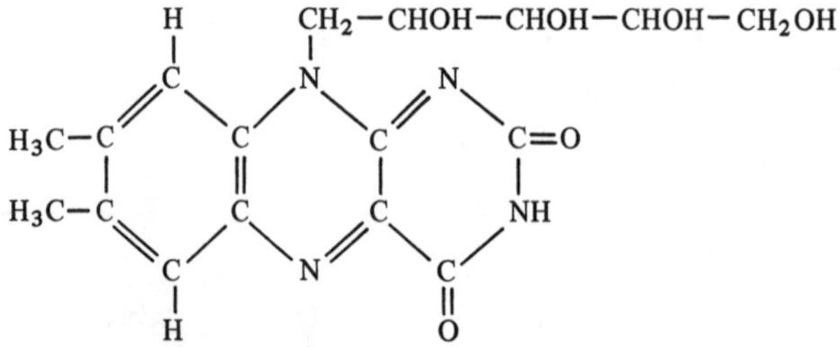

25. Which of the following compounds is tetrahydrofuran?

(a)

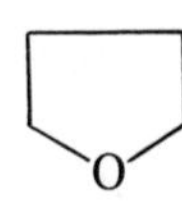


(c)

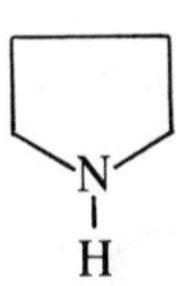


(e)

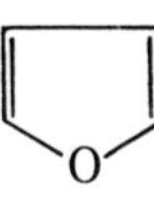


(b)

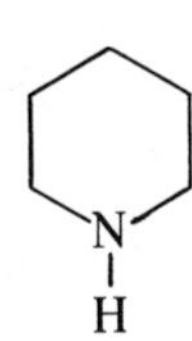


(d)

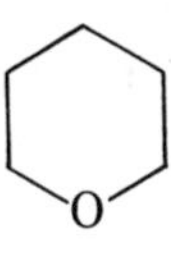


26. Which of the following structural units is absent in the nicotinamide-adenine-dinucleotide (NAD)?

(a) a pyridine nucleus
(b) a diphosphate unit
(c) a pentose unit
(d) an isoquinoline nucleus
(e) a purine nucleus

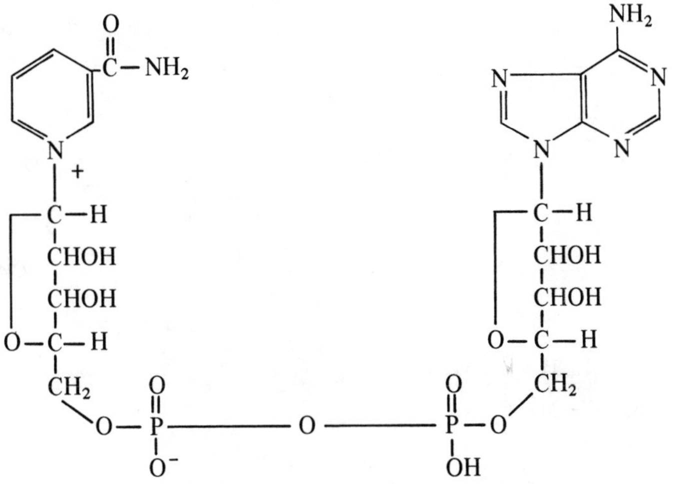

NAD

27. Which one of the following structures represents a synthetic female hormone?

(a)

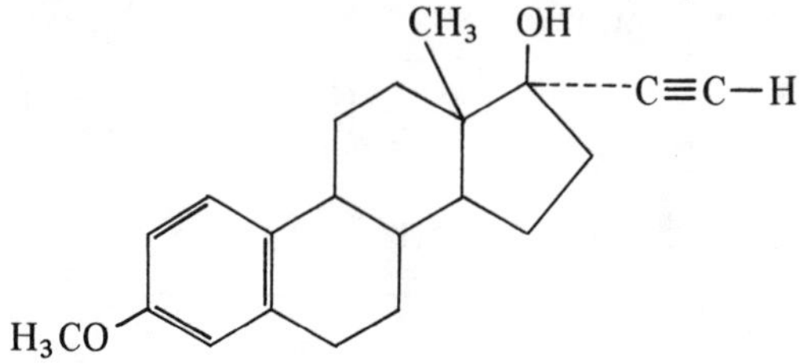

$CH_3$ OH

C≡C—H

$H_3CO$

(b) HO

$CH_2CH_2NH_2$

N—H

(c)

H—N

H

H

O

C

OH

N

$CH_3$

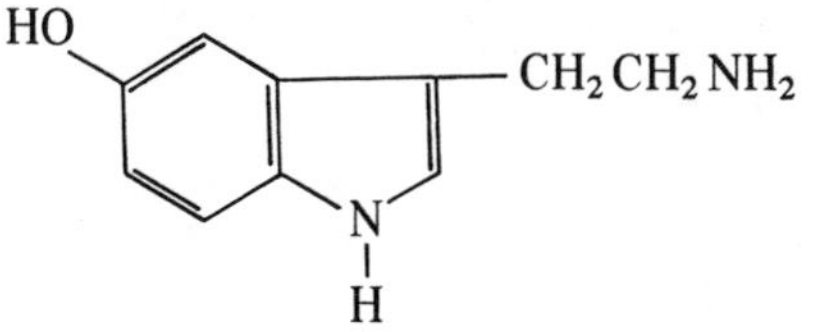

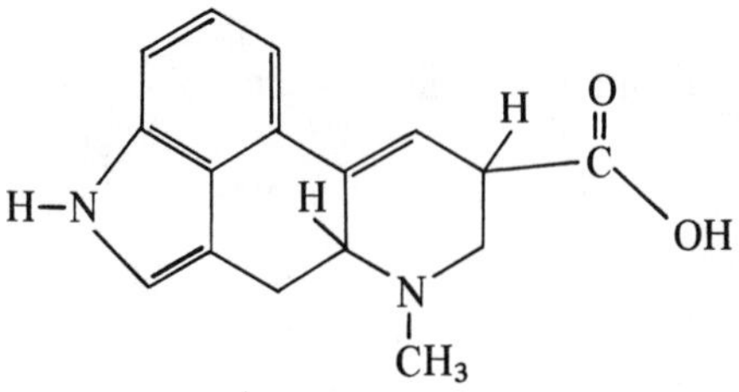

(d)

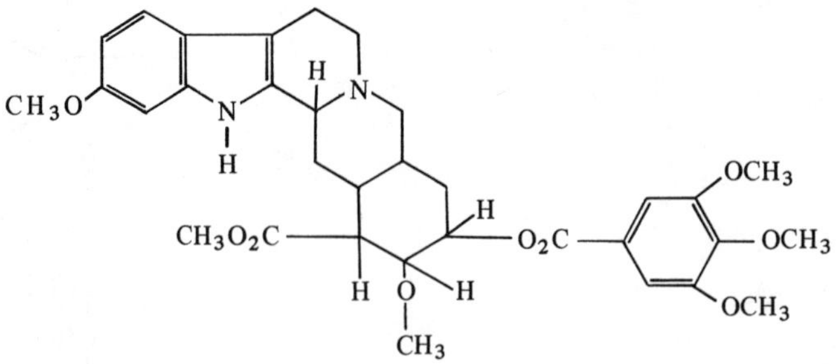

$CH_3O$

N—H

H

N

$CH_3O_2C$

H

$O_2C$

$OCH_3$

$OCH_3$

$OCH_3$

H O H

$CH_3$

(e)

O H H H

R—C—N—C—C—S

C—N

O

$CH_3$

C

$CH_3$

C—COOH

H

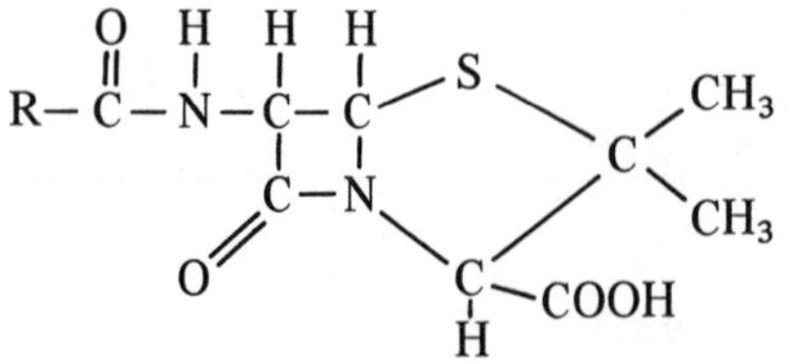

28. Which class of compounds is a necessary animal nutrient but required only in trace quantities?

(a) fats
(b) vitamins
(c) proteins
(d) carbohydrates

**Completion**

29. Cyclic compounds with only carbon atoms joined together in the ring are called ________________ compounds.
30. Cyclic compounds which have, in addition to carbon, one or more atoms of other elements in the ring are called ________________ compounds.
31. Nitrogen-containing plant products that possess marked physiological properties are called ________________.
32. Two rings which have one or more bonds in common are called ________________ ________________.
33. Abundant, potent, naturally produced substances that the animal organism requires, but usually is incapable of synthesizing, are called ________________.
34. Vitamins A, D, E, and K are classified as ____________ - ____________ vitamins; vitamin C and the B vitamins are classified as ____________ - ____________ vitamins.
35. Steroids containing hydroxyl groups in their structures may be called ____________.
36. Penicillin, streptomycin, aueromycin, and terramycin are commonly called ________________.
37. The general term used to describe a large class of natural products with carbon skeletons that are multiples of the $C_5$ isoprene unit is ________________.
38. The knowledge that the isoprene unit often appears in natural products in a regular head-to-tail sequence established a guide to the elucidation of the structures of a number of plant products. This guide is referred to as the ____________ ____________.
39. The most common hetero atoms found in heterocyclic compounds are ____________, ____________, and ____________.
40. Cross-linking of adjacent chains of natural rubber with sulfur is called ________________.

**Matching** (Match each compound to the appropriate classification or description.)

(a)

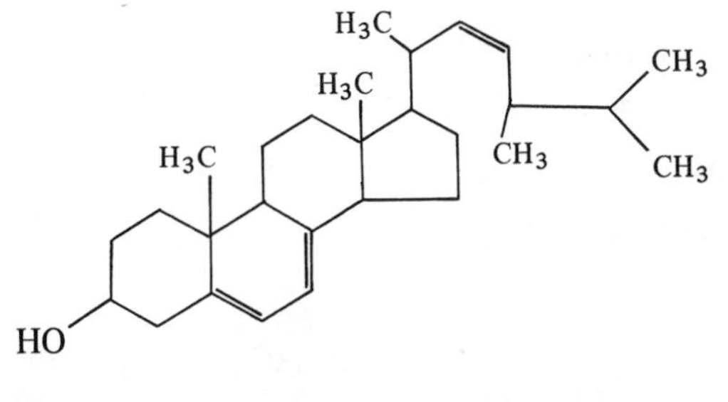

(b)

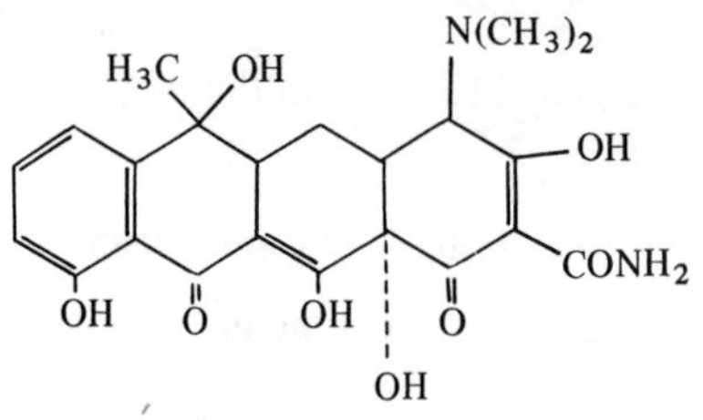

$N(CH_3)_2$, $H_3C$, OH, OH, $CONH_2$, OH, O, OH, O, OH

(c)

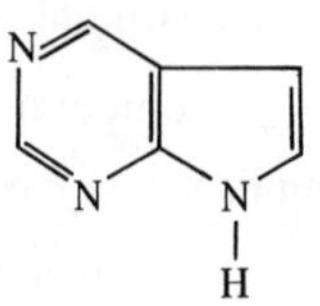

N, N, N, H

(d)

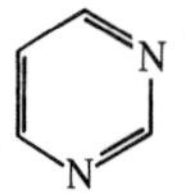

N, N

(e)

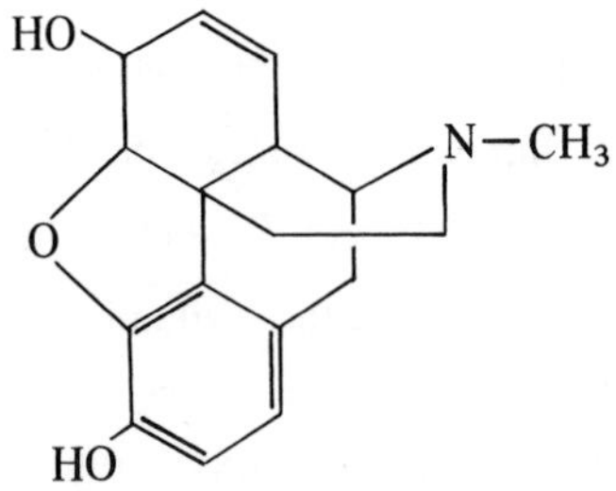

HO, $N-CH_3$, O, HO

(f)

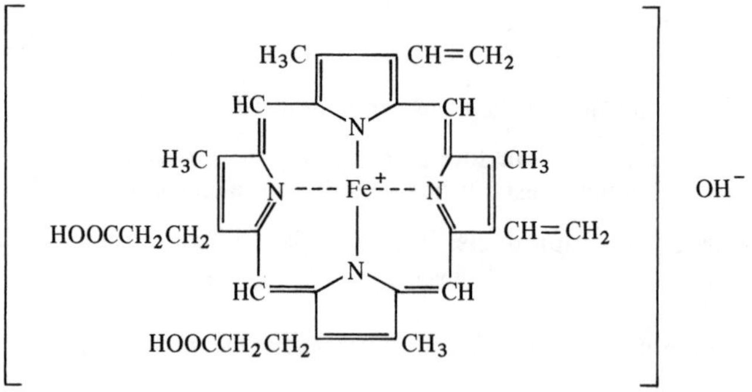

$H_3C$, $CH=CH_2$, HC, CH, $H_3C$, $CH_3$, N, $Fe^{+}$, N, N, N, $HOOCCH_2CH_2$, $CH=CH_2$, HC, CH, $HOOCCH_2CH_2$, $CH_3$, $OH^-$

(g)

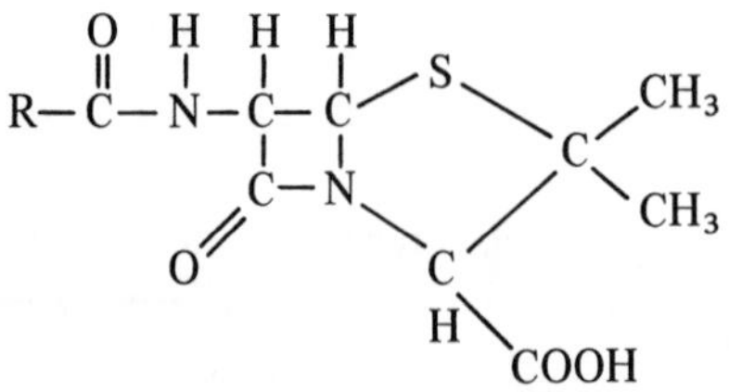

O H H H, R—C—N—C—C, S, $CH_3$, C, C—N, $CH_3$, O, C, H, COOH

(h)

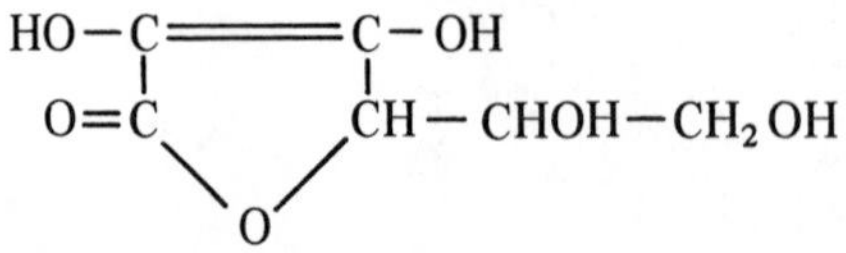

HO—C══C—OH, O═C, CH—CHOH—$CH_2OH$, O

(i) O ‖ —C—OH on a pyridine ring (N)

(j) $CH_2{=}C(CH_3){-}CH{=}CH_2$

(k) $\cdots{-}CH_2{-}C(CH_3){=}CH{-}CH_2{-}\left(CH_2{-}C(CH_3){=}CH{-}CH_2\right)_n{-}CH_2{-}C(CH_3){=}CH{-}CH_2{-}\cdots$

41. _____ a porphyrin
42. _____ a purine
43. _____ a pyrimidine
44. _____ a steroid
45. _____ a penicillin
46. _____ a tetracycline
47. _____ a terpene
48. _____ isoprene
49. _____ one of the B vitamins
50. _____ vitamin C or ascorbic acid
51. _____ the alkaloid known as morphine

**Miscellaneous**

52. Vitamin A is a terpene. Draw broken lines through segments of the structure of vitamin A to separate and identify the isoprene units. Does vitamin A exemplify the isoprene rule?

Ring: $C(CH_3)_2$, $H_2C$, $H_2C$, $CH_2$, $C{-}CH_3$, C, with side chain:

$C{-}CH{=}CH{-}C(CH_3){=}CH{-}CH{=}CH{-}C(CH_3){=}CH{-}CH_2OH$

Vitamin A

## ANSWERS

**True–False**

1. False. Heterocyclic compounds are very abundant in the plant and animal kingdoms.

2. True

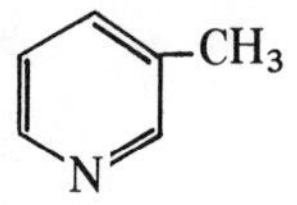

3-Methylpyridine

3. True. Substitution of the ring in each case takes place preferentially at the 2,5-positions.
4. True
5. True
6. True
7. False. A union of four pyrrole rings makes up the porphyrin structure.
8. True
9. True
10. False. Methadone is addicting, but does not produce the same mental and physical deterioration as other addicting drugs such as heroin.
11. True. Xylocaine and mepivacaine are also local anesthetics that have certain structural similarities to cocaine.
12. True
13. True
14. True
15. True
16. False. Cholesterol is found in almost all animal tissue and is particularly abundant in the brain, spinal cord, and in gallstones.
17. True

**Multiple Choice**

18. (e) Tetracycline does not have a ring containing a hetero atom.
19. (b)
20. (b) The structure of vitamin C, ascorbic acid, does not contain the characteristic steroid ring system shown below.

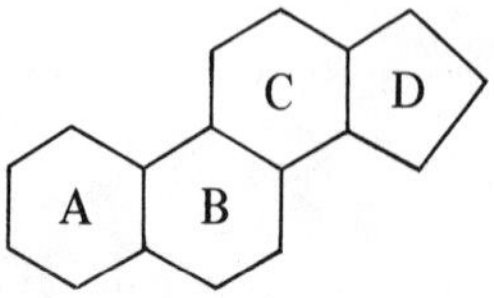

21. (b) A monoterpene contains two isoprene units.

22. (b) Alkaloids are naturally occurring nitrogenous substances of plant origin that possess marked physiological properties.
23. (a) The structure of nicotine contains a pyrrolidine ring and a pyridine ring, but not a piperidine ring.
24. (c) Skatole, 3-methylindole, is a degradation product of the α-amino acid tryptophan.
25. (a)
26. (d)
27. (a)
28. (b)

**Completion**

29. carbocyclic
30. heterocyclic
31. alkaloids
32. fused rings
33. vitamins
34. fat-soluble, water-soluble
35. sterols
36. antibiotics
37. terpenes
38. isoprene rule
39. nitrogen, oxygen, and sulfur
40. vulcanization

**Matching**

41. (f)
42. (c)
43. (d)
44. (a)
45. (g)
46. (b)
47. (k)
48. (j)
49. (i)
50. (h)
51. (e)

**Miscellaneous**

52. $H_3C$ $CH_3$ C $H_2C$ $H_2C$ C $H_2$ (ring) $C-CH=CH-C(CH_3)=CH-CH=CH-C(CH_3)=CH-CH_2OH$ ; $C-CH_3$

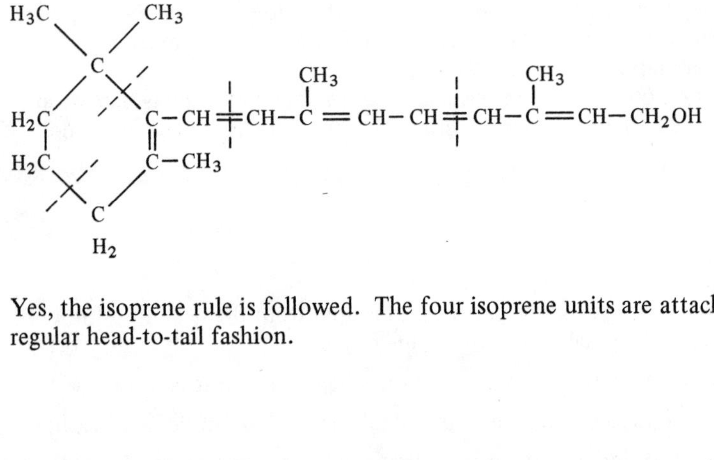

Yes, the isoprene rule is followed. The four isoprene units are attached in a regular head-to-tail fashion.

# 18

# *Natural Gas, Petroleum, and Petrochemicals*

## INTRODUCTION

This chapter is primarily a descriptive chapter designed to increase our knowledge of petroleum, its composition, and some of its many uses. Gasoline, its manufacture, and how its quality as a motor fuel may be improved through the use of additives is discussed in some detail. Also methods for the conversion of high molecular weight hydrocarbons into suitable motor fuels are outlined.

Some of the terms relative to the production of motor fuels that we will want to become familiar with are: **octane number**, **catalytic cracking**, **reforming** or **isomerization**, **alkylation**, and **aromatization**.

A short section on petrochemicals deals with petroleum products that we are inclined to overlook when we consider crude oil only as an abundant fossil fuel.

## PROBLEMS

### True–False

1. T F There is evidence that petroleum resulted from the decomposition of plant and animal matter of marine origin.
2. T F Natural gas has the same composition regardless of its source.
3. T F Crude oil is a complex mixture of alkanes and aromatic hydrocarbons.
4. T F The maximum octane number that can be assigned to a fuel is 100.

5. T F The law now requires the elimination of lead alkyls, such as tetraethyl lead, from automobile fuel. Removing these additives will require either engines capable of using lower octane fuels, the inclusion of more aromatics which possess high octane numbers, or the addition of new antiknock agents.
6. T F A low quality gasoline does not burn smoothly and results in knocking and a loss of power.
7. T F Higher molecular weight petroleum fractions unsuitable as motor fuel can be broken down by heat and the fragments converted to high quality gasoline.
8. T F Increased branching in alkanes results in a lower octane gasoline.
9. T F Cyclopentane is a better gasoline than *n*-pentane.
10. T F Reforming, alkylation, and aromatization reactions are reactions used to improve the octane rating of hydrocarbons.
11. T F Gasoline can be produced synthetically from coal as economically as it can be produced from petroleum.
12. T F Ethylene chloride and ethylene bromide are used in "ethyl" gasoline as scavengers for the lead, which results from the decomposition of tetraethyl lead.

**Completion**

13. The most abundant alkane present in natural gas is _______________.
14. The principal component of L.P. gas is _______________.
15. A measure of the ability of a gasoline to deliver maximum power when burned in an internal combustion engine is the _______________ _______________.
16. An octane number of 100 is assigned to ____________________ and a value of zero is assigned to ____________________.
17. The process by which large molecular weight compounds are broken down into smaller molecules by the application of heat in the presence of a catalyst is called _______________.
18.

$$CH_3-CH_2-CH_2-CH_2-CH_3 \xrightarrow[\text{heat}]{\text{catalyst,}} CH_3-\underset{\underset{CH_3}{|}}{\overset{\overset{CH_3}{|}}{C}}-CH_3$$

The above reaction is an example of _______________.

19. Liquid fuels can be produced from coal. The best known methods are the _______________-_______________ synthesis and the _______________ process.
20.

$$CH_3-\underset{\underset{CH_3}{|}}{\overset{\overset{CH_3}{|}}{C}}-H + H_2C=\overset{\overset{CH_3}{|}}{C}-CH_3 \xrightarrow[\text{heat}]{\text{catalyst,}} CH_3-\underset{\underset{CH_3}{|}}{\overset{\overset{CH_3}{|}}{C}}-CH_2-\underset{\underset{H}{|}}{\overset{\overset{CH_3}{|}}{C}}-CH_3$$

The reaction above can be used to produce high octane gasoline. It is called an ______________ reaction.

21. Chemicals, produced from petroleum, that are used in the manufacture of products other than fuels are called ______________.

**Matching** (Match the proper petroleum fractions with their principal uses.)

(a) Gas
(b) Gas-oil
(c) Gasoline
(d) Paraffin wax
(e) Kerosine
(f) Residuum
(g) Wax-oil

22. ____ Jet fuel, fuel oil, diesel fuel
23. ____ Packaging (wax paper), candles
24. ____ Fuel, such as natural gas and L.P. gas
25. ____ Lubricants, mineral oils, cracking stock
26. ____ Motor fuel, solvent naphtha
27. ____ Roofing, water-proofing, road building materials
28. ____ Diesel fuel, fuel oil, cracking stock

## ANSWERS

**True–False**

1. True
2. False. Some natural gases have a high carbon dioxide content and others contain high percentages of nitrogen, hydrogen sulfide, and some helium.
3. True
4. False. An octane number of 100 is assigned to 2,2,4-trimethylpentane, but there are other gasolines which perform better in an internal combustion engine and would, therefore, have higher octane numbers.
5. True
6. True. Low quality gasolines have low octane numbers.
7. True. High molecular weight hydrocarbons (above $C_9$) not suitable for use as gasoline are catalytically "cracked" or broken into smaller molecules. The latter are rearranged (isomerized), cyclized and dehydrogenated (aromatized), or combined (alkylated) to form compounds of high "octane number".
8. False. The octane numbers of hydrocarbons increase with chain branching, unsaturation, and cyclization.
9. True
10. True

11. False. The Fischer-Tropsch synthesis and the Bergius process are two developments for the production of liquid fuel from coal. The cost of gasoline when produced by these processes is not competitive with that produced from petroleum.
12. True. The lead dichloride ($PbCl_2$) and dibromide ($PbBr_2$) formed at the operating temperature of the engine are carried away with the exhaust. This is a source of lead in our atmosphere and poses a problem which concerns all of us.

**Completions**

13. methane
14. propane
15. octane number
16. 2,2,4-trimethylpentane, *n*-heptane
17. catalytic cracking
18. isomerization or reforming
19. Fischer-Tropsch, Bergius
20. alkylation
21. petrochemicals

**Matching**

22. (e)
23. (d)
24. (a)
25. (g)
26. (c)
27. (f)
28. (b)

# 19

# *Color in Organic Compounds, Dyes*

## INTRODUCTION

This chapter relates the phenomenon of color to the absorption of electromagnetic radiation in the visible region. The color produced is a function of the molecular structure. Colored organic molecules usually have a rather extensive $\pi$-system involving a number of conjugated double bonds and one or more structural groups capable of undergoing a $\pi \rightarrow \pi^*$ or $n \rightarrow \pi^*$ electronic transition. Color producing groups are called **chromophores**. Certain electron-pair donor groups, called **auxochromes**, intensify color when they are present in a molecule with a chromophore. The color perceived by the eye, the **complimentary color**, is a composite of all wavelengths not absorbed.

Dyes, of course, are colored organic molecules, but in order to be useful as a dye the molecule must have in its structure parts other than those which are responsible for its color. A dye molecule must have the ability to attach itself firmly to whatever is to be colored, and it must be reasonably permanent–i.e., wash-fast and light-fast. In this chapter we shall study the various structures of dyes and classify them according to structure and method of application.

## PROBLEMS

### True-False

1. T F White light is a mixture of all wavelengths within the visible region.

2. T F The color produced by a compound and perceived by the eye is the complimentary color of the wavelengths absorbed by the compound.
3. T F A dye must possess the ability to attach itself to whatever is being colored.
4. T F The formation of an azo dye can be carried out by a coupling reaction directly within the fabric.
5. T F Most dyes were isolated from a natural source in ancient times, but most modern dyes are synthetic.
6. T F The nature of the material to be dyed determines the method of dye application.
7. T F If the energy necessary to bring about a $\pi \rightarrow \pi^*$ or $n \rightarrow \pi^*$ transition in an organic compound corresponds to radiation in the ultraviolet region the compound will appear colorless.
8. T F The $\pi$-system in a colored compound is usually rather extensive.

**Multiple Choice**

9. Which group is not considered to be a chromophore?
   (a) $-N{=}O$
   (b) $-NH_2$
   (c) 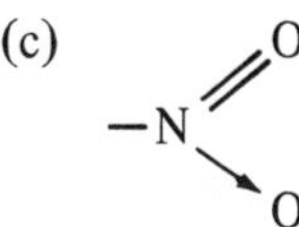
   (d) $-N{=}N-$
   (e) $-\overset{O}{\overset{\|}{C}}-\left(\underset{H}{\overset{}{C}}{=}\underset{H}{\overset{}{C}}\right)_n-\overset{O}{\overset{\|}{C}}-$

10. Which group is least likely to be listed as an auxochrome?
    (a) $-OH$
    (b) $-OCH_3$
    (c) $-CH_3$
    (d) $-NH_2$
    (e) $-\underset{H}{\underset{|}{N}}-CH_3$

**Completion**

11. The multiple-bonded, electron-attracting groups which are capable of absorbing radiation in the visible region and thus producing color are called ____________.
12. Groups which themselves are not responsible for the formation of color, but which enhance the colors produced by the chromophores are called ____________.
13. Silk and wool are animal fibers and classified as ____________. Cotton and linen are of vegetable origin and classified as ____________.

**Matching** (Match each compound to the appropriate description.)

(a) OH, $NO_2$, $NO_2$

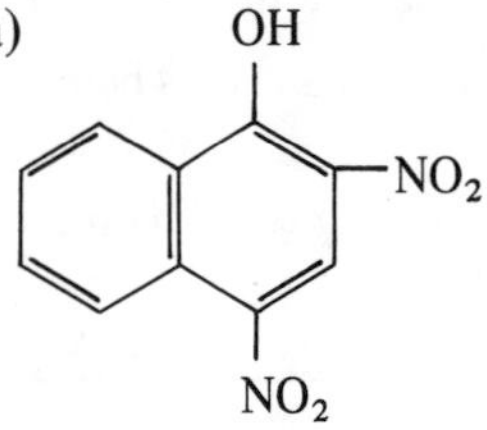

(b) NO, OH

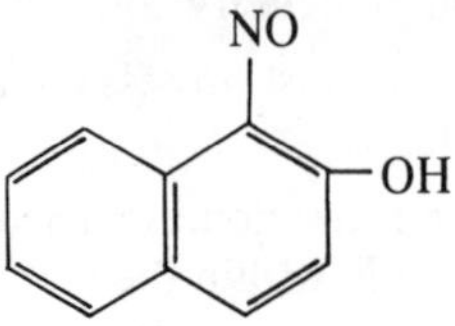

(c) N=N, $SO_3H$, OH

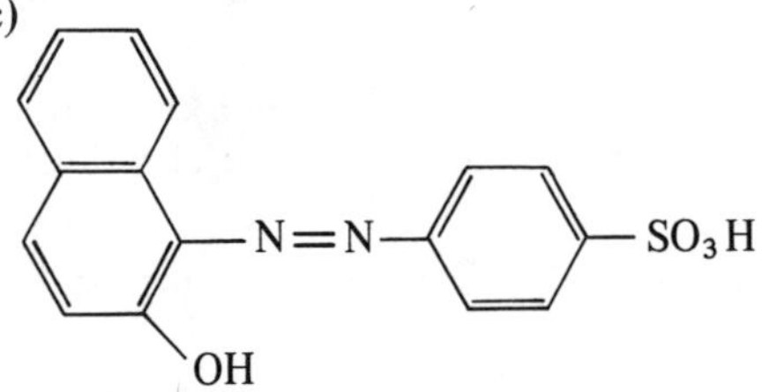

(d) O, OH, OH, O

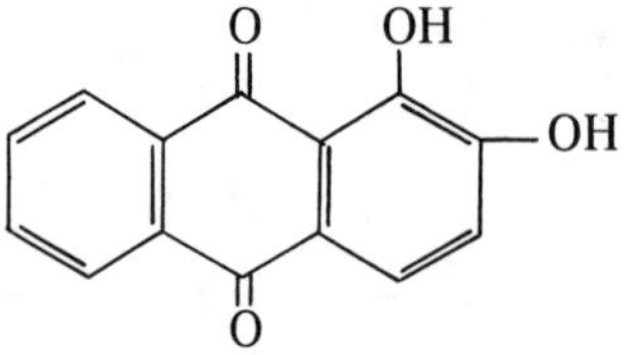

(e) O, C, H, N, C=C, N, H, C, O

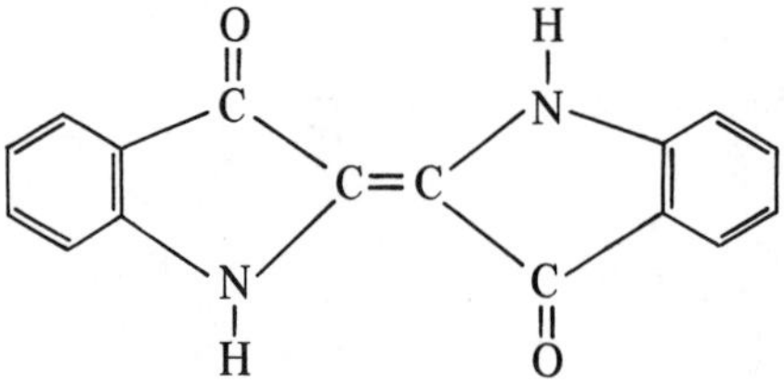

(f) $N(CH_3)_2$, C, $N(CH_3)_2$, +, $Cl^-$

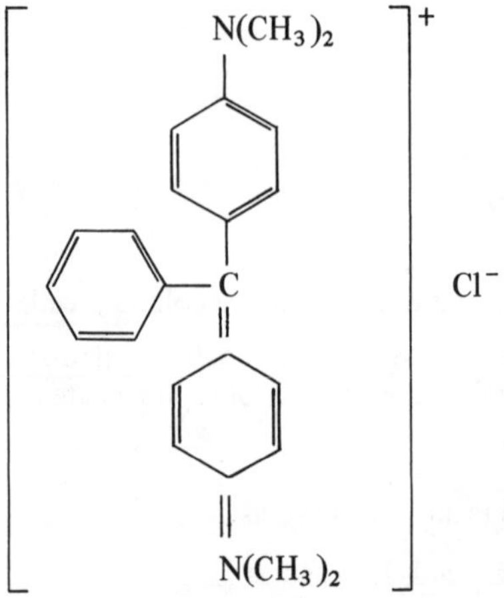

14. ____ An azo dye
15. ____ A nitro dye
16. ____ An anthraquinone dye
17. ____ A triphenylmethane dye
18. ____ A nitroso dye
19. ____ An indigoid dye

(Match each class of dye to the appropriate description.)

(g) Direct or substantive dyes
(h) Mordant dyes
(i) Vat dyes
(j) Ingrain dyes
(k) Disperse dyes

20. ____ Dye which uses a metallic oxide or hydroxide to form a link between the fabric and the dye. Useful for silk, wool, or cotton.
21. ____ Water-insoluble, fiber-soluble dyes used largely for fibers which lack polar groups.
22. ____ Dyes which are formed directly in the fiber. Good for silk and wool.
23. ____ Dyes that contain polar groups capable of combining with polar groups of the fiber. Good for silk, wool, and nylon.
24. ____ Dyes which are introduced into a fiber in a soluble and reduced form, but are then converted to an insoluble oxidized form by chemical reaction outside the dyeing vat.

## ANSWERS

### True–False

1. True
2. True. If a material absorbs light in the blue region of the visible spectrum, the color of the material is a composite of all wavelengths not absorbed and appears yellow.
3. True
4. True
5. True
6. True. Different types of groups serving as points of attachment for a dye are present in different types of materials. Different dyeing techniques, therefore, are required.
7. True. The eye perceives radiation only in the visible region of the electromagnetic spectrum.
8. True

## Multiple Choice

9. (b) Chromophores are usually multiple-bonded, electron-attracting structures.
10. (c) Auxochromes are usually electron-releasing groups which can enter into resonance with the chromophore to extend the conjugated system. The methyl group is not such a group.

## Completion

11. chromophores
12. auxochromes
13. proteins, carbohydrates

## Matching

14. (c)
15. (a)
16. (d)
17. (f)
18. (b)
19. (e)
20. (h)
21. (k)
22. (j)
23. (g)
24. (i)

# ANSWERS TO EXERCISES IN ORGANIC CHEMISTRY: A BRIEF COURSE

*by Linstromberg and Baumgarten*

## CHAPTER 1

1.1

| Na | Mg | Al | Si | P | S | Cl | Ar |
|---|---|---|---|---|---|---|---|
| 2-8-1 | 2-8-2 | 2-8-3 | 2-8-4 | 2-8-5 | 2-8-6 | 2-8-7 | 2-8-8 |

1.2 H :O: / :O: H — Hydrogen peroxide

Cl : Cl — Chlorine

O / Cl : C : Cl — Phosgene

1.3

H : C : H (with H above), H : C (with H above and below) : N : O, H : C : H (with H below)

• = electrons from carbon

· = electrons from hydrogen

○ = electrons from oxygen

* = electrons from nitrogen

Yes, the N–O bond is of the coordinate covalent type.

1.4 Bond angle = 120°. Boron trifluoride is a planar molecule.

1.5 Hydrogen fluoride is most polar; hydrogen iodide least polar. *Reason*: Fluorine is more electronegative than iodine, and the bond between the hydrogen and the fluorine atom is slightly more ionic than covalent.

1.6 Only one structure is possible for dichloromethane. Notice that all the following are actually the same structure.

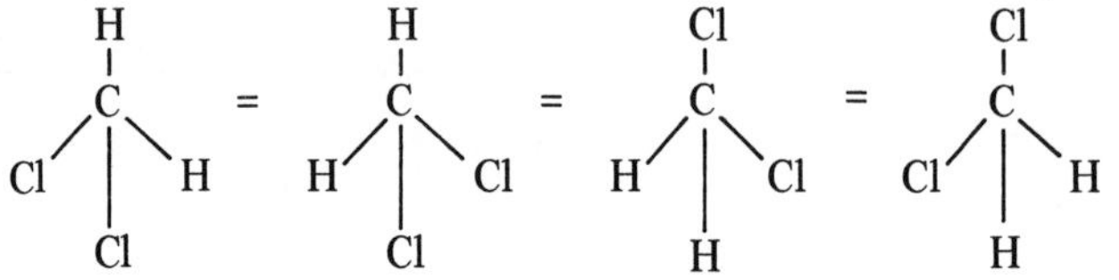

If carbon occupied the apex of a pyramid, *cis* and *trans* structures would be possible. These are unknown.

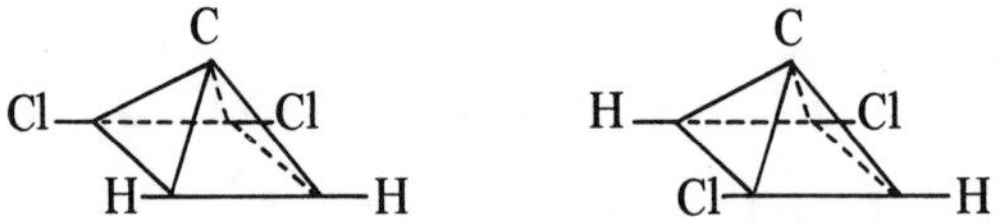

*cis* *trans*

1.7 Four alcohol structures and three ether structures can be drawn for $C_4H_{10}O$.

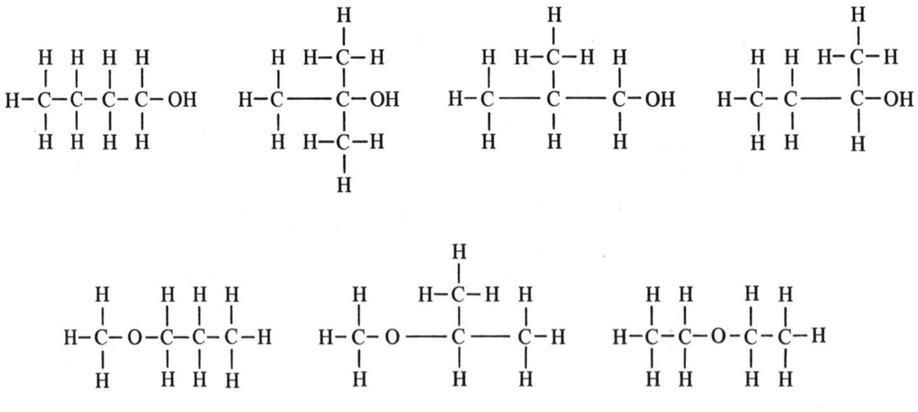

1.8

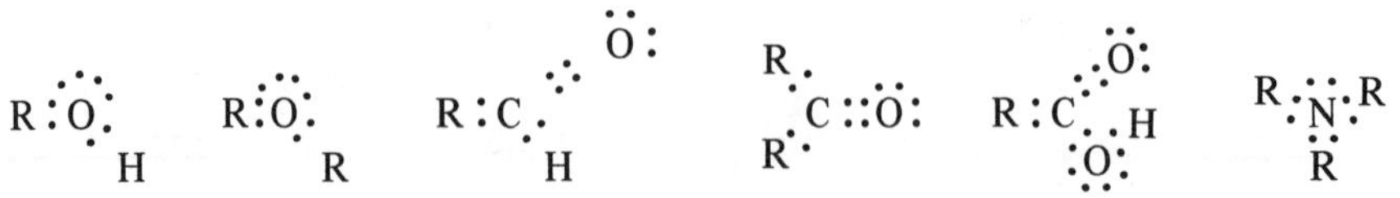

1.9

1.10 (1) base; (2) base; (3) acid; (4) base (can donate two of the electrons in the double bond); (5) acid (Mg can accept an electron pair).

1.11

```
                                              H  H                    H
     ..                                       *. o*                  .*  oo
(a) H*S:          (b) H*C :8 N8        (c) H*C o C * H       (d) H * C o O o
       *H                                     *. o*                  .*   o*
                                              H  H                    H     H

     H                   oo                  H                  H
                        oCl o                                    
      *.                 .o    oo           *.   oo            *.   oo
(e)    C :8 O8     (f) H * C o Cl8    (g) H*C oCl8     (h) H*C o O8
      *.                 .o    oo           *.   oo            .*   o   .H
     H                  oCl o                H                  H     .C*
                         oo                                          H*  .H

(i) .F:    .F:          (j)     H
     .*   *.                   *.  ooo H
       B                     H*C o N*
      .*                       *.   o  H
     :F:                        H
```

| 1.12 | | | | |
|---|---|---|---|---|
| NaCl | Very sol. in $H_2O$ | Insol. in hydrocarbons | Strong electrolyte | Nonflammable |
| $C_{10}H_8$ | Insol. in $H_2O$ | Sol. in hydrocarbons | Nonelectrolyte | Flammable |

| | | |
|---|---|---|
| NaCl | High m.p. (800°) | Vapor pres. not measurable at room temperature. |
| $C_{10}H_8$ | Low m.p. (80°) | Vapor pres. measurable at room tempature.* |

*Approximately 7 x $10^{-4}$ mm at room temperature.

The vapor pressure of naphthalene when it is kept in a confined space with woolens is sufficiently great to be lethal to moth larvae.

1.13 (a) The ability of carbon to bond covalently and (b) its ability to bond to other carbon atoms makes possible the great number of organic compounds.

1.14 (a)

```
   H  H  H
   |  |  |
H—C—C—C—H
   |  |  |
   H  H  H
```

Only one structure possible for $C_3H_8$, propane.

(b)

```
   H  H  H                      H  H  H
   |  |  |                      |  |  |
H—C—C—C—Cl       and       H—C—C—C—H
   |  |  |                      |  |  |
   H  H  H                      H  Cl H
```

1-Chloropropane and 2-Chloropropane

(c)

```
    H  H  H              H  H  H                  H     H  H
    |  |  |              |  |  |                  |     |  |
 H- C- C- C -O-H ,   H- C- C- C -H ,  and  H- C -O- C - C -H
    |  |  |              |  |  |                  |     |  |
    H  H  H              H  O  H                  H     H  H
                            |
                            H
```

*n*-Propyl alcohol — Isopropyl alcohol — Methyl ethyl ether

(d)

```
    H  H                        H  H
    |  |                        |  |
 H- C- C -Cl       and     Cl- C- C - Cl
    |  |                        |  |
    H  Cl                       H  H
```

1,1-Dichloroethane — 1,2-Dichloroethane

(e)

```
                                   H                        H
                                   |                        |
   H  H  H  H  H           H  H- C -H  H  H          H  H- C -H  H
   |  |  |  |  |           |     |     |  |          |     |     |
H- C- C- C- C- C -H     H- C --- C --- C- C -H    H- C --- C --- C -H
   |  |  |  |  |           |     |     |  |          |     |     |
   H  H  H  H  H           H     H     H  H          H  H- C -H  H
                                                           |
                                                           H
```

*n*-Pentane — Isopentane — Neopentane

(f)

```
          H                O                   O                   H
          |                ||                 / \                  |
 CH3CH2 - C=O      CH3 - C - CH3      CH3 - C --- CH2       H2C=C - O - CH3
                                            |
                                            H
```

Propionaldehyde — Acetone — Propylene oxide — Methyl vinyl ether

```
        H                                H2C - O
        |                                 |    |
 H2C= C - CH2 - OH                       H2C - CH2
```

Allyl alcohol — Oxetane (Trimethylene oxide)

(g)

```
    H                  H                           H
    |                  |    ..                     |   +     -
 H- C - C≡N         H- C - N=C:   <-------->   H- C - N≡C:
    |                  |                           |
    H                  H                           H
```

Methyl cyanide (Acetonitrile) — Methyl isocyanide

(h)

Allylamine

1-Methylaziridine

2-Methylaziridine

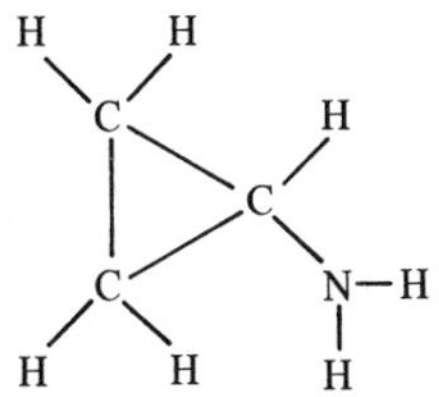

Cyclopropylamine

$CH_3-\overset{H}{C}=N-CH_3$

Ethylidene-methylamine

Azetidine

1.15 Three

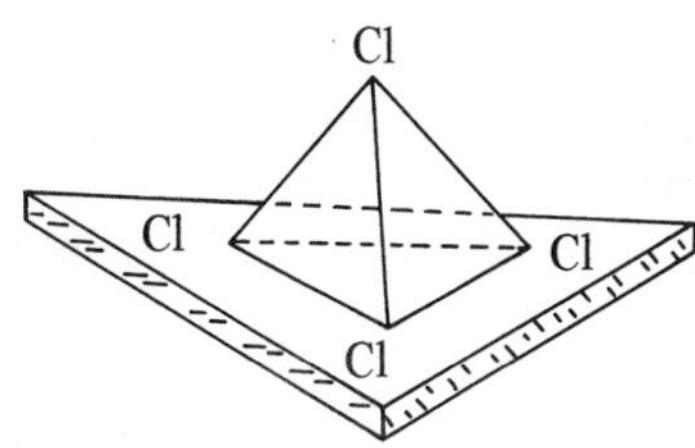


1.16 Polar bonds: (b), (c), (d), (e), (g), (h), (i), (j)
Polar molecules: (b), (d), (g), (i), (j)

1.17 (a) C=37.5%; H=12.5%; 0=50%
(b) C=39%; H=16%; N=45%
(c) C=8.5%; H=2.1%; I=89.4%
(d) C=20%; H=6.67%; N=46.67%; O=26.67%
(e) C=42%; H=6.45%; O=51.5%

1.18 $$\frac{12/44 \times 0.044\text{g}}{0.060\text{ g}} \times 100 = \text{percent of C in sample} = 20\%$$

$$\frac{2/18 \times 0.036\text{ g}}{0.060\text{ g}} \times 100 = \text{percent H in sample} = 6.67\%$$

The molecular weight of nitrogen, $N_2$, is 28 g/mole

22,400 ml = 1 molar volume of gas under standard conditions.

22.4 ml $N_2$ = 0.001 mole = 0.028 g of nitrogen

$$\frac{0.028}{0.060} \times 100 = \text{percent of N in sample} = 46.67\%$$

100 - (20 + 6.67 + 46.67) = 26.67% oxygen in sample

$$C_{20/12}H_{6.67/1}N_{46.67/14}O_{26.67/16} = C_{1.67}H_{6.67}N_{3.33}O_{1.67} = CH_4$$

```
H        O        H
 \       ||      /
  N  —   C   —  N
 /               \
H                 H
```

Urea

1.19 C = 73.35%; H = 3.8%; N = 10.7%; O = 12.15%

$$C_{73.35/12.01}H_{3.8/1.008}N_{10.7/14.01}O_{12.15/15.99} = C_{6.11}H_{3.77}N_{0.764}O_{0.760}$$

$$= C_{8.04}H_{4.96}N_{1.005}O_1$$

Thus *empirical* formula is $C_8H_5NO$. Molecular weight of $C_8H_5NO$ = 131; whereas, molecular weight of compound = 250–275. Therefore, *molecular* formula must be $C_{16}H_{10}N_2O_2$ (molecular weight = 262).

1.20

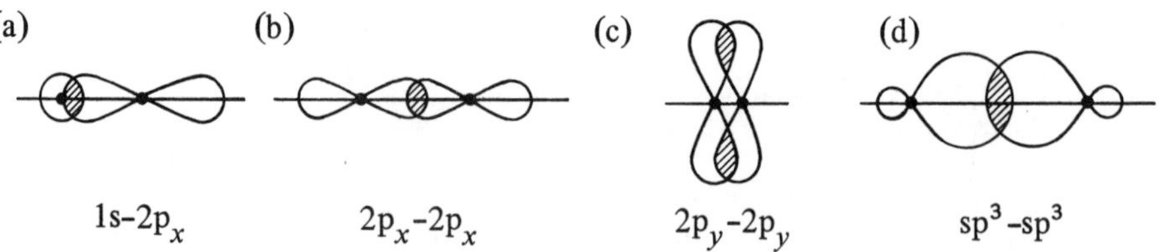


1.21 (a) isomers (differ in position of an H atom); (b) resonance forms; (c) resonance forms; (d) resonance forms; (e) resonance forms; (f) isomers (differ in position of an H atom); (g) isomers (differ in the positions of a C and an N atom); (h) isomers (differ in the positions of a C and an N atom).

1.22

| (a) | | (b) | | (c) | | (d) | |
|---|---|---|---|---|---|---|---|
| H— | 0 | —C— | 0 | H— | 0 | :O: | -1 |
| —O— | 0 | H— | 0 | —C— | 0 | :O:: | +1 |
| —N= | +1 | —N≡ | +1 | —S— | +1 | ::O: | 0 |
| =O: | 0 | ≡C— | 0 | —O: | -1 | | |
| —O: | -1 | —O: | -1 | | | | |

## CHAPTER 2

2.1

```
     H  H  H  H  H  H
     |  |  |  |  |  |
  H- C— C— C— C— C— C— H
     |  |  |  |  |  |
     H  H  H  H  H  H
```

*n*-Hexane

```
             H
             |
     H  H— C—H  H  H  H
     |       |     |  |  |
  H— C—————C—————C— C— C— H
     |       |     |  |  |
     H       H     H  H  H
```

2-Methylpentane
(Isohexane)

```
     H  H        H        H  H
     |  |        |        |  |
  H— C— C————————C————————C— C— H
     |  |        |        |  |
     H  H    H— C— H      H  H
                 |
                 H
```

3-Methylpentane

```
              H
              |
     H    H— C— H   H  H
     |        |       |  |
  H— C————————C———————C— C— H
     |        |       |  |
     H    H— C— H   H  H
              |
              H
```

2,2-Dimethylbutane

```
              H
              |
     H    H— C— H   H          H
     |        |       |          |
  H— C————————C———————C——————————C— H
     |        |       |          |
     H        H   H— C— H      H
                      |
                      H
```

2,3-Dimethylbutane

2.2

```
            CH3
  CH3 — C — CH2 — Cl
            |
          CH2CH3
```

The longest continuous chain of carbon atoms was not correctly determined by the student. The correct name of the compound is

1-Chloro-2,2-dimethylbutane

2.3

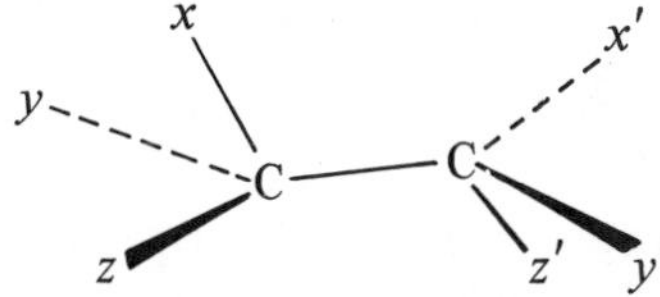


(a) An anti conformation of *n*-butane may have methyl groups at $x$ and $z'$, or at $y$ and $y'$, or at $z$ and $x'$.

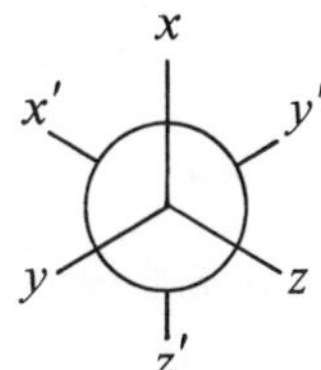


(b) Gauche conformations of *n*-butane may have methyl groups at $x$ and $x'$ or at $x$ and $y'$; at $y$ and $x'$ or at $y$ and $z'$; at $z$ and $z'$ or at $z$ and $y'$.

An anti conformation for *n*-butane is most likely in order to allow less crowding of methyl groups, thus producing greater stability of the molecule.

2.4 Since *n*-pentane is a liquid, the heat of combustion will be the total heat less the number of kilocalories equivalent to the heat of vaporization.

2.6 $CH_3{-}H + Cl\cdot \rightarrow CH_3\cdot + H{-}Cl$

104 103

$\Delta H = -(103\text{–}104) = +1$ kcal/mole

$CH_3\cdot + Cl{-}Cl \rightarrow CH_3{-}Cl + Cl\cdot$

58 83.5

$\Delta H = -(83.5 - 58) = -25.5$ kcal/mole

2.7 Five cyclic structures may be drawn for $C_5H_{10}$

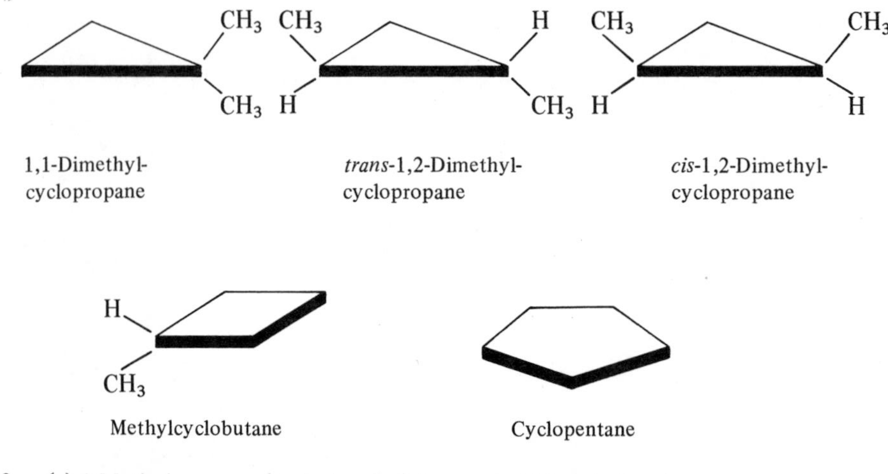


2.8 (a) 2-Methylpropane (Isobutane); (b) *trans*-1,2-Dimethylcyclopentane; (c) 2-Methylpentane (Isohexane); (d) 2-Chlorobutane (*sec*-Butyl chloride); (e) 1-Chloro-2,2-dimethylpropane (Neopentyl chloride); (f) 2,2,4-Trimethyl-pentane (Isooctane*).

*Isooctane is a common name, but not correct for an isoalkane. (See text, Table 18.4.)

2.9 Structures (a) and (c) are those of 2,4-dimethylpentane; structures (b) and (d) are those of 3-methylpentane; (e) 2,2-Dimethylbutane; (f) 2,3-Dimethyl-butane; (g) 2-Methylpentane; (h) 3,3-Dimethylpentane.

2.10 (a) The longest carbon chain was not named as the parent compound. Correct name: 2,3-Dimethylpentane.
(b) The substituent was not located by the smallest numbers. Correct name: 2-Methylpentane.
(c) Substituents were not located by their smallest numbers. Correct name: 2,2,4-Trimethylhexane.

(d) This name is redundant. Systematic (IUPAC) and trivial names should not be mixed. Correct name: either 2-Chloropropane or Isopropyl chloride.

(e) Substituents were not located by smallest numbers. Correct name: 2-Chloro-3-methylpentane.

(f) Substituents were not located by smallest numbers. Further, since cyclopropane is a planar structure the methyl groups may be either on the same side of the plane (*cis*) or on opposite sides of the plane (*trans*). Correct names: *cis*-1,2-Dimethylcyclopropane or *trans*-1,2-Dimethylcyclopropane. (See text, Section 2.12.)

2.11 (a) $CH_3-CH_2-CH(CH_3)-CH_3$

(b) $CH_3CH_2-CH(CH_3)-CH_2-C(CH_3)_2-CH_3$

(c) $CH_3-C(CH_3)_2-CH_2-Cl$

1-Chloro-2,2-dimethylpropane

(d) $CH_3-C(CH_3)_2-Cl$

2-Chloro-2-methylpropane

(e) $CH_3-CH(CH_3)-CH_2-Cl$

1-Chloro-2-methylpropane

(f) $H-CCl_3$

Trichloromethane

(g) $Br-CH_2-CH_2-CH_2-Br$

(h) $CH_3-CH(CH_3)-CH(CH_2-CH_3)-CH_2-CH_3$

2.12

$CH_3-CH(CH_3)-CH_2-CH_3$ (positions: ⑥ the two $CH_3$ groups on C1 side, ① CH, ② $CH_2$, ③ $CH_3$) $\xrightarrow[h\nu]{Cl_2}$

$Cl-CH_2-CH(CH_3)-CH_2-CH_3$ $\quad \dfrac{6 \times 1}{22.4} \times 100 = 26.8\%$

$CH_3-CCl(CH_3)-CH_2-CH_3$ $\quad \dfrac{1 \times 5.8}{22.4} \times 100 = 25.9\%$

$CH_3-CH(CH_3)-CHCl-CH_3$ $\quad \dfrac{2 \times 3.8}{22.4} \times 100 = 33.9\%$

$CH_3-CH(CH_3)-CH_2-CH_2Cl$ $\quad \dfrac{3 \times 1}{22.4} \times 100 = 13.4\%$

Denominator = $(6 \times 1) + (1 \times 5.8)$
$+ (2 \times 3.8)$
$+ (3 \times 1)$
$= 22.4$

2.13

$Cl-CH_2-CH_2-CH_2-Cl$

1,3-Dichloropropane

$CH_3-CH(Cl)-CH_2-Cl$

1,2-Dichloropropane

$CH_3-CH_2-CHCl_2$

1,1-Dichloropropane

$CH_3-CCl_2-CH_3$

2,2-Dichloropropane

2.14 (a) $C_2H_5MgBr$

(b)

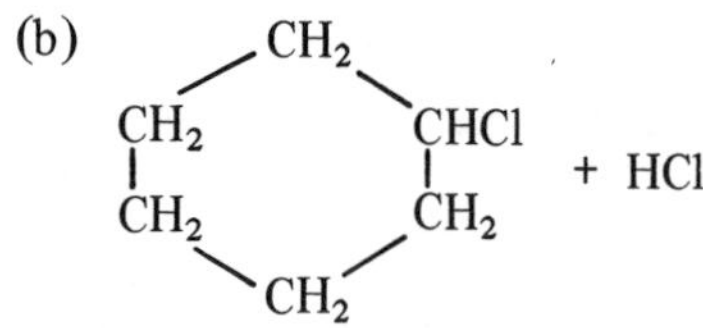


(c) $CH_4 + MgI(OH)$

(d) $3\ CO_2 + 4\ H_2O$

(e) (mixtures of all chlorinated methanes) Possibilities:

$$CH_4 + Cl_2 \xrightarrow{uv} CH_3Cl + HCl$$

$$4\ CH_4 + 5\ Cl_2 \rightarrow 3\ CH_3Cl + CH_2Cl_2 + 5\ HCl$$

or $$4\ CH_4 + 10\ Cl_2 \rightarrow CH_3Cl + CH_2Cl_2 + CHCl_3 + CCl_4 + 10\ HCl$$

2.15 (1) $\underset{104}{CH_3-H} + Br\cdot \rightarrow CH_3\cdot + \underset{87}{H-Br}$ $\qquad \Delta H = -(87-104) = +17$ kcal/mole

(2) $CH_3\cdot + \underset{46}{Br-Br} \rightarrow \underset{70}{CH_3-Br} + Br\cdot$ $\qquad \Delta H = -(70-46) = -24$ kcal/mole

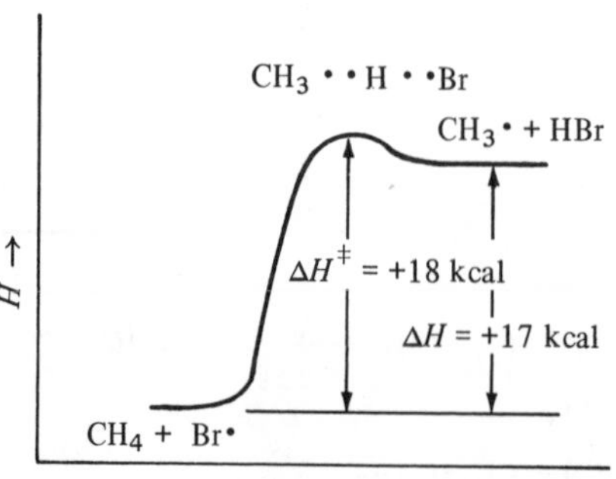


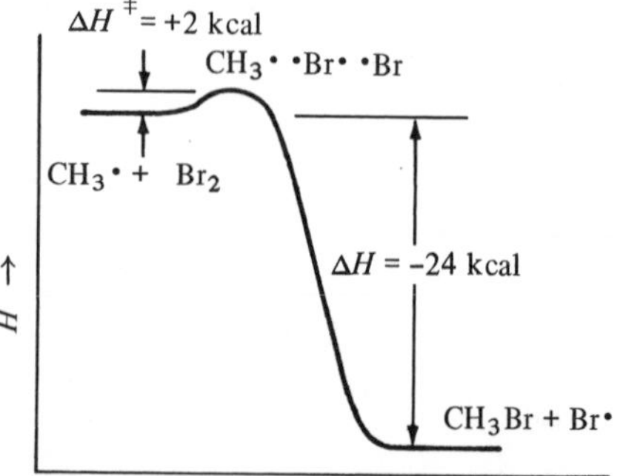


Step (1) is rate determining.

2.16 38,126 liters of air at (STP) or 1,352 cu. ft. of air; 6.12 liters of liquid water.

2.17 L.P., or liquified petroleum, gas is completely combustible. Natural gas, on the hand, contains a number of nonfuel components.

2.18 (a) $2\ H_2 + O_2 \rightarrow 2\ H_2O$

$\Delta H = 2 \times 104$ kcal + 119 kcal - 4(111 kcal) = -117 kcal

(b) $C_2H_6 + Cl_2 \rightarrow C_2H_5Cl + HCl$

$\Delta H$ = 98 kcal + 58 kcal - 81.5 kcal - 103 kcal = -28.5 kcal

(c) $C_3H_8 + Cl_2 \rightarrow CH_3\underset{\displaystyle Cl}{\underset{|}{C}}HCH_3 + HCl$

$\Delta H$ = 98 kcal + 58 kcal - 81 kcal - 103 kcal = -28 kcal

2.19

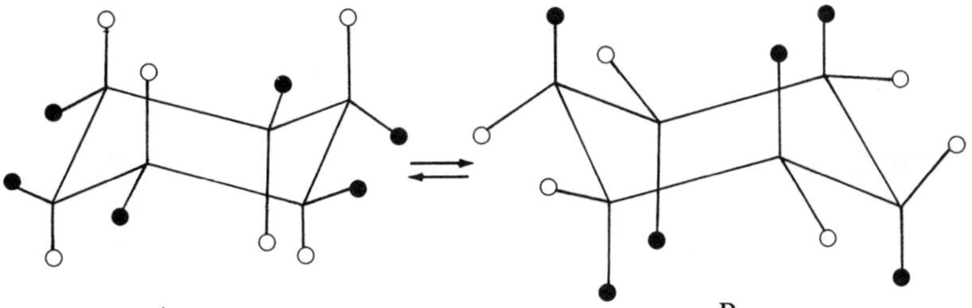

A

○ axial

● equatorial

B

○ equatorial

● axial

2.20

$$CH_3Br \xrightarrow[2.\ HOH]{1.\ Mg} CH_4$$

$$CH_3Br \xrightarrow[2.\ CuI]{1.\ Li} (CH_3)_2CuLi \xrightarrow{CH_3Br} CH_3{-}CH_3$$

$$CH_3CH_2Br \xrightarrow[2.\ HOH]{1.\ Mg} CH_3{-}CH_3$$

$$(CH_3)_2CuLi + CH_3CH_2Br \longrightarrow CH_3CH_2CH_3$$

$$CH_3CH_2Br \xrightarrow[2.\ CuI]{1.\ Li} (CH_3CH_2)_2CuLi \xrightarrow{CH_3Br} CH_3CH_2CH_3$$

$$CH_3CH_2CH_2CH_2Br \xrightarrow[2.\ HOH]{1.\ Mg} CH_3CH_2CH_2CH_3$$

$(CH_3CH_2)_2CuLi + CH_3CH_2Br \longrightarrow CH_3CH_2CH_2CH_3$

$CH_3CH_2CH_2CH_2Br \xrightarrow[2.\ CuI]{1.\ Li} (CH_3CH_2CH_2CH_2)_2CuLi \xrightarrow{CH_3Br} CH_3CH_2CH_2CH_2CH_3$

$(CH_3CH_2CH_2CH_2)\ CuLi + CH_3CH_2Br \longrightarrow CH_3CH_2CH_2CH_2CH_2CH_3$

$(CH_3CH_2CH_2CH_2)_2CuLi + CH_3CH_2CH_2CH_2Br \longrightarrow CH_3CH_2CH_2CH_2CH_2CH_2CH_2CH_3$

2.21 $CH_3-CH(CH_3)-CH(Br)-CH_3 \xrightarrow[2.\ HOH]{1.\ Mg} CH_3-CH(CH_3)-CH_2-CH_3$

A

A $\xrightarrow{(CH_3)_2CuLi}$ $CH_3-CH(CH_3)-CH(CH_3)-CH_3$

2.22 $(CH_2)_4 + 6\ O_2 \rightarrow 4\ CO_2 + 4\ H_2O \qquad \Delta H = -655.86$ kcal/mole

For strain-free cyclohexane with 6 $CH_2$ groups:

$$\frac{-944.80}{6} = -157.47 \text{ kcal/}CH_2 \text{ group}$$

For strain-free cyclobutane (hypothetical):

$$\Delta H_{calc} = 4 \times (-157.47) = -629.87 \text{ kcal/mole}$$

For actual cyclobutane

$$\Delta H_{expt} = \quad -655.86 \text{ kcal/mole}$$

$$\text{Strain energy} \quad = \quad 25.99 \text{ kcal/mole}$$

2.23

A: Newman projection — front carbon: $CH_3$ (top), H, H; back carbon: $CH_3$ (bottom), H, H.
Most stable conformer
Let $E = 0$

B: Newman projection — front carbon: $CH_3$ (top), H, H; back carbon: $CH_3$ (upper right), H, H.
$E = 0.9$

C: Newman projection — front carbon: $CH_3$ (top), H, H; back carbon: $CH_3$ (upper left), H, H.
$E = 0.9$

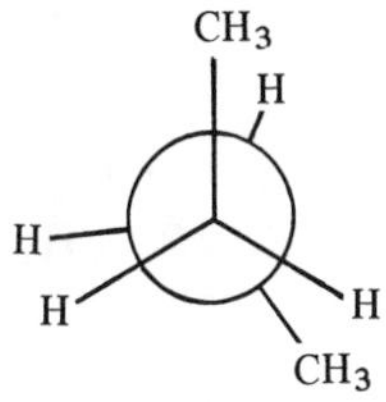

$E = 3.8$ D

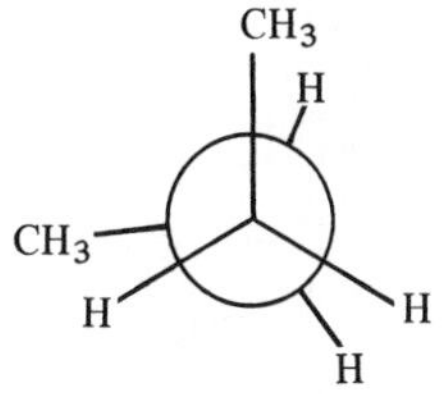

$E = 3.8$ E

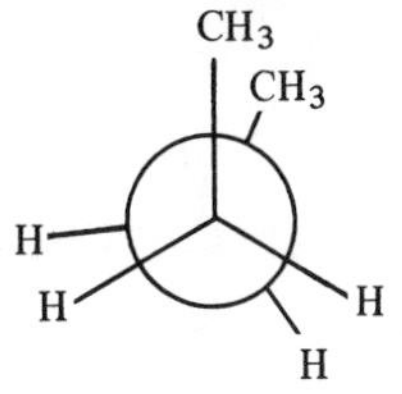

$E = 4.5$ F

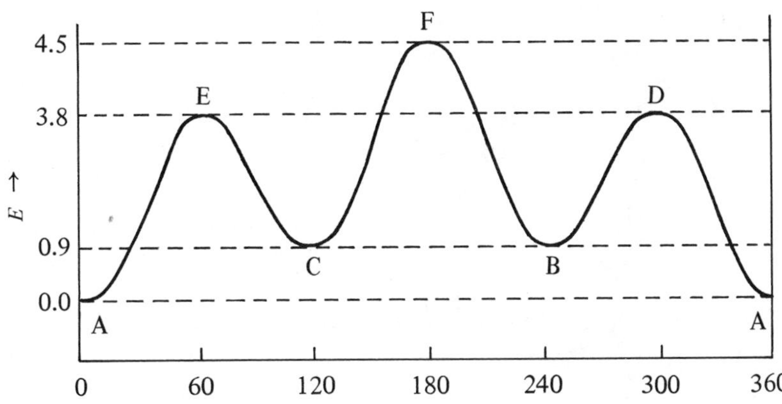


2.24 (a)

$$^{13}CH_3Br \xrightarrow[\text{2. HOH}]{\text{1. Mg}} {}^{13}CH_4$$

(b)

$$CH_3Br \xrightarrow[\text{2. DOD}]{\text{1. Mg}} CH_3D$$

(c)

$$^{13}CH_3Br \xrightarrow[\text{2. CuI}]{\text{1. Li}} ({}^{13}CH_3)_2CuLi \xrightarrow{{}^{13}CH_3Br} {}^{13}CH_3{-}{}^{13}CH_3$$

(d)

$$CH_3CH_2CH_2CH_2CH_2Br \xrightarrow[\text{2. DOD}]{\text{1. Mg}} CH_3CH_2CH_2CH_2CH_2D$$

## CHAPTER 3

3.1

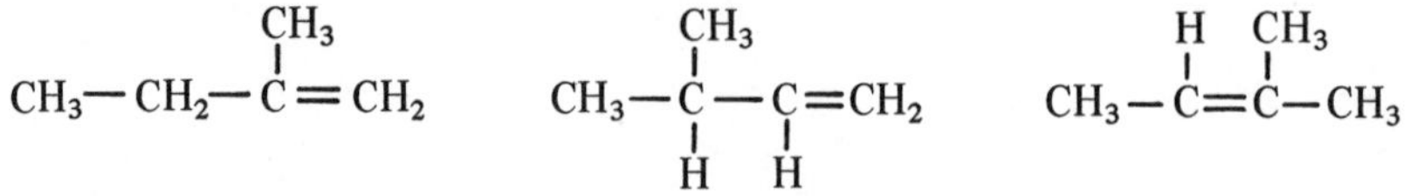

$$CH_3-CH_2-\underset{}{\overset{CH_3}{\overset{|}{C}}}=CH_2$$

2-Methyl-1-butene

$$CH_3-\underset{H}{\underset{|}{\overset{CH_3}{\overset{|}{C}}}}-\underset{H}{\underset{|}{C}}=CH_2$$

3-Methyl-1-butene

$$CH_3-\overset{H}{\overset{|}{C}}=\overset{CH_3}{\overset{|}{C}}-CH_3$$

2-Methyl-2-butene

3.2 $CH_3CH_2CH_2CH{=}CH_2$

1-Pentene

$$\begin{matrix} CH_3CH_2 & & H \\ & C{=}C & \\ H & & CH_3 \end{matrix}$$

*trans*-2-Pentene

$$\begin{matrix} CH_3CH_2 & & CH_3 \\ & C{=}C & \\ H & & H \end{matrix}$$

*cis*-2-Pentene

3.3

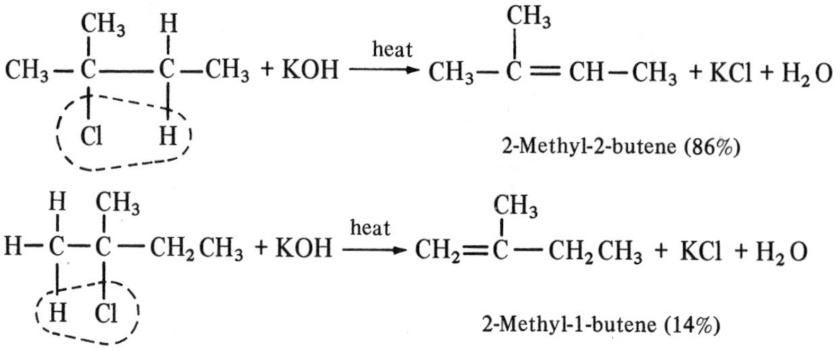

$$CH_3-\underset{Cl}{\underset{|}{\overset{CH_3}{\overset{|}{C}}}}-\underset{H}{\underset{|}{\overset{H}{\overset{|}{C}}}}-CH_3 + KOH \xrightarrow{\text{heat}} CH_3-\overset{CH_3}{\overset{|}{C}}=CH-CH_3 + KCl + H_2O$$

2-Methyl-2-butene (86%)

$$H-\underset{H}{\underset{|}{\overset{H}{\overset{|}{C}}}}-\underset{Cl}{\underset{|}{\overset{CH_3}{\overset{|}{C}}}}-CH_2CH_3 + KOH \xrightarrow{\text{heat}} CH_2{=}\overset{CH_3}{\overset{|}{C}}-CH_2CH_3 + KCl + H_2O$$

2-Methyl-1-butene (14%)

3.4 Cycloalkanes

3.5

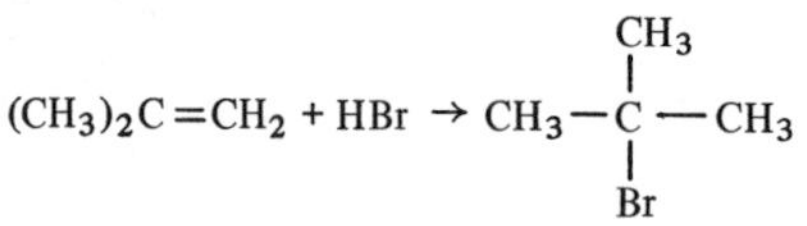

$$(CH_3)_2C{=}CH_2 + HBr \rightarrow CH_3-\underset{Br}{\underset{|}{\overset{CH_3}{\overset{|}{C}}}}-CH_3$$

The predominant product is 2-bromo-2-methylpropane because a more stable intermediate tertiary carbonium ion is formed more rapidly than a primary carbonium ion.

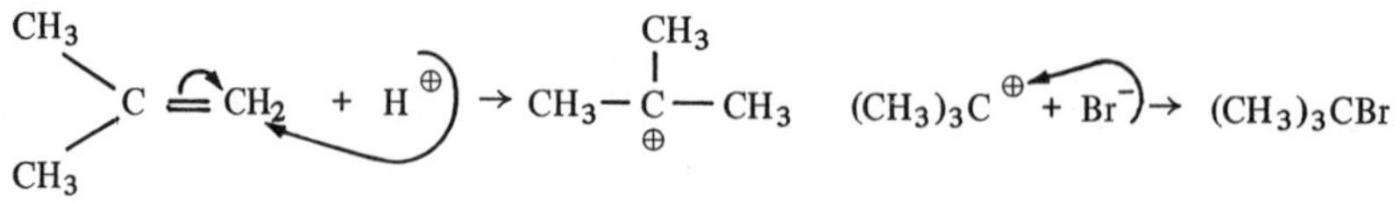

$$(CH_3)_2C{=}CH_2 + H^{\oplus} \rightarrow CH_3-\underset{\oplus}{\overset{CH_3}{\overset{|}{C}}}-CH_3 \qquad (CH_3)_3C^{\oplus} + Br^- \rightarrow (CH_3)_3CBr$$

3.6 $CH_3{-}CH{=}CH_2 + 0.1\%\ KMnO_4 \rightarrow CH_3-\underset{OH}{\underset{|}{CH}}-CH_2OH$ (Propylene glycol)

$$CH_3{-}CH{=}CH_2 + KMnO_4 \xrightarrow{\text{heat}} CH_3-C\begin{matrix}{\nearrow}O \\ {\searrow}O^-K^+\end{matrix} + H-C\begin{matrix}{\nearrow}O \\ {\searrow}O^-K^+\end{matrix}$$

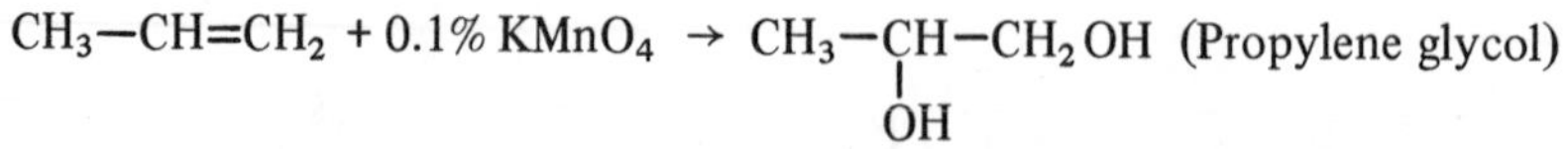

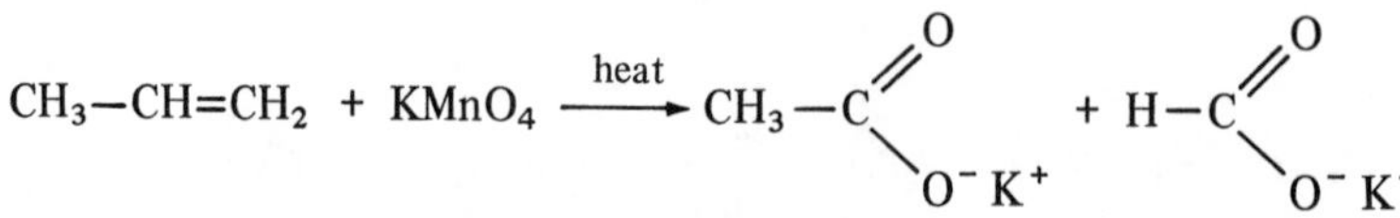

Potassium acetate

Potassium formate

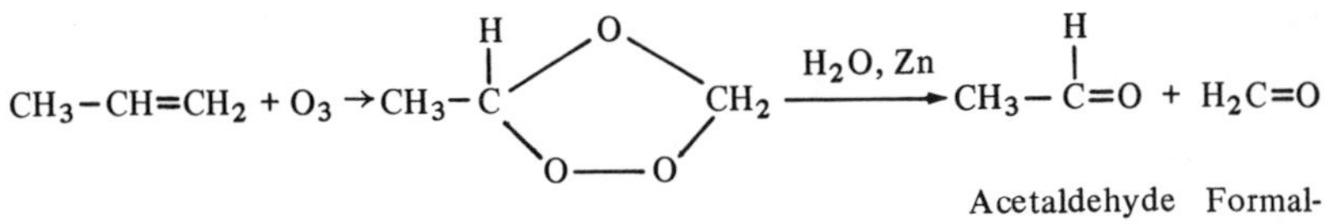

$$CH_3-CH=CH_2 + O_3 \rightarrow CH_3-\overset{H}{\underset{|}{C}}\!\overset{O}{\underset{O-O}{\frown}}\!CH_2 \xrightarrow{H_2O,\ Zn} CH_3-\overset{H}{\underset{|}{C}}=O + H_2C=O$$

Acetaldehyde Formaldehyde

3.7 $n$=approximately 892

3.8 $CH_2=CH-CH=CH_2 + R\cdot \rightarrow [R-CH_2-\dot{C}H-CH=CH_2 \leftrightarrow R-CH_2-CH=CH-\dot{C}H_2]$

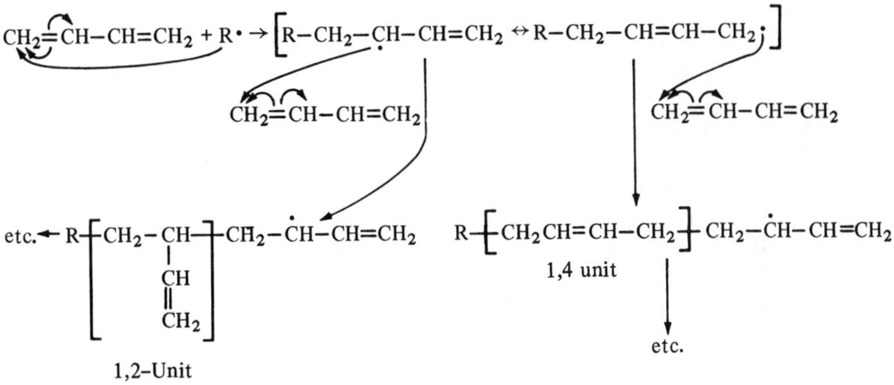

$CH_2=CH-CH=CH_2$ ... $CH_2=CH-CH=CH_2$

etc. ← $R-[CH_2-CH(CH=CH_2)]-CH_2-\dot{C}H-CH=CH_2$ $R-[CH_2CH=CH-CH_2]-CH_2-\dot{C}H-CH=CH_2$

1,2-Unit 1,4 unit

etc.

3.9 (a) $:\ddot{S}::C::\ddot{S}:$ (b) $H:C:::N:$ (c) $:\ddot{Cl}:Be:\ddot{Cl}:$

All three structures, like acetylene, exhibit *sp* hybridization and are linear molecules.

3.10 The sigma bond represented by (a) is formed by $sp^2$-$sp^2$ overlap; that by (b) is formed by $sp^2$ - $sp$ overlap; that by (c) is formed by $sp$-$sp$ overlap; and that by (d) is formed by $sp$-$s$ overlap. The value of $\theta$ is approximately $120°$; $\phi = 180°$

3.11

(a) $CH_3-CH=CH_2$ Propene

(b) $Br-CH_2CH_2-Br$ 1,2-Dibromoethane

(c) $CH_2=C(CH_3)-CH=CH_2$ 2-Methyl-1,3-butadiene

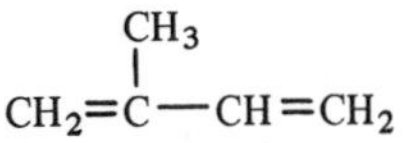

(d) $CH_3-C(CH_3)=CH_2$ 2-Methylpropene

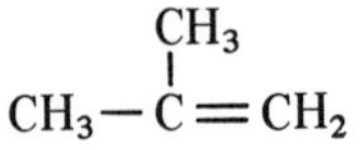

(e) $CH_3CH_2-C\equiv C-CH_2CH_2CH_3$ Ethyl-*n*-propylacetylene

3.12 (a) $CH_3-CH(CH_3)-CH=CH-CH(CH_3)-CH_3$

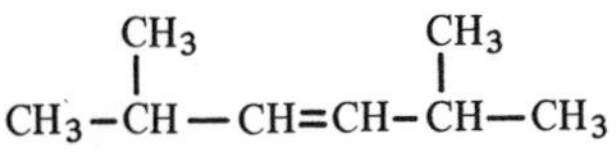

(b) $CH_2=C(CH_2-CH_3)-CH(CH_3)-CH_2-CH_3$

(c) [cyclohexene ring with Cl, H, H, Cl substituents]

Cl H H Cl

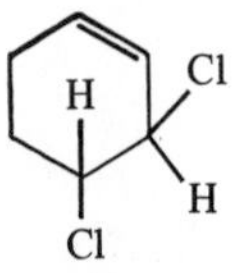

(d) [1,3-cyclohexadiene ring]

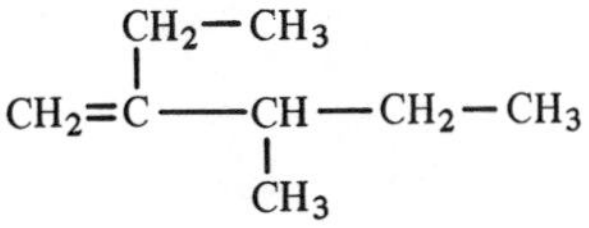

(e) $CH_2{=}CH{-}CH_2{-}C{\equiv}C{-}CH_2{-}CH{=}CH_2$

(f) $(CH_3)HC{=}CH{-}CH(CH_3){-}CH_3$ (H and H on the same side of C=C; $CH_3$ and $CH(CH_3)_2$ on the same side)

3.13 (a) 2-Hexene (b) 4-Chloro-2-pentene (c) 2-Pentyne (d) 2,3-Dibromo-2-butene (e) 2-Bromo-3-methyl-2-butene (f) 3-Methyl-1-butene (g) cyclopentene (h) 1,2-Dimethylcyclopentane

3.14 (a), (b), (d), (h)

3.15 (a) $CH_3CH_2OH \xrightarrow{H_2SO_4\text{,heat}} H_2C{=}CH_2 + H_2O$

(b) $H_2C{=}CH_2$ (from a) + HBr $\rightarrow CH_3CH_2Br$

(c) $CH_3CH_2Br$ (from b) + $Na{-}C{\equiv}C{-}H$ (from f) $\rightarrow CH_3CH_2{-}C{\equiv}C{-}H + NaBr$

(d) $C_2H_5Br$ (from b) + Mg $\xrightarrow{\text{anhydrous ether}} C_2H_5MgBr$

(e) $C_2H_5MgBr$ (from d) + $H_2O \rightarrow C_2H_6 + Mg(OH)Br$ or

$H_2C{=}CH_2$ (from a) + $H_2 \xrightarrow{\text{Ni, pressure}} C_2H_6$

(f) $C_2H_4$ (from a) + $Br_2 \rightarrow BrCH_2CH_2Br$

$BrCH_2CH_2Br + KOH \xrightarrow{\text{alcoholic solution}} CH_2{=}CHBr + KBr + H_2O$

$CH_2{=}CHBr + NaNH_2 \rightarrow H{-}C{\equiv}C{-}H + NaBr + NH_3$

(g) $C_2H_2$ (from f) + HCl $\longrightarrow CH_2{=}CH{-}Cl$

$CH_2{=}CH{-}Cl + Cl_2 \longrightarrow ClCH_2{-}CHCl_2$

(h) $C_2H_4$ (from a) + $Br_2 \rightarrow BrCH_2CH_2Br$

(i) $C_2H_2$ (from f) + HCN $\rightarrow H_2C{=}CH{-}CN$

(j) $2\,C_2H_2 \xrightarrow{CuCl,\ NH_4Cl} H{-}C{\equiv}C{-}CH{=}CH_2$

(k) $C_2H_2 + HCl \rightarrow H_2C{=}CHCl$

(l)

$H_2C{=}CH_2$ (from a) $\xrightarrow[Zn(Cu)]{CH_2I_2}$ cyclopropane ($CH_2$—$CH_2$—$CH_2$ ring)

3.16 (a) Propene; (b) HBr; (c) $H_2O$; (d) A peroxide; (e) Alcoholic KOH (f) hot $KMnO_4$, then HCl; (g) 2 $CH_3-CHO$; (h) $C_2H_4$; (i) $C_2H_2$; (j) *cis*-1,2-Dimethylcyclopentane; (k) *trans*-1,2-Dibromocyclopentane; (l) $Cl-CH_2CH(OH)CH{=}CH_2$ and $Cl-CH_2-CH{=}CH-CH_2OH$

3.17 (a) $CH_3CH(X)CH_3$ (I) and $CH_3CH_2CH_2-X$ (II) $\xrightarrow{\text{alcoholic KOH}}$ $CH_3CH{=}CH_2$

Compound I is more reactive toward HX elimination.

(b) $CH_3-C(CH_3)(X)CH_2CH_3$ (III) and $CH_3-CH(CH_3)-CH(X)-CH_3$ (IV) $\xrightarrow{\text{alcoholic KOH}}$ $(CH_3)_2C{=}CHCH_3$

Compound III is more reactive toward HX elimination.

(c)

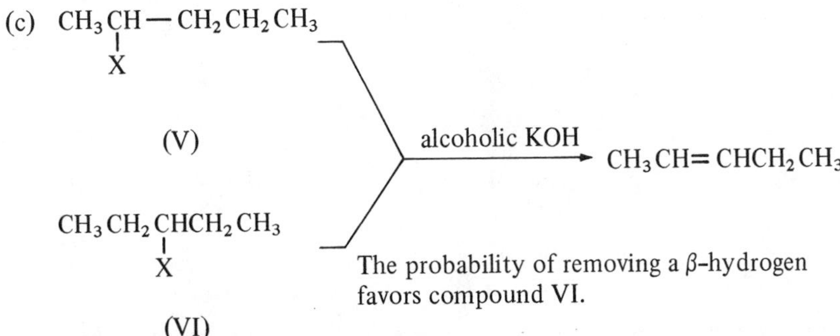


The probability of removing a $\beta$-hydrogen favors compound VI.

The order of ease of removal of hydrogen halide from alkyl halides is: tertiary > secondary > primary.

3.18 Cyclobutane

3.19 (A) may be either $(CH_3)_2CH-C{\equiv}C-H$ (3-Methyl-1-butyne) or $CH_3(CH_2)_2C{\equiv}C-H$ (1-Pentyne)

(B) is

$$CH_3-\overset{H}{\overset{|}{C}}=\overset{H}{\overset{|}{C}}-\overset{H}{\overset{|}{C}}=CH_2$$

1,3-Pentadiene

3.20 (a) OH, $CH_3$, H, C, C, OH, H, $CH_3$

(b)

$$CH_3-\overset{CH_3}{\overset{|}{\underset{CH_3}{\underset{|}{C}}}}-OH$$

(c) $CH_3$, Br

(d) Br, $CH_3$, H, C, C, Br, $CH_3CH_2$, H + Br, H, $CH_3$, C, C, Br, H, $CH_2CH_3$

(These are enantiomers, see Section 5.2.)

(e) C, Cl, Cl

3.21 (a) $CH_2$, $CO_2CH_3$, CH, C, H, H, CH, C, $CH_2$, $CO_2CH_3$

(b) $CH_2$, $CO_2CH_3$, CH, C, CH, C, $CH_2$, $CO_2CH_3$

(c) $CH_2$, CH, CH, $CH-CO_2CH_3$, CH, CH, $CH_2$

3.22

R–•–•–•–•· ≡ R(H) → R–•–•–•–•–H

$CH_2{=}CH_2$

Growing radical chain

Abstraction of H-atom from 5th C-atom from end of chain

Polymer grows from new new radical site

3.23

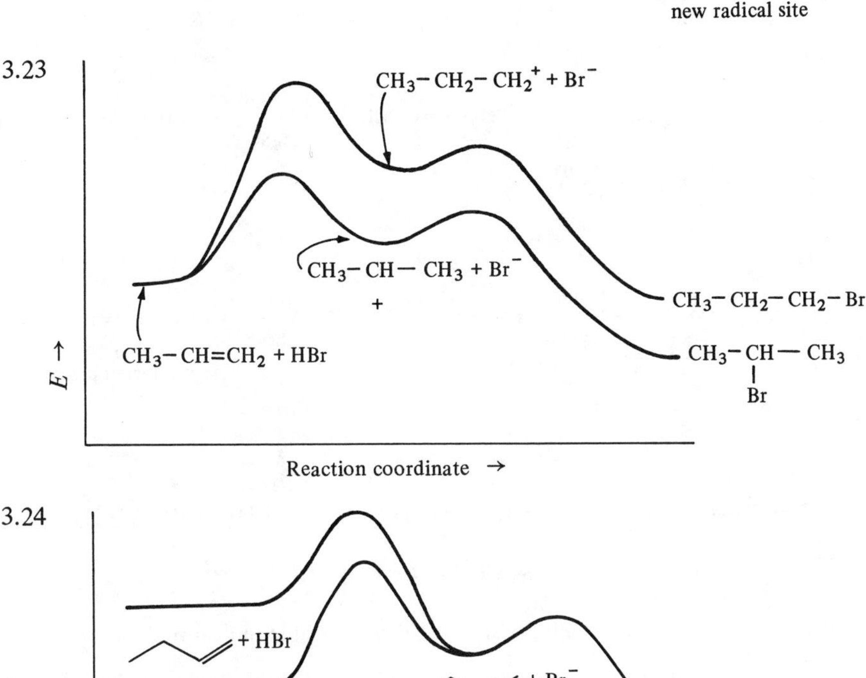


3.24

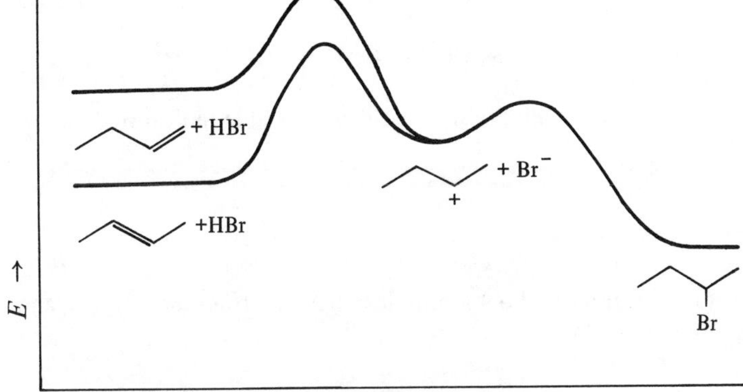


2-Butene is more stable than 1-butene.

3.25

$$Br{-}CH_2{-}CH_2{-}CH_2{-}Br + Zn \rightarrow Br{-}CH_2{-}CH_2{-}CH_2{:}ZnBr \rightarrow \text{cyclopropane } (CH_2)_3 + ZnBr_2$$

3.26 (a)

$$\text{bromocyclohexane} \xrightarrow[\text{KOH}]{\text{alcoholic}} \text{cyclohexene} \xrightarrow[\text{HOH (cold)}]{KMnO_4} \text{cyclohexane-1,2-diol (OH, OH)}$$

(b)

$$H{-}C{\equiv}C{-}H \xrightarrow[2.\ CH_3Br]{1.\ NaNH_2} H{-}C{\equiv}C{-}CH_3 \xrightarrow[2.\ CH_3Br]{1.\ NaNH_2} CH_3{-}C{\equiv}C{-}CH_3 \xrightarrow[Ni]{H_2} CH_3CH_2CH_2CH_3$$

(c)

$$CH_3{-}CH{=}CH{-}CH_3 \xrightarrow[\text{2. alcoholic KOH; 3. } NaNH_2]{1.\ Br_2} CH_3{-}C{\equiv}C{-}CH_3 \xrightarrow[HgSO_4]{H_2O,\ H_2SO_4} CH_3{-}\overset{O}{\overset{\|}{C}}{-}CH_2{-}CH_3$$

(d)

$$CH_3{-}CH_2{-}CH{=}CH_2 \xrightarrow[H_2SO_4]{H_2O} CH_3{-}CH_2{-}\overset{OH}{\overset{|}{C}H}{-}CH_3 \xrightarrow[\text{heat}]{H_2SO_4} CH_3{-}CH{=}CH{-}CH_3$$

(preferred product)

3.27

$$Cl_2 \xrightarrow{\text{light}} 2\ Cl\cdot$$

$$CH_2{=}CH{-}CH_3 + Cl\cdot \longrightarrow [CH_2{=}CH{-}CH_2\cdot \leftrightarrow \cdot CH_2{-}CH{=}CH_2] + HCl$$

$$CH_2{=}CH{-}CH_2\cdot + Cl_2 \longrightarrow CH_2{=}CH{-}CH_2{-}Cl + Cl\cdot$$

The resonance-stabilized allyl radical is the most stable radical that can form from propene.

## CHAPTER 4

4.1 Structure (a), (b), and (d) show aromatic behavior; structures (c), (e), and (f) are nonaromatic.

4.2

$$\text{o-xylene } (CH_3, CH_3) \xrightarrow{Br_2,\ Fe} \text{3-bromo-o-xylene } (CH_3, CH_3, Br) + \text{4-bromo-o-xylene } (CH_3, CH_3, Br)$$

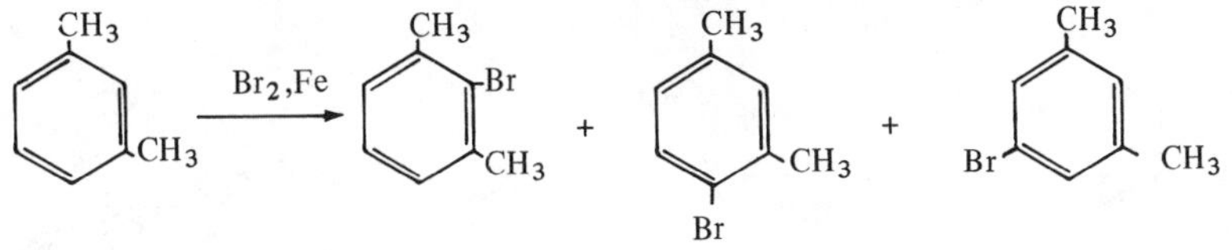

$CH_3$, $CH_3$ $\xrightarrow{Br_2,Fe}$ $CH_3$, Br, $CH_3$ + $CH_3$, $CH_3$, Br + $CH_3$, Br, $CH_3$

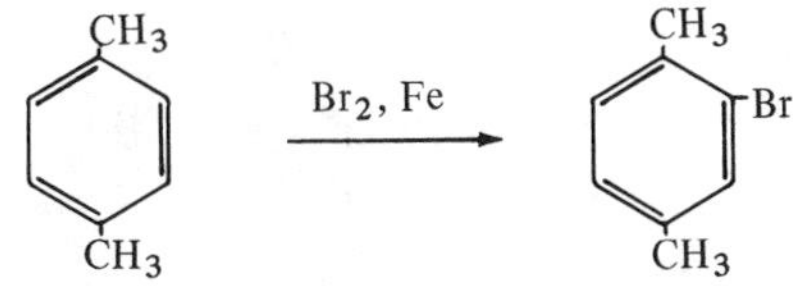

$CH_3$, $CH_3$ $\xrightarrow{Br_2, Fe}$ $CH_3$, Br, $CH_3$

4.3

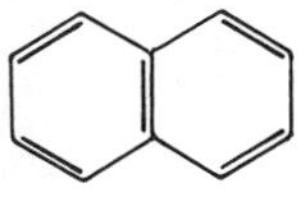

(a)

(shown in text)

Structure (a) is the most important contributor to the resonance hybrid.

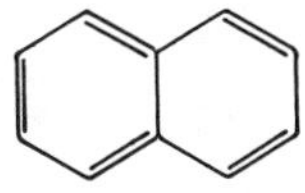

(b) and (c)

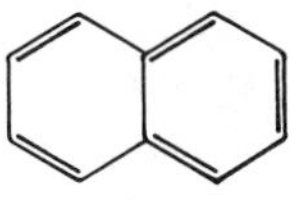

4.4

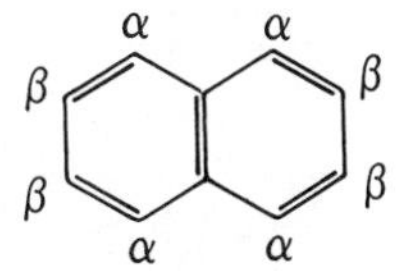

α α β β β β α α

Only two mono-substituted products are possible for naphthalene—α and β. Ten disubstituted products are possible.

4.5

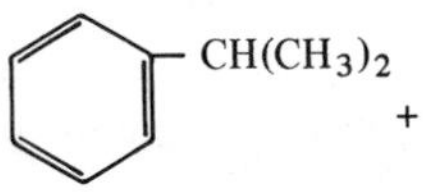

$CH(CH_3)_2$ + $3K_2Cr_2O_7$ + $12H_2SO_4$ → COOH + $2CO_2$ + $3Cr_2(SO_4)_3$ + $3K_2SO_4$ + $15H_2O$

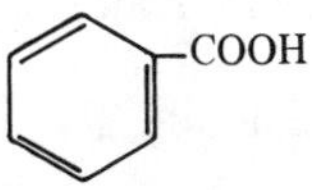

4.6 (a) 2,4,6-Tribromophenol, (b) 1,3-Dinitrobenzene (*m*-Dinitrobenzene), (c) Styrene (Vinyl benzene), (d) Benzyl cyanide (Phenylacetonitrile), (e) 4,4′-Dibromobiphenyl, (f) Triphenylmethane, (g) β-Naphthalenesulfonic acid.

4.7

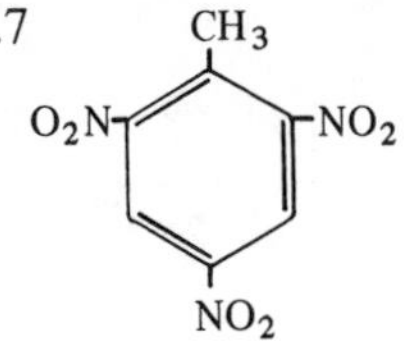

(a) 2,4,6-trinitro structure: $CH_3$, $O_2N$, $NO_2$, $NO_2$

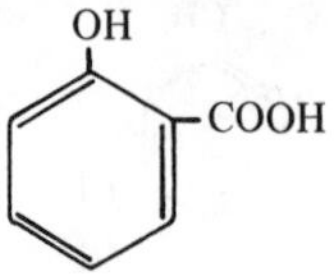

(b) OH, COOH

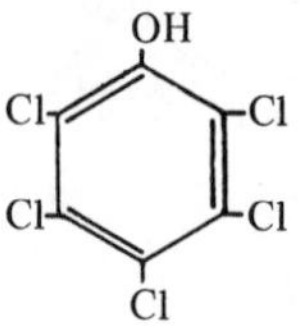

(c) OH, Cl, Cl, Cl, Cl, Cl

(d) Cl, Cl

4.8

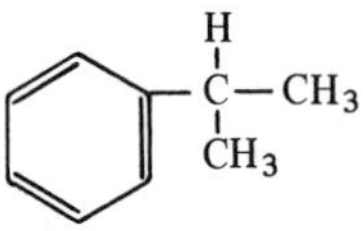

H, C—$CH_3$, $CH_3$

Isopropylbenzene

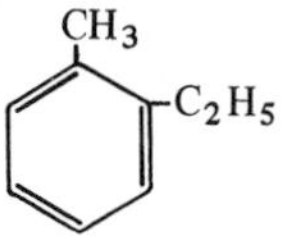

$CH_3$, $C_2H_5$

2-Ethyltoluene
(*o*-Ethyltoluene)

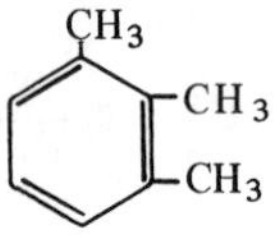

$CH_3$, $CH_3$, $CH_3$

1,2,3-Trimethyl-
benzene

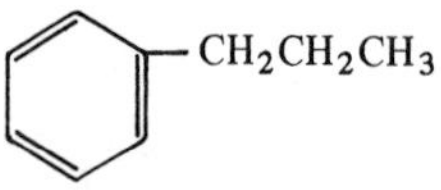

$CH_2CH_2CH_3$

*n*-Propylbenzene

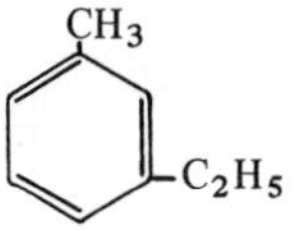

$CH_3$, $C_2H_5$

3-Ethyltoluene
(*m*-Ethyltoluene)

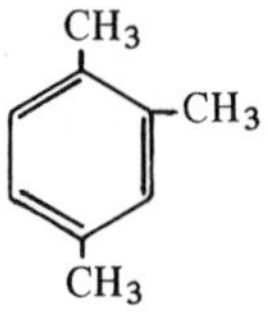

$CH_3$, $CH_3$, $CH_3$

1,2,4-Trimethyl-
benzene

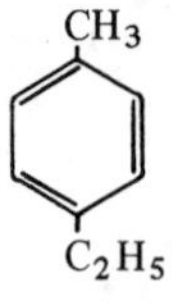

$CH_3$, $C_2H_5$

4-Ethyltoluene
(*p*-Ethyltoluene)

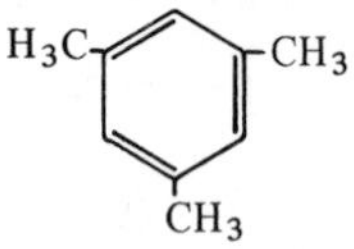

$H_3C$, $CH_3$, $CH_3$

1,3,5-Trimethylbenzene

4.9 The presence of $FeBr_3$ as a Lewis acid (presumably formed before the ring is attacked) would cause a polarization of the bromine molecule to make one bromine atom of the molecular pair electrophilic.

Aluminum trichloride, also a Lewis acid, would accomplish the same purpose.

4.10

(a) $CH_3$, $CH_3$

(b) $CH_3$, $H_3C$, $CH_3$

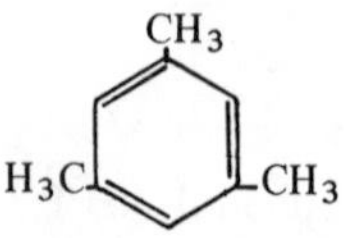

4.11 Bromine in carbon tetrachloride solution when added (in the absence of light) to a sample of the unknown will be discolored if the unknown is cyclohexene. Should this test be negative, then expose the reaction mixture to bright light and note if hydrogen bromide gas is being produced by blowing the breath across the mouth of the test tube. If this test is negative, then the sample is benzene; if positive, the sample is cyclohexane.

4.12 (a)

$C_6H_5CH_3 \xrightarrow[H_2SO_4]{HNO_3}$ o-$CH_3C_6H_4NO_2 \xrightarrow[H_2SO_4]{K_2Cr_2O_7}$ o-$O_2NC_6H_4C(=O)-OH$

(b)

$C_6H_6 \xrightarrow[H_2SO_4]{HNO_3} C_6H_5NO_2 \xrightarrow[FeBr_3]{Br_2}$ m-$BrC_6H_4NO_2$

(c)

$C_6H_5CH_3 \xrightarrow[FeBr_3]{Br_2}$ p-$BrC_6H_4CH_3 \xrightarrow[H_2SO_4]{K_2Cr_2O_7}$ p-$BrC_6H_4C(=O)OH$

(d)

$C_6H_5CH_3 \xrightarrow{H_2SO_4}$ p-$HO_3SC_6H_4CH_3$

(e)

$C_6H_6 \xrightarrow[AlCl_3]{CH_3-CHCl-CH_3} C_6H_5CH(CH_3)_2$

(f)

$C_6H_5CH_3 \xrightarrow[light]{Cl_2} C_6H_5CH_2Cl$

(g)

$C_6H_6 \xrightarrow[FeBr_3]{Br_2} C_6H_5Br \xrightarrow[H_2SO_4]{HNO_3}$ p-$O_2NC_6H_4Br$

4.13 (a) OH $CH_3$ $NO_2$ and $NO_2$ OH $CH_3$

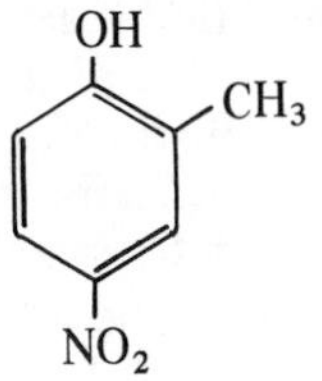

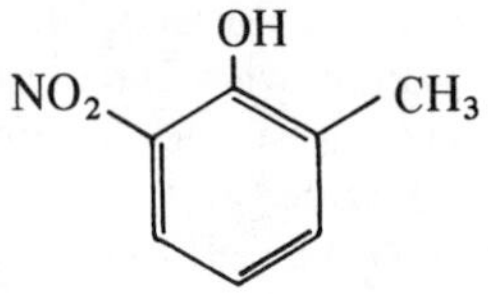

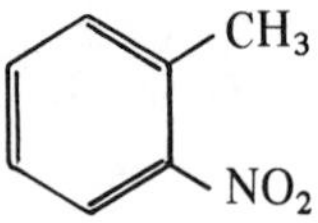

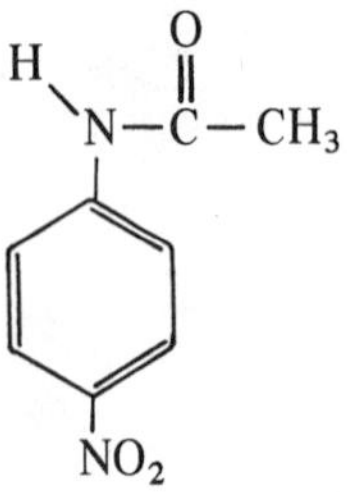

(b) $CH_3$ $NO_2$ and $CH_3$ $O_2N$

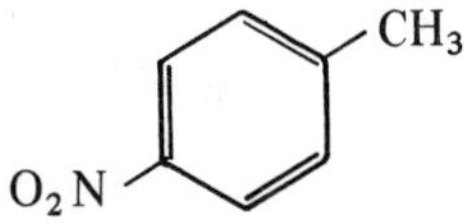

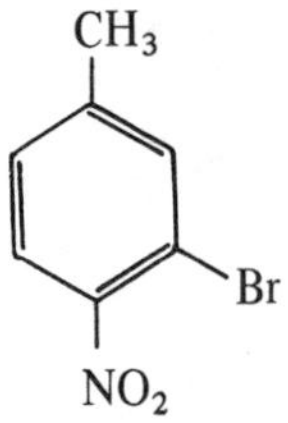

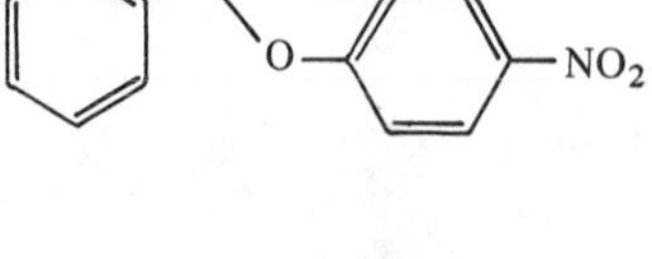

(c) H O N—C—$CH_3$ $NO_2$ and H O N—C—$CH_3$ $NO_2$

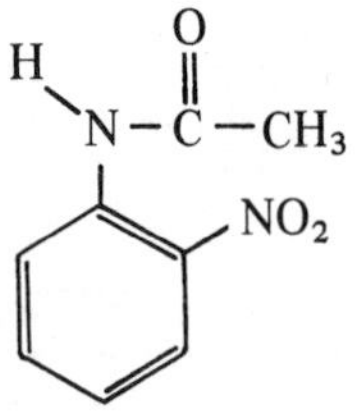

(d) $CH_3$ Br $NO_2$ and $CH_3$ $NO_2$ Br

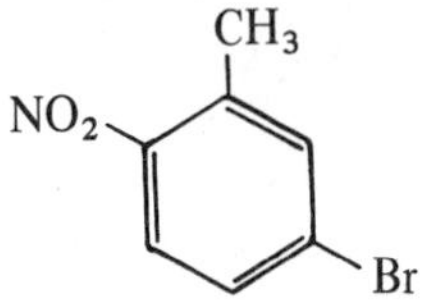

(e) O C OH $NO_2$

(f) O C O $NO_2$ and O C O $NO_2$

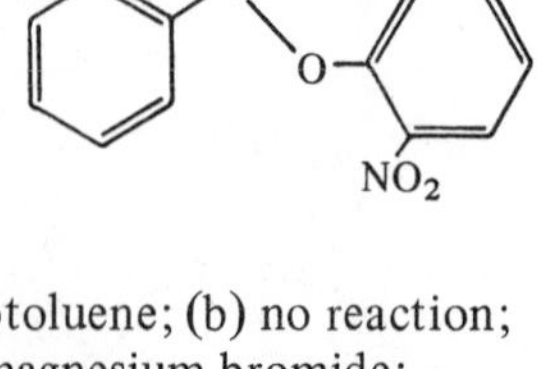

4.14 (a) 4-Bromo-2-nitrotoluene and 2-bromo-6-nitrotoluene; (b) no reaction; (c) Benzyl chloride; (d) no reaction; (e) Phenylmagnesium bromide; (f) Benzenesulfonic acid.

4.15 *o*-Xylene,

4.16

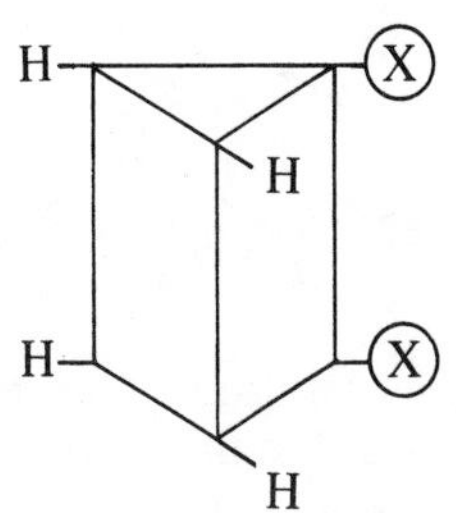

*ortho* *meta* *para*

4.17 (a)

(b)

(c)

(d)

4.18

Both benzyl ion and benzyl radical are resonance stabilized and should be more stable than primary alkyl radicals and cations which lack such resonance stabilization.

4.19

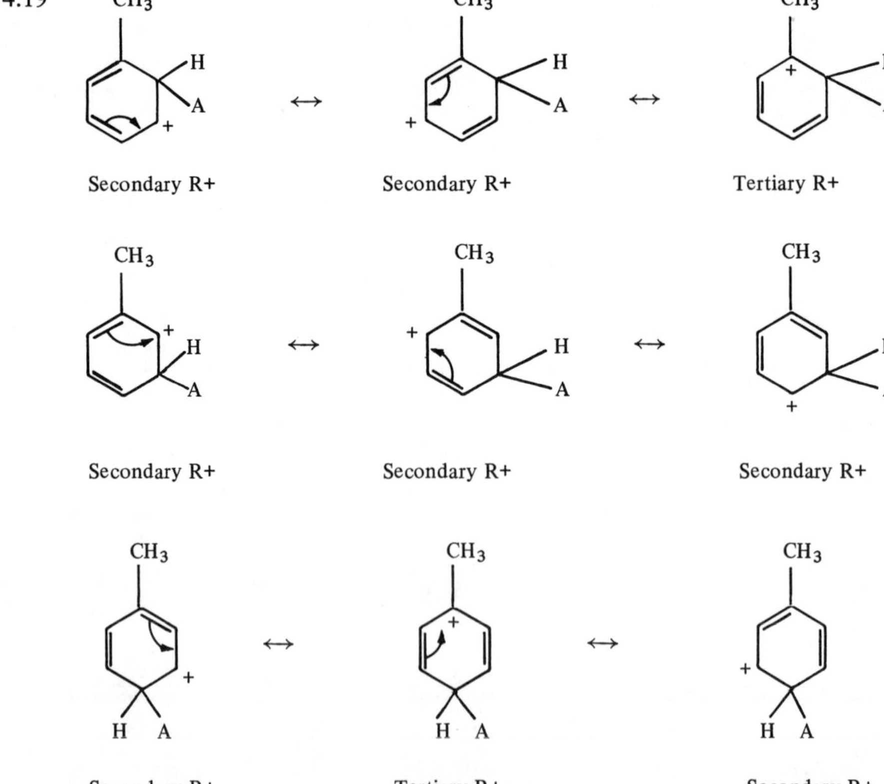


One explanation of the *o,p*-orientation of the methyl group is based on the relative stabilities of the carbonium ion intermediates in *o*–, *m*–, and *p*-substitution. Both the *o*- and *p*-intermediates are hybrids of two secondary and one tertiary carbonium ion structures, whereas the *m*-intermediate is a hybrid of three secondary carbonium ion structures. On this basis the *o*- and *p*-intermediates should be somewhat more stable than the *m*-intermediate.

4.20

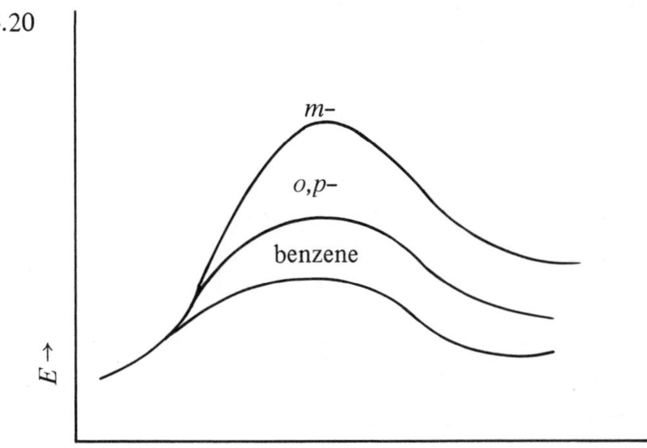


4.21 (a)

$$C_6H_5CH_3 \xrightarrow[\text{light}]{Br_2} C_6H_5CH_2Br \xrightarrow[\text{2. } D_2O]{\text{1. Mg}} C_6H_5CH_2D$$

(b)

$$CH_3C_6H_5 \xrightarrow[FeBr_3]{Br_2} p\text{-}CH_3C_6H_4Br \xrightarrow[\text{2. } D_2O]{\text{1. Mg}} p\text{-}CH_3C_6H_4D$$

(c)

$$C_6H_5CH_3 \xrightarrow[\text{light}]{Br_2} C_6H_5CH_2Br \xrightarrow{(CH_3CH_2CH_2)_2CuLi} C_6H_5CH_2CH_2CH_2CH_3$$

(d)

$$C_6H_5CH_2CH_3 \xrightarrow[\text{light}]{Br_2} C_6H_5\text{-}CH(Br)\text{-}CH_3 \xrightarrow{\text{alcoholic KOH}} C_6H_5\text{-}CH{=}CH_2$$

(e)

$$C_6H_5CH_2Br \xrightarrow[\text{2. CuI}]{\text{1. Li}} [C_6H_5CH_2]_2CuLi \xrightarrow[\text{(from (a))}]{C_6H_5CH_2Br} C_6H_5CH_2CH_2C_6H_5$$

From (a) or (c)

(f)

$$C_6H_5-CH{=}CH_2 \xrightarrow[\text{2. alcoholic KOH; 3. } NaNH_2]{\text{1. } Br_2} C_6H_5-C{\equiv}CH$$

From (d)

(g)

$$C_6H_5CH_3 \xrightarrow[\text{Pt, high temp. and pressure}]{3\ H_3} C_6H_{11}CH_3$$

(h)

$$C_6H_5-CH{=}CH_2 \xrightarrow[Zn(Cu)]{CH_2I_2} C_6H_5-CH(CH_2)CH_2 \text{ (cyclopropyl)}$$

From (d)

4.22

$$Br_2 \xrightarrow[\text{light}]{} 2\ Br\cdot$$

$$C_6H_5-CH_2CH_3 + Br\cdot \rightarrow C_6H_5-\dot{C}H-CH_3 \quad \text{or} \quad C_6H_5-CH_2CH_2\cdot + HBr$$

A benzyl radical, stabilized by resonance — A primary alkyl radical

In the bromination of alkanes this step is endothermic (see Exercise 2.18) although the overall bromination reaction sequence is exothermic (see Exercise 2.5). Any electronic effect that stabilizes the intermediate free radical will tend to lower the activation energy for the endothermic step. Thus, bromination is a very selective reaction, showing preference for bromination at the carbon atom giving the most stable free radical. Chlorination is much less selective, and chlorination of ethylbenzene gives only slightly more of the 1-chloro-1-phenylethane than of the 2-chloro-1-phenylethane.

$$C_6H_5-\dot{C}H-CH_3 + Br_2 \rightarrow C_6H_5-CH(Br)-CH_3 + Br\cdot$$

4.23 (a)

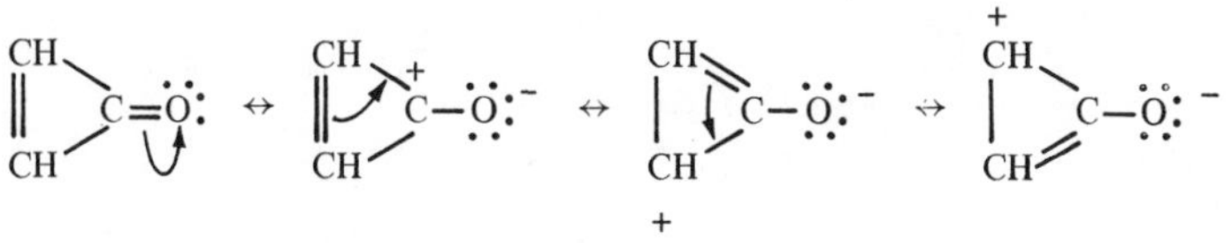

(b)

(c)

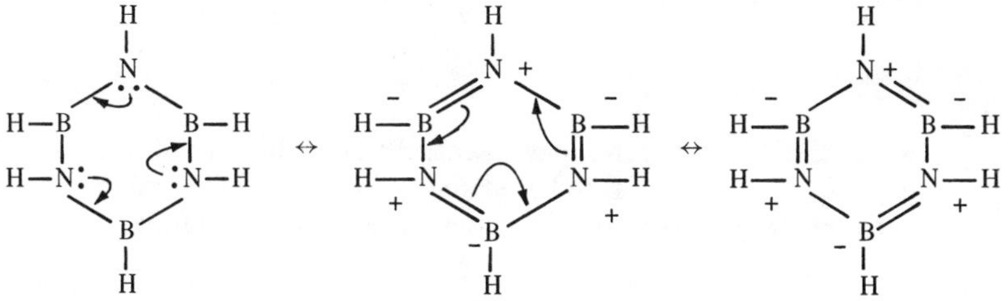

(d)

(e)

(f)

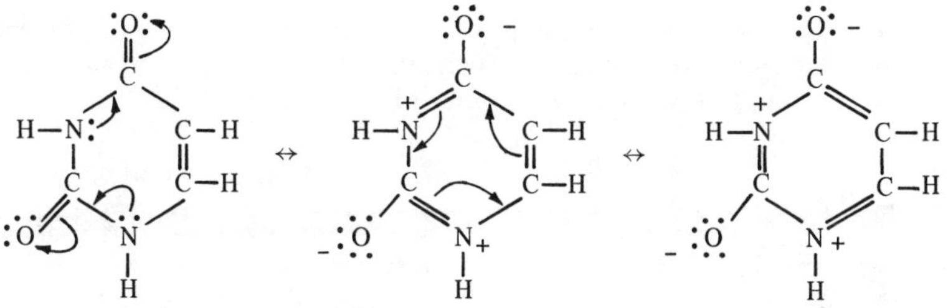

## CHAPTER 5

5.1 Objects (a), (b), (e), (f), and (g) may exist in enantiomeric forms.

5.2

CHO
H—+—OH —rotate 90°→ HOCH₂—⊕—CHO (H top, OH bottom)
CH₂OH

(*R*)-Glyceraldehyde (*S*)-Glyceraldehyde

↓ rotate 180°

CH₂OH
HO—+—H —exchange H and CHO→ HO—+—CHO (CH₂OH top, H bottom) —exchange HO and CH₂OH→ HOCH₂—⊕—CHO (OH top, H bottom)
CHO

H not at top or bottom (*R*)-Glyceraldehyde

Rule: Fischer structures may be rotated 180°, but not 90°, in the plane of the paper if configuration is to be retained. (Rotation of a structure having only one asymmetric C-atom by 90° converts that structure into its enantiomer.)

5.3

H, $C_2H_5$ on C3; $CH_3$ (1), H on C2: C3=C2

*trans*-2-Pentene

H, $C_2H_5$ on C3; H, $CH_3$ (1) on C2: C3=C2

*cis*-2-Pentene

Hydrogen bromide addition to either compound gives largely a racemic mixture of 2-bromopentane. Should hydrogen bromide add in such manner as to place bromine on carbon atom 3, then the product is optically inactive.

5.4 Since the lowest priority group (H) is neither on the top nor on the bottom, two exchanges are made on *each* asymmetric C-atom in such a way as to bring one H-atom to the top and the other to the bottom.

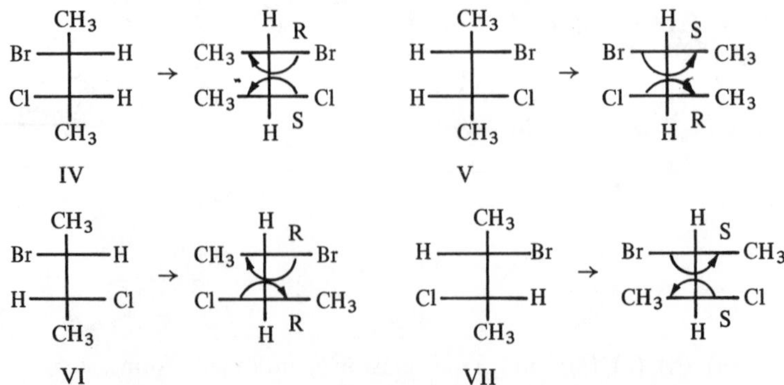

5.5 Both substituted carbon atoms of *cis*-1,2-dimethylcyclopentane are bonded to four unlike groups but the molecule represents a *meso* form and is optically inactive. One half of the molecule is a mirror image of the other.

5.6

(a) (b)

Structure (a) is that of natural rubber and (b) that of gutta percha.

5.7 (a) two: *cis* and *trans*; (b) three: *cis* (*meso*) and *trans* (d, 1) (c) three: d, 1, and *meso*; (d) two: d, 1; (e) sixteen: 8 D-forms, 8 L-forms.

5.8

(a) Epinephrine

(b) Limonene

(c) Aureomycin

(d) Camphor

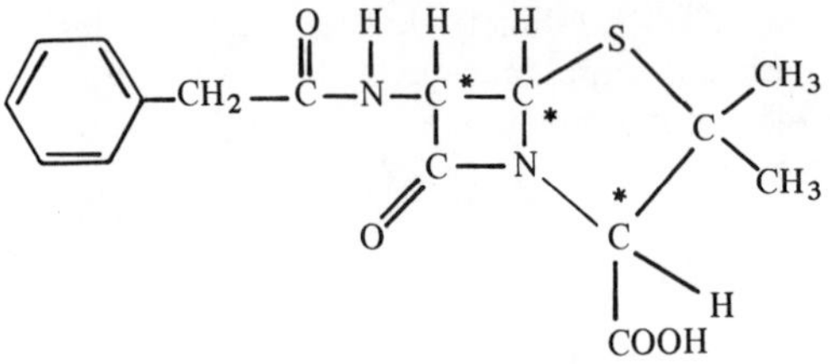

(e) Penicillin G

$$CH_3-(CH_2)_5-\overset{*}{C}H(OH)-CH_2-CH=CH-(CH_2)_7COOH$$

(f) Ricinoleic acid

5.9 (a) Yes; (b) No; (c) Yes; (d) No, a *meso* form; (e) Yes.

5.10

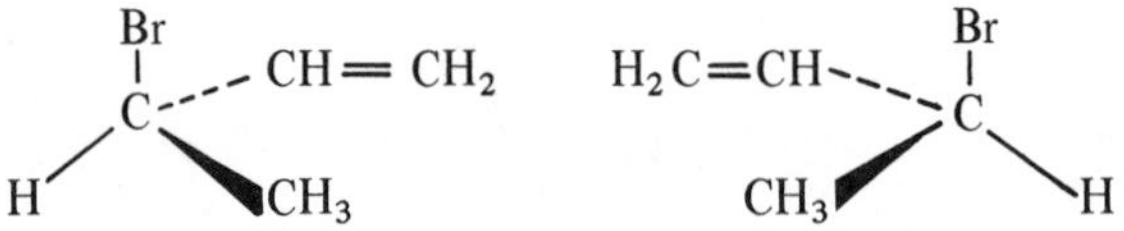

(*R*) (*S*)

5.11

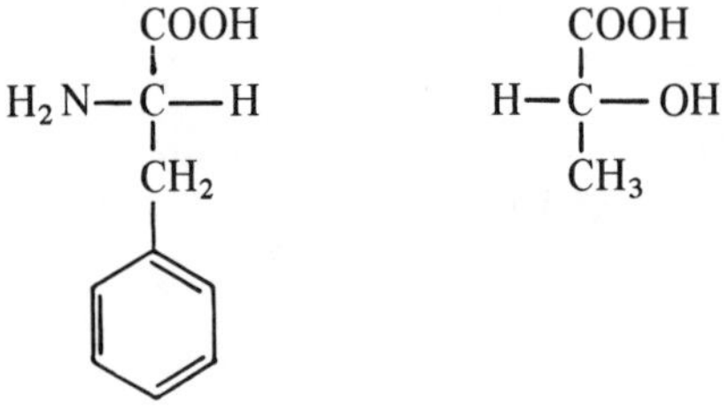

L-Phenylalanine (*S*)

D-Lactic acid (*R*)

5.12 (a) $H_2C{=}CH-CH_2-CH(Br)-CH_3$ (b) $CH_3CH_2CH(Br)-CH{=}CH_2$

(c) $CH_3-CH{=}CH-CH(Br)-CH_3$ (c) may also exist in *cis* and *trans* forms.

5.13 (a) 1-Butene + HBr → $CH_3CH_2CH(Br)CH_3$ (Resolvable into two optical forms.)

(b) 2-Butene + HBr → $CH_3CH(Br)CH_2CH_3$ (Same as a.)

(c) Cyclopentene + HBr → Bromocyclopentane (Not optically active.)

(d) Cyclopentene + $Br_2$ → *trans*-1,2-Dibromocyclopentane (Resolvable into two optical forms.)

(e) 2-Butene + HOCl → $CH_3CH(OH)CHCH_3$ with Cl on the third carbon (Resolvable into four optical forms.)

(f) Propylene + HOCl → $CH_3CHCH_2Cl$ with OH on the second carbon (Resolvable into two optical forms.)

(g) Propylene + $KMnO_4$ → $CH_3CH(OH)CH_2OH$ (Resolvable into two optical forms.)

5.14

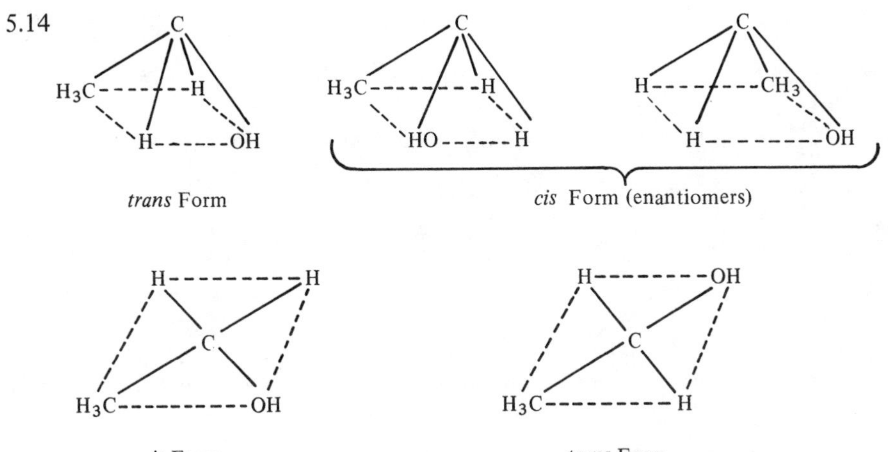


*trans* Form — *cis* Form (enantiomers)

*cis* Form — *trans* Form

Only one structure for ethyl alcohol is known because of the tetrahedral nature of the carbon atom.

5.15 The mode of addition of bromine to an olefin is *trans*. Thus the *trans* addition of bromine to *trans*-2-butene gives *meso*-2,3-dibromobutane. On the other hand, the *trans* addition of bromine to *cis*-2-butene will result in enantiomers.

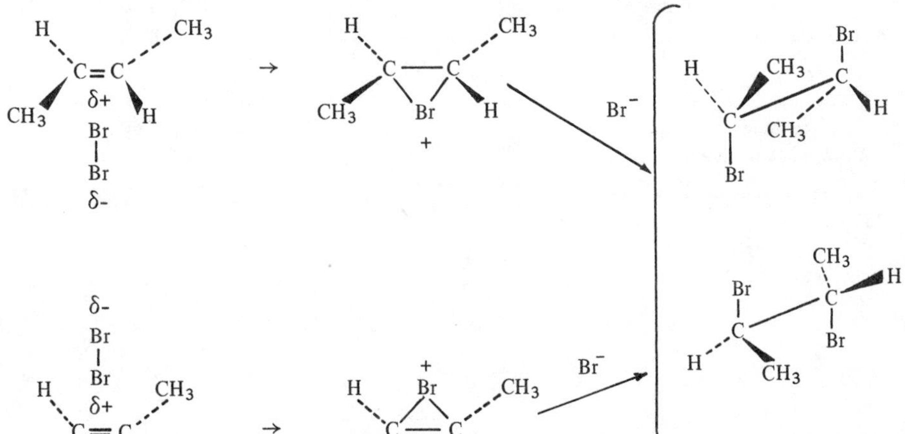

(These are identical.)

*meso* form

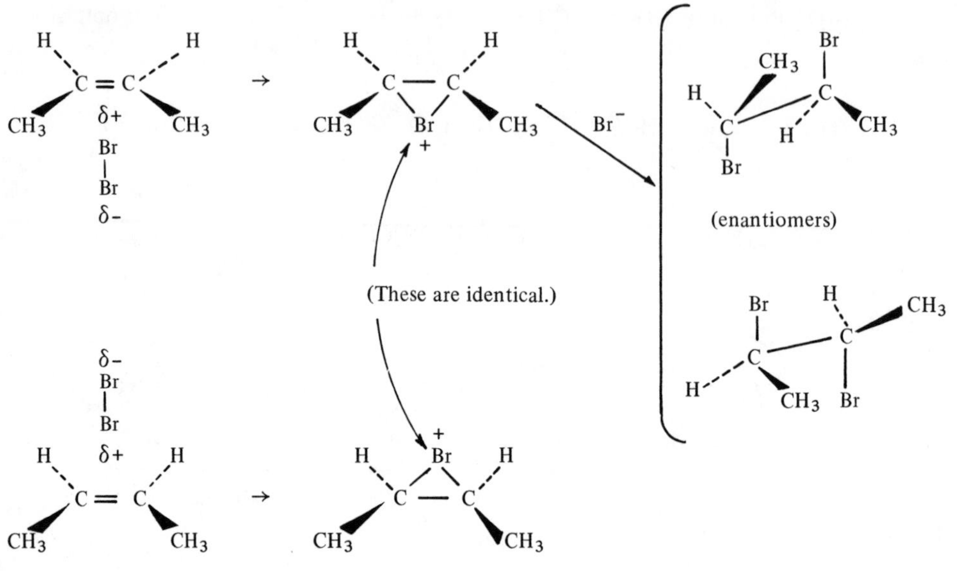

5.16 $CH_3CH_2\overset{*}{C}HCH_3$ (Cl) $\xrightarrow{-(HCl)}$ *trans*-2-Butene $\xrightarrow{Br_2}$ *meso*-2,3-Dibromobutane

(A) (B) (C)

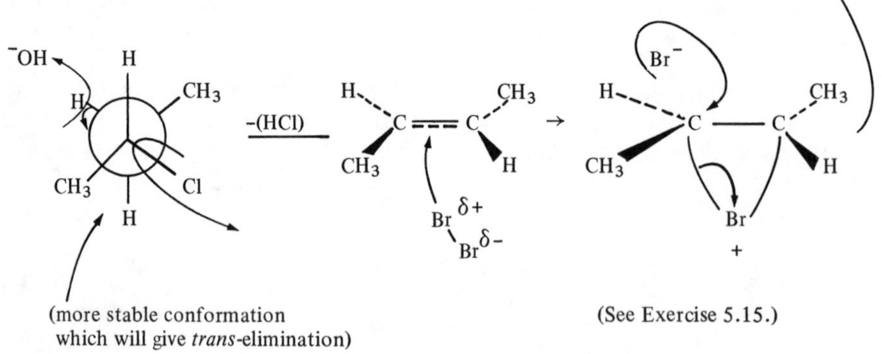

(more stable conformation which will give *trans*-elimination)

(See Exercise 5.15.)

5.17

C; Y, X; A, B; Z $\equiv$ C, Z; X, A; B; Y $\equiv$

| | C | |
|---|---|---|
| A | | B |
| X | | Y |
| | Z | |

5.18 (a)

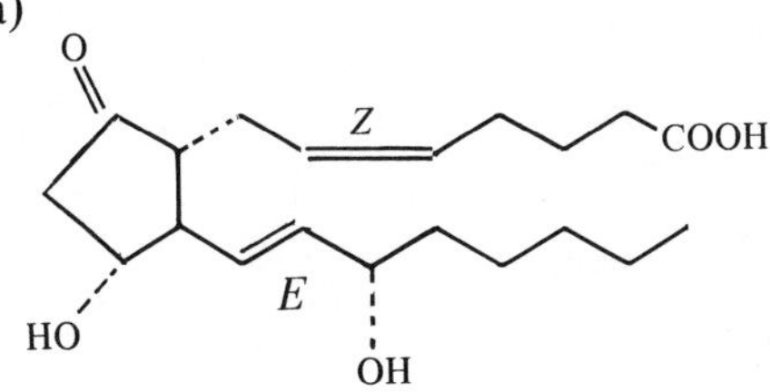


(b) $CH_3(CH_2)_3$–CH$\overset{E}{=}$CH–CH$\overset{E}{=}$CH–CH$\overset{Z}{=}$CH–$(CH_2)_7CO_2H$

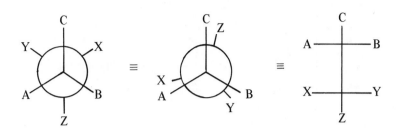

5.19 First make the exchanges necessary to get like groups at the top and bottom.

(a)

| | $CH_3$ | |
|---|---|---|
| HO | | H |
| HO | | H |
| | $CH_3$ | |

*meso*

(b)

| | $CH_3$ | |
|---|---|---|
| Cl | | H |
| H | | Cl |
| | $CH_3$ | |

not *meso*

(c) *meso*

(d)

| | $CH_3$ | |
|---|---|---|
| H | | OH |
| HO | | H |
| H | | OH |
| | $CH_3$ | |

*meso*

5.20 First make the exchanges necessary to get the lowest priority groups in the top and/or bottom positions.

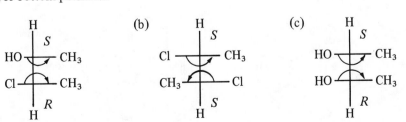

(a)

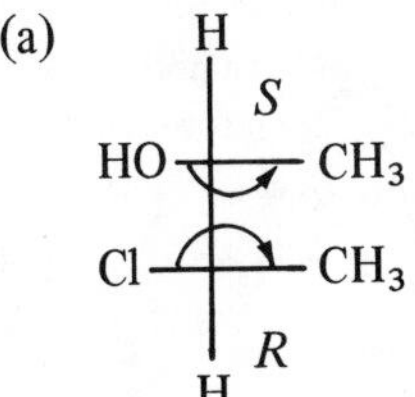


(b)

| | H | |
|---|---|---|
| | *S* | |
| Cl | | $CH_3$ |
| $CH_3$ | | Cl |
| | *S* | |
| | H | |

(c)

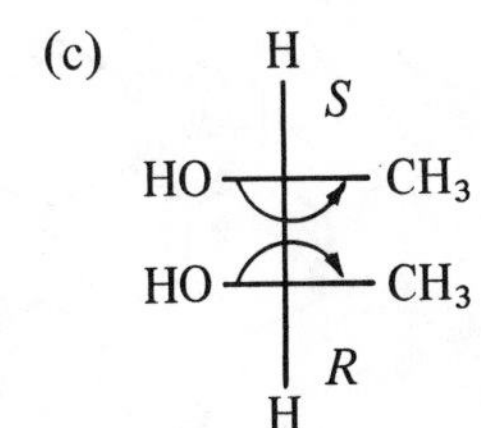

(d)

H, *S*; HO—CH$_3$; H—Cl; HO—CH$_3$; H, *R* ≡ *S*; H—Cl; *R* ≡ H; Cl—*S* (s); *R*

(e)

CHO; H—OH; H—OH; H—OH; H—OH; $CH_2OH$ ≡ H; HO—CHO (*R*); H—OH; H—OH; HO—$CH_2OH$ (*R*); H ≡ H; HO—CHOH—CHO (*R*); HO—CHOH—$CH_2OH$ (*R*); H

5.21 (a) (*S*)-; (b) (*R*)-; (c) (*S*)-; (d) (*R*)-; (e) (*R*)-; (f) (*R*)-

5.22

$$[\alpha]_D^{20} = \frac{+1.91^\circ}{1 \times (0.90/50)} = +106^\circ$$

5.23 (a)

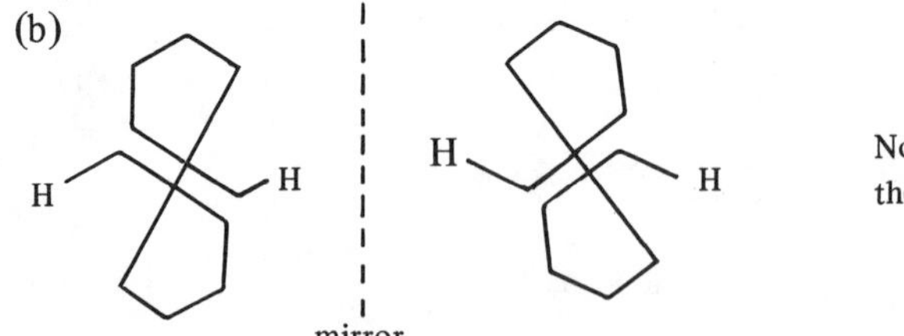


Nonsuperimposable; therefore, enantiomers

(b)

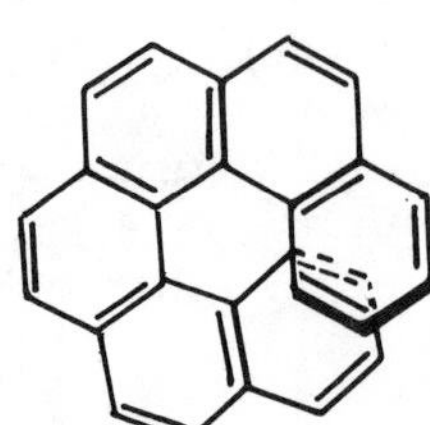


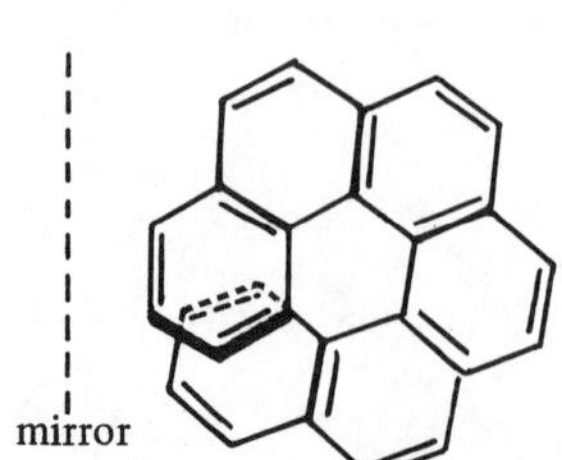


mirror

Nonsuperimposable; therefore, enantiomers

(c)

mirror

Nonsuperimposable; therefore, enantiomers

(d) $CH_3$ H $C{=}C{=}C$ H $CH_3$ | H $CH_3$ $C{=}C{=}C$ $CH_3$ H

mirror

Nonsuperimposable; therefore, enantiomers

## CHAPTER 6

6.1

$$\nu = \frac{\gamma H_0}{2\pi} = 42.57602 \times H_0 \quad (\text{for } \nu \text{ in MHz and } H_0 \text{ in T; } 1\ \text{T} = 10^4 \text{ gauss})$$

$$H = \frac{220}{42.57602} \times 10^4 \text{ gauss}$$

$$= 51{,}672 \text{ gauss}$$

A simple alternative solution is:

$$\frac{H_{220}}{\nu_{220}} = \frac{H_{60}}{\nu_{60}}$$

$$H_{220} = \frac{\nu_{220}}{\nu_{60}} \times H = \frac{220}{60} \times 14{,}092.44 = 51{,}672 \text{ gauss}$$

6.2

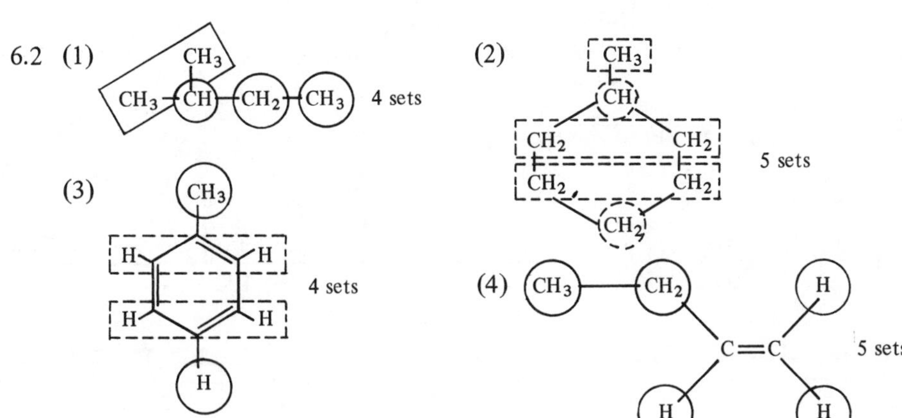


6.3

$$\frac{\gamma}{2\pi} H_1(1 - \sigma_1) = 60.000000 \text{ where } \gamma/2\pi = 42.57602 \times 10^{-4} \text{ (for } H_1 \text{ in gauss and } \nu \text{ in MHz)}$$

$$\frac{\gamma}{2\pi} H_2(1 - \sigma_2) = 60.000000$$

$$\text{Therefore, } H_1 = \frac{60}{42.57602 \times 10^{-4} \times (1 - 5 \times 10^{-6})} = 14{,}092.51029 \text{ gauss}$$

$$H_2 = \frac{60}{42.57602 \times 10^{-4} \times (1 - 7 \times 10^{-6})} = 14{,}092.53848 \text{ gauss}$$

$$\Delta H_{12} = H_2 - H_1 = 0.02819 \text{ gauss}$$

$$\Delta\nu_{12} = 4257.602 \times 0.02819 = 120 \text{ Hz}$$

Note that $(\sigma_2 - \sigma_1) \times \nu = (7 \times 10^{-6} - 5 \times 10^{-6}) \times 60 \times 10^6 = 120$ Hz

6.4 $\delta_1 = 1.45$ $\delta_2 = 1.55$

**At 60 MHz:**

$$(\Delta\nu_1)_{TMS} = 1.45 \times 60 = 87 \text{ Hz}$$

$$(\Delta\nu_2)_{TMS} = 1.55 \times 60 = 93 \text{ Hz}$$

$$\Delta\nu_{12} = 93 - 87 = 6 \text{ Hz}$$

**At 220 MHz:**

$$(\Delta\nu_1)_{TMS} = 1.45 \times 220 = 319 \text{ Hz}$$

$$(\Delta\nu_2)_{TMS} = 1.55 \times 220 = 341 \text{ Hz}$$

$$\Delta\nu_{12} = 341 - 319 = 22 \text{ Hz}$$

Note: $(220/60) \times 6 \text{ Hz} = 22 \text{ Hz}$

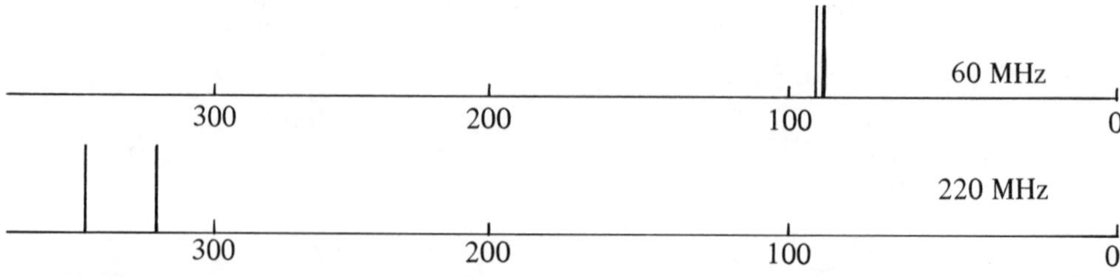


Thus, increasing the spectrometer frequency increases the separation in Hz between peaks caused by chemical shift differences between nuclei.

6.5 Where the exactly analogous structure is not shown in the chart, similar structures may be studied to estimate chemical shifts in $\delta$ values.

(a) $C_6H_5—CH_3$
7.17 2.5?

The closest structure for the $CH_3$ group is $CH_2Ph$ at δ 2.87. Examination of differences between $CH_3$ and $CH_2$ for a variety of structures shows differences ranging from about δ 0.2 to 1.0, the higher values for $CH_3$ and $CH_2$ attached to electronegative atoms (Br, Cl, I). A reasonable estimate for toluene would be a difference of δ 0.3 to 0.4 (the actual value for toluene is δ 2.32, a difference of 2.87 - 2.32 = 0.55; our error is only δ 0.18).

(b) $CH_3—CH_2—C(=O)—H$
1.10 2.20 9.58

(c) 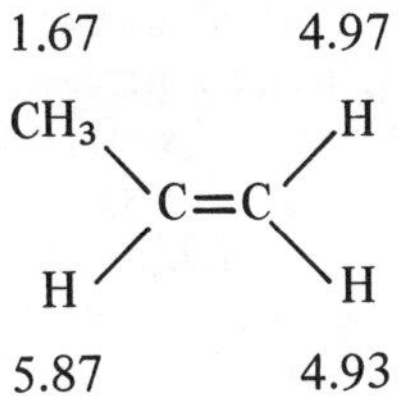 $CH_3(H)C=CH_2$: $CH_3$ 1.67; H (on same carbon as $CH_3$) 5.87; =CH₂ H's 4.97 and 4.93

(d) $CH_3—CH_2—OH$
1.18 3.55 variable

(e) $CH_3—C(CH_3)_2—CH_3$
all at 0.90

6.6 1 6 15 20 15 6 1

6.7 (a) $CH_3-CH_2-Cl$ A 3-proton 1:2:1 triplet at δ 1.48 and a 2-proton 1:3:3:1 quartet at δ 3.57. $J = 7$ Hz.

(b) $CH_3-C{\equiv}C-H$ The predicted spectrum would consist of a 3-proton and a 2-proton singlet in the δ 1.7 to 2.6 region (exact values can't be predicted from the chart and must be estimated from Table 6.1). By one of those quirks of nature the actual spectrum consists of a single peak at δ 1.80 (that is, by coincidence both types of proton in the molecule have the same chemical shift in $CDCl_3$).

(c) $Cl-CH_2—CH_2—CH_2—Cl$

A 2-proton 1:4:6:4:1 quintet at δ 2.20 and a 4-proton 1:2:1 triplet at δ 3.70. $J = 7$ Hz.

(d) $CH_3—CHBr_2$ No closely analogous compound can be found in Fig. 6.6. Therefore, we must estimate the chemical shifts on the basis of what is available. From Fig. 6.6 we get the following:

| | |
|---|---|
| $C\underline{H}_3CCH_2$ | 0.90 |
| $CH_2Br$ | 3.30 |
| $CH_2CBr$ | 1.75 |

The chemical shift of a proton on a C-atom attached to two Br-atoms should be somewhat less than twice that of a proton on a C-atom attached to one Br-atom. To estimate such a shift we note that substituting one Br-atom on an ethane molecule (δ 0.9 approximately) gives ethyl bromide (δ 3.30 approximately (δ 3.43 actual value)). Therefore, replacing one H-atom by one Br-atom increased the chemical shift by δ 2.40 (3.30 - 0.90). We assume that replacing a second H-atom by a Br-atom would also increase the chemical shift by about an additional δ 2.40. Therefore, we estimate the shift for $CHBr_2$ to be 3.30 + 2.40 = 5.70 (actual value = δ 5.86—not a bad estimate; if we had used the actual value for ethyl bromide as a basis for our estimate, our calculated value would be 3.43 + 2.53 = 5.96). To estimate the chemical shift for the $CH_3$ group we note that replacing one H-atom of ethane by a Br-atom should increase the shift of the $CH_3$ group from 0.9 to 1.75 (δ 1.67 actual value), an increase of 0.85. Therefore, we estimate the chemical shift of $CH_3CBr_2$ to be 1.75 + 0.85 = 2.60 (actual value = δ 2.47; if we had used the actual value for ethyl bromide as a basis for our estimate, our calculated value would have been 1.67 + 0.77 = 2.44!). Thus, we predict a 3-proton doublet at δ 2.6 and a 1-proton quartet at δ 5.7. J = 7 Hz.

(e) Cyclohexane (six $H_2$ ring carbons): A singlet at δ 1.34 (actual value δ 1.43).

6.8 (a) Carbon-carbon double bond: R—CH=CH—R′ (R ≠ R′).

(b) 1,1-Disubstituted carbon-carbon double bond: $R_2C{=}CH_2$

(c) Carbon-carbon triple bond: $R{-}C{\equiv}C{-}H \xrightarrow[H_2SO_4,\ HgSO_4]{H_2O} R{-}\underset{\underset{O}{\|}}{C}{-}CH_3$

6.9 Phenylalanine (257 nm); tyrosine (275 nm); tryptophan (280 nm).

6.10

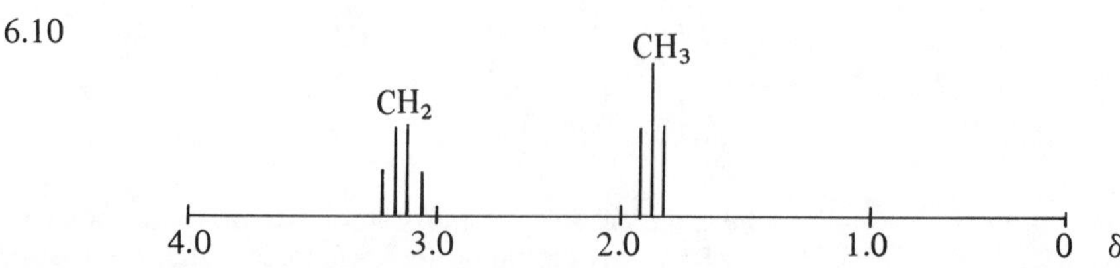

(a) three; (b) four; (c) δ 1.83; (d) δ 3.20; (e) 3.20 × 60 = 192 Hz; (f) 3.20 × 100 = 320 Hz; (g) 7 Hz (J is independent of the spectrometer frequency).

6.11 (a) 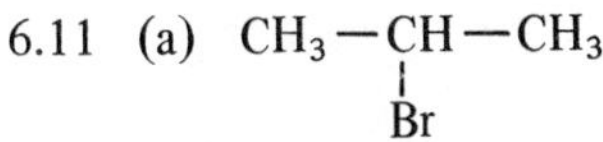

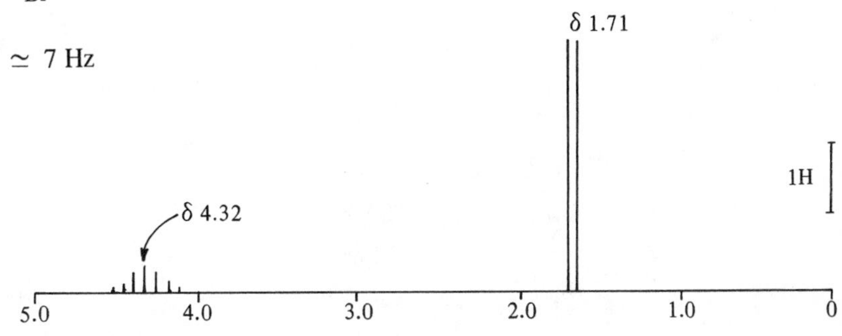


(b) $CH_3-O-CH_2-CH_3$

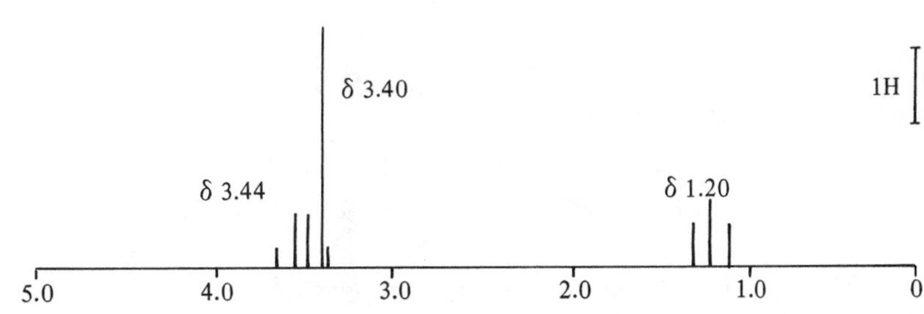


(c) 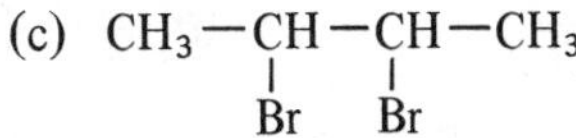

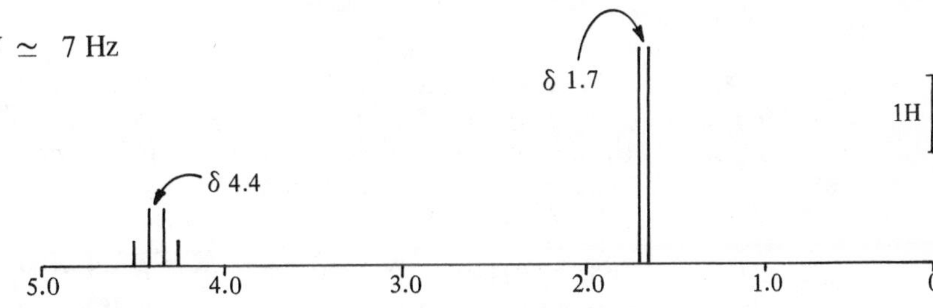


(d) 

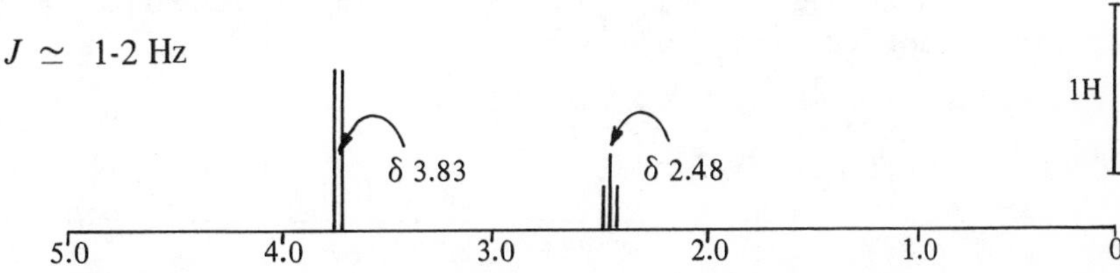

(e) $Cl{-}CH_2{-}CH_2{-}CH_2{-}Cl$

$J \simeq 7$Hz

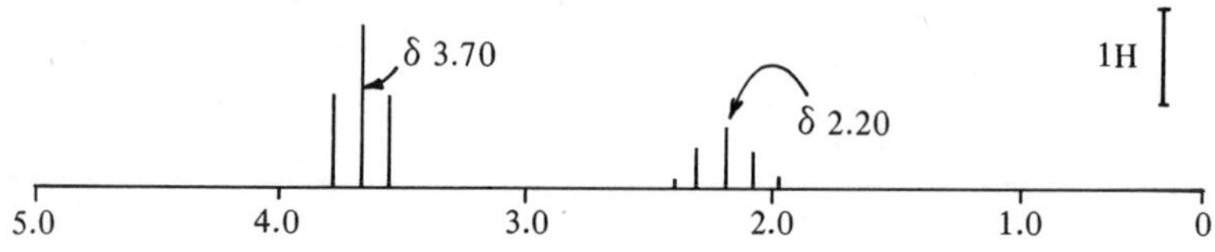


(f) 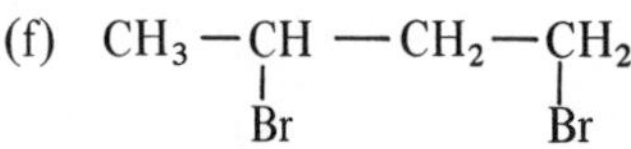


$J \simeq 7$ Hz

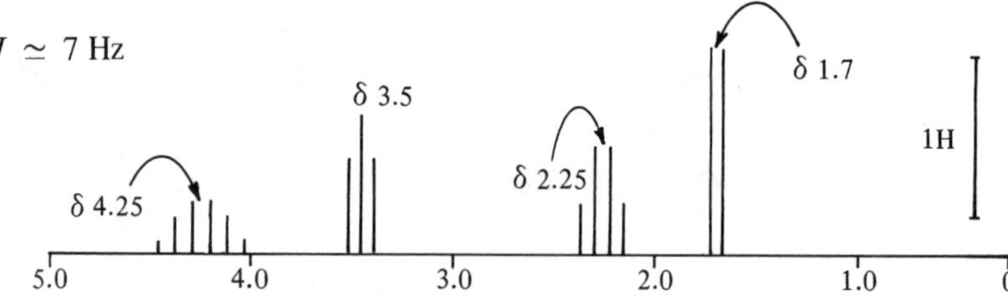


6.12 (a) $H_a$: 343 Hz (δ 5.72); $H_b$: 463 Hz (δ 7.72); $J$ = 7 Hz. (Note: shift values are pH dependent.)

(b) $CH_3$: 203 Hz (δ 3.38); $CH_2$: 212 Hz (δ 3.53).

(c) $CH_3$: 73 Hz (δ 1.22); $CH_2$: 222 Hz (δ 3.70); OH: 176 Hz (δ 2.93) (pH and medium dependent); $J$ = 7 Hz.

(d) According to the computer printout there are peaks at: 1800 (δ 72.0), 1852 (δ 74.1), 2025 (δ 81.0), 2070 (δ 82.8), 2105 (δ 84.2), 2110 (δ 84.4), 2147 (δ 85.9), 2207 (δ 88.3), 2330 (δ 93.2), 2600 (δ 104.0), and 2890 (δ 115.6) Hz.

(e) $CH_3$: 70 Hz (δ 1.17); CH: 208 (δ 3.47); $J$ = 7 Hz.

6.13 3200 $cm^{-1}$ (phenyl CH stretch); 2950–2800 $cm^{-1}$ (CH stretch of CH and $CH_2$); four bands between 1960 and 1740 $cm^{-1}$ (overtones of monosubstituted phenyl); 1618 and 1508 $cm^{-1}$ (phenyl ring stretch); 1470 $cm^{-1}$ ($CH_2$ bending); 770 and 715 $cm^{-1}$ (monosubstituted phenyl bending).

6.14 3100 $cm^{-1}$ (phenyl CH stretch); 3000–2900 $cm^{-1}$ (CH stretch of CH and $CH_3$); four bands between 1980 and 1790 $cm^{-1}$ (overtones of monosubstituted phenyl); 1695 $cm^{-1}$ (C=O stretch); 1605 and 1490 $cm^{-1}$ (phenyl ring stretch); 1460, 1360, and 1295 $cm^{-1}$ ($(CH_3)_2C\lt$ bending); 800 and 660 $cm^{-1}$ (monosubstituted phenyl bending).

6.15 At pH 7:

$$\epsilon_{max} = \frac{A}{c \times 1} = \frac{0.668}{8.44 \times 10^{-5} \times 1.0} = 7920$$

At pH 10:

$$\epsilon_{max} = \frac{0.450}{8.44 \times 10^{-5} \times 1.0} = 5330$$

6.16 For band at 240 nm, $\epsilon_{max} = \frac{0.728}{5.6 \times 10^{-5} \times 1.0} = 13{,}000$

For band at 280 nm, $\epsilon_{max} = \frac{0.083}{5.6 \times 10^{-5} \times 1.0} = 1480$

For band at 317 nm, $\epsilon_{max} = \frac{0.266}{5.0 \times 10^{-3} \times 1.0} = 53$

(Handbook values are 12,900 and 1230 at 240 and 280 nm, respectively.)

6.17 (a)

$$\tilde{\nu} = 1303 \sqrt{\frac{5\,(12 + 12)}{12 \times 12}} = 1189 \text{ cm}^{-1}$$

(b) 1682 $cm^{-1}$; (c) 2059 $cm^{-1}$; (d) 1985 $cm^{-1}$; (e) 3033 $cm^{-1}$; (f) 3003 $cm^{-1}$; (g) 1574 $cm^{-1}$; (h) 973 $cm^{-1}$.

Note: another worked example, (d): $\tilde{\nu} = 1303\sqrt{\frac{15\,(12 + 14)}{12 \times 14}}$

$$= 1985 \text{ cm}^{-1}$$

6.18

$$\tilde{\nu}_H = 1303 \sqrt{\frac{k\,(m_H + m_C)}{m_H \times m_C}} \quad \text{and} \quad \tilde{\nu}_D = 1303 \sqrt{\frac{k\,(m_D + m_C)}{m_D \times m_C}}$$

$$\frac{\tilde{\nu}_H}{\tilde{\nu}_D} = \sqrt{\frac{k\,\frac{(m_H + m_C)}{(m_H \times m_C)}}{k\,\frac{(m_D + m_C)}{(m_D \times m_C)}}} = \sqrt{\frac{(m_H + m_C)\,(m_D \times m_C)}{(m_D + m_C)\,(m_H \times m_C)}} = \sqrt{\frac{(1 + 12)\,(2 \times 12)}{(2 + 12)\,(1 \times 12)}}$$

$$= 1.363$$

Therefore, $\tilde{\nu}_D = \dfrac{\tilde{\nu}_H}{1.363} = \dfrac{2900}{1.363} = 2128\ \text{cm}^{-1}$

6.19 $Cl-CH_2-CH_2-C{\equiv}N$

6.20

$$C_6H_5(H)C{=}C(H)CO_2H$$

6.21

$$CH_3-C(CH_3)(Cl)-CH_2-CH_3$$

6.22

$$Cl-CH_2-CH(O-CH_3)_2$$

6.23

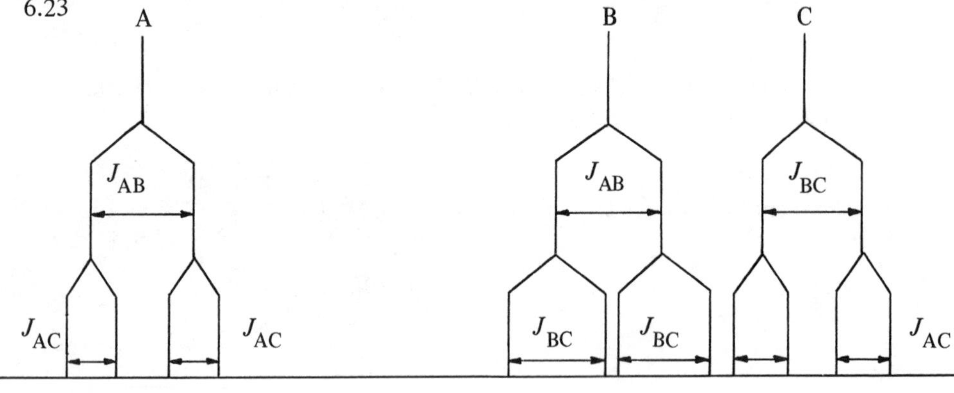


6.24

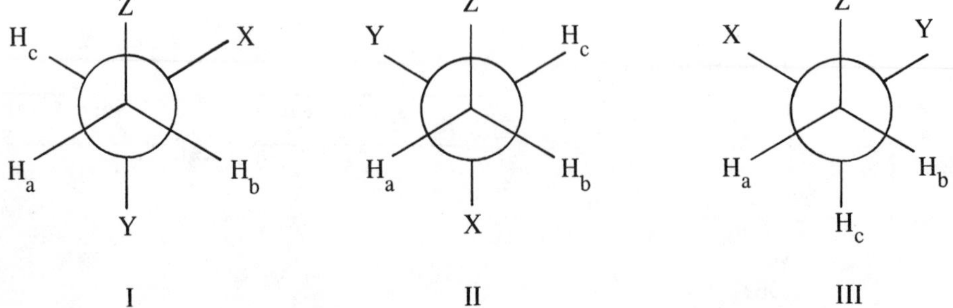

Compare I and II. In I $H_b$ is between X and Y, and in II $H_a$ is between X and Y. However, in I the order of groups about $H_b$ is X, Z, $H_c$, $H_a$, Y (going from X to Y); whereas, in II the order of groups about $H_a$ is X, $H_b$, $H_c$, Z, Y (going from X to Y). Thus, $H_b$ in I is not in the same environment as $H_a$ in II. Similar arguments can be used to show that in none of the three forms is $H_a$ ever in an environment exactly like that occupied by $H_b$ in any of the three forms. The relationship between $H_a$ and $H_b$ is somewhat similar to that of diastereomers, and $H_a$ and $H_b$ are said to be *diastereotopic* groups. To see the relationship, draw the molecule as in the exercise but replace $H_a$ by D (deuterium). Now draw it again but replace $H_b$ by D. The two structures you have drawn are diastereomers. Incidentally, molecules such as

```
      X
      |
H_a --+-- H_b
      |
      Y
```

are said to be *enantiotopic* (from the word, enantiomers). Can you see why? Draw the two structures obtained by replacing $H_a$ by D in one structure and $H_b$ by D in the other, and compare the structures. Certain enzymes will attack only one of a pair of enantiotopic H-atoms.

## CHAPTER 7

7.1 $CH_3CH_2CH_2CH_2CH_2Cl$

1-Chloropentane (1°)**

$CH_3CH_2CH_2CH(Cl)CH_3$

2-Chloropentane (2°)

$CH_3CH_2CH(Cl)CH_2CH_3$

3-Chloropentane (2°)

$(CH_3)_2CHCH_2CH_2Cl$

1-Chloro-3-methylbutane (1°)

$(CH_3)_2CHCH(Cl)CH_3$

2-Chloro-3-methylbutane (2°)

$CH_3C(CH_3)(Cl)CH_2CH_3$

2-Chloro-2-methylbutane (3°)

$Cl{-}CH_2CH(CH_3)CH_2CH_3$

1-Chloro-2-methylbutane (1°)

$CH_3C(CH_3)_2CH_2Cl$

1-Chloro-2,2-dimethylpropane (1°)

**(1° = primary, 2° = secondary, 3° = tertiary)

7.2 $CH_2\cdot$ ↔ $CH_2\cdot$ ↔ $CH_2$ ↔ $CH_2$ ↔ $CH_2$

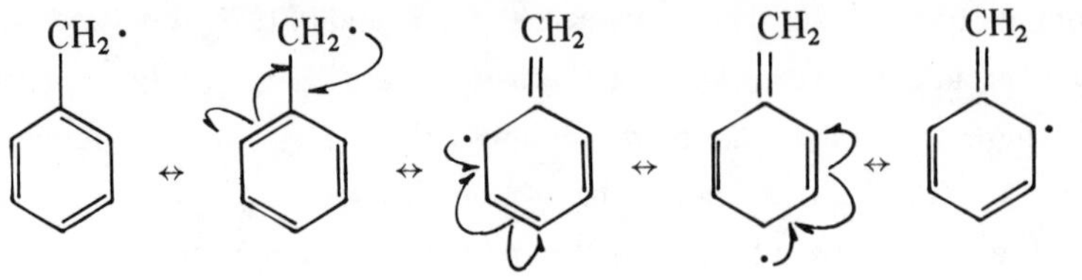

7.3 (a)

$$CH_3CH=CH_2 + HCl \longrightarrow CH_3-\underset{}{\overset{Cl}{\overset{|}{CH}}}-CH_3$$

(b) $CH_3CH=CH_2 + HBr \xrightarrow{\text{peroxides}} CH_3CH_2CH_2Br$

(c) $CH_3CH_2CH_2Br + NaI \xrightarrow{\text{acetone}} CH_3CH_2CH_2I + NaBr$

(d)

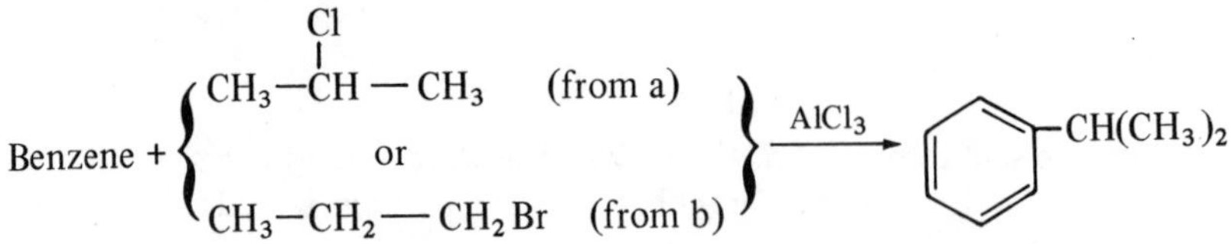

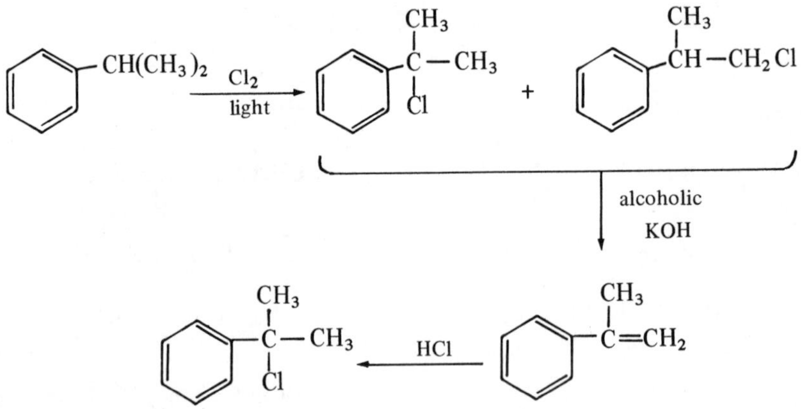

$$\text{Benzene} + \left\{\begin{matrix} CH_3-\overset{Cl}{\overset{|}{CH}}-CH_3 & \text{(from a)} \\ \text{or} & \\ CH_3-CH_2-CH_2Br & \text{(from b)} \end{matrix}\right\} \xrightarrow{AlCl_3} C_6H_5-CH(CH_3)_2$$

$$C_6H_5-CH(CH_3)_2 \xrightarrow[\text{light}]{Cl_2} C_6H_5-\underset{Cl}{\underset{|}{\overset{CH_3}{\overset{|}{C}}}}-CH_3 + C_6H_5-\overset{CH_3}{\overset{|}{CH}}-CH_2Cl$$

$$\xrightarrow{\text{alcoholic KOH}} C_6H_5-\overset{CH_3}{\overset{|}{C}}=CH_2 \xrightarrow{HCl} C_6H_5-\underset{Cl}{\underset{|}{\overset{CH_3}{\overset{|}{C}}}}-CH_3$$

7.4 Transition state for E2 elimination in *sec*-butyl chloride:

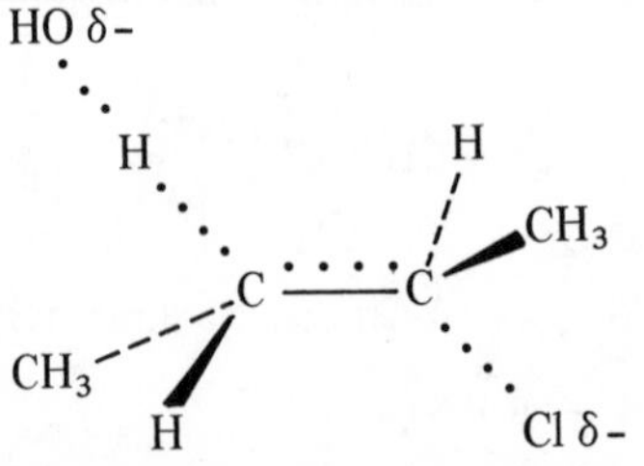

HO δ-

H H

$CH_3$

C ---- C

$CH_3$

H Cl δ-

7.5

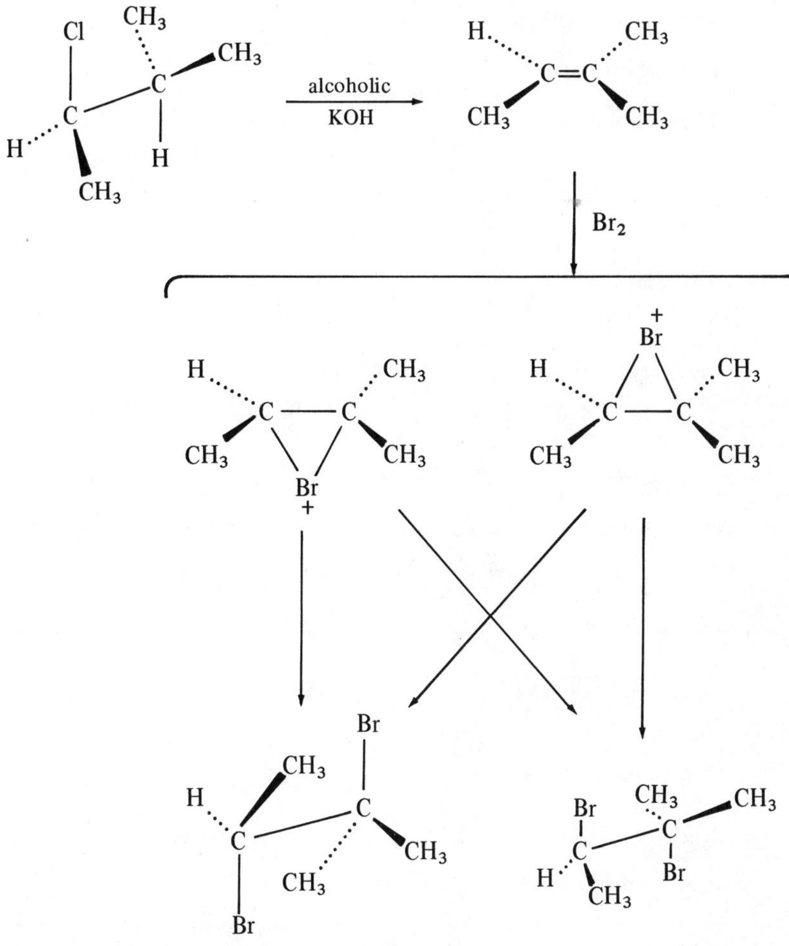

Cl, CH3, CH3, C, C, H, H, CH3 — alcoholic KOH → H, CH3, C=C, CH3, CH3

↓ $Br_2$

(S)-2,3-Dibromo-2-methylbutane

(R)-2,3-Dibromo-2-methylbutane

7.6 (a) $S_N1$; (b) E1; (c) $S_N2$; (d) E2

7.7 (a) Allyl bromide, 3-Bromo-1-propene
(b) Methylmagnesium iodide
(c) Isopropyl bromide, 2-Bromopropane
(d) Iodoform
(e) 2,4-Dinitrochlorobenzene
(f) 1,4-Dichlorobenzene, *p*-Dichlorobenzene
(g) 4-Chlorobenzyl chloride, *p*-Chlorobenzyl chloride
(h) Octafluorocyclobutane
(i) 1-Chloro-1-phenylpropane

7.8 (a)

$CH_3CH_2CH_2CH_2Br$
*n*-Butyl bromide
(1-Bromobutane)

$CH_3\underset{H}{\overset{CH_3}{C}}CH_2Br$
Isobutyl bromide
(1-Bromo-2-methylpropane)

$CH_3CH_2\underset{Br}{\overset{*}{C}}HCH_3$
*sec*-Butyl bromide
(2-Bromobutane)

$CH_3\underset{Br}{\overset{CH_3}{C}}CH_3$
*tert*-Butyl bromide
(2-Bromo-2-methylpropane)

( * denotes asymmetric carbon)

(b) Nine isomers are possible, three of which could exist as enantiomorphs.

$CH_3CH_2CH_2CHCl_2$
1,1-Dichlorobutane

$CH_3CH_2CCl_2CH_3$
2,2-Dichlorobutane

$CH_3CH_2\underset{Cl}{\overset{*}{C}}HCH_2Cl$
1,2-Dichlorobutane

$CH_3\underset{Cl}{\overset{*}{C}}HCH_2CH_2Cl$
1,3-Dichlorobutane

$ClCH_2CH_2CH_2CH_2Cl$
1,4-Dichlorobutane

$CH_3\underset{Cl}{\overset{*}{C}}H\underset{Cl}{\overset{*}{C}}HCH_3$
2,3-Dichlorobutane

(Optically active if configuration is *R,R* or *S,S*. If configuration is *R,S* or *S,R* it is a *meso* form.)

$Cl_2CH\underset{H}{\overset{CH_3}{C}}CH_3$
1,1-Dichloro-2-methylpropane

$ClCH_2\underset{H}{\overset{CH_3}{C}}CH_2Cl$
1,3-Dichloro-2-methylpropane

$ClCH_2\underset{Cl}{\overset{CH_3}{C}}CH_3$
1,2-Dichloro-2-methylpropane

7.9 (a) *tert*-Butyl chloride (b) 1-Bromobutane (c) Phenylmagnesium bromide (d) Isobutylene (e) 1-butanol (f) no reaction (g) no reaction (h) 3-Bromobenzoic acid (*m*-Bromobenzoic acid) (i) racemic 2,3-Dibromobutane (j) 2,4-Dinitroaniline

(k) *n*-Butyl iodide (1) 1-Chloro-1-phenylethane and 1-chloro-2-phenylethane (m) *sec*-Butylbenzene (n) 1-Butyne (o) Methyl ethyl ether (p) (*S*)-2-Butanol (q) 2-Chloro-2-phenylpropane and 1-chloro-2-phenylpropane (r) 2-Bromo-2-methylbutane (s) 1-Isopropyl-4-methylcyclohexene (major product) and 3-isopropyl-6-methylcyclohexene

7.10 (a) Isopropyl alcohol $\xrightarrow{P,Br_2}$ Isopropyl bromide $\xrightarrow[2.\ H_2O]{1.\ Mg,\ ether}$ $C_3H_8$

(b) Benzene $\xrightarrow{C_2H_5Br,\ AlCl_3}$ Ethylbenzene $\xrightarrow[heat]{Al_2O_3}$ Styrene

(c) Isopropyl bromide $\xrightarrow{\text{alcoholic KOH}}$ $CH_3CH{=}CH_2$ $\xrightarrow{HBr,\ H_2O_2}$ $CH_3CH_2CH_2Br$

(d) 1,2-Dibromopropane $\xrightarrow[2.\ NaNH_2]{1.\ \text{alcoholic KOH}}$ Propyne $\xrightarrow{2HBr}$ $CH_3CBr_2CH_3$

(e) Toluene $\xrightarrow{Br_2,\ Fe}$ *p*-Bromotoluene $\xrightarrow{[O]}$ *p*-Bromobenzoic acid

7.11 (a) $CH_3{-}CH_2{-}CH_2{-}OH$

(b) $CH_3{-}CH_2{-}CH_2{-}MgCl$

(c) no reaction

(d) $C_6H_5{-}CH(CH_3)_2$ (benzene ring–CH with two $CH_3$ groups)

(e) $CH_3{-}CH_2{-}CH_2{-}I$

(f) $CH_3{-}C{\equiv}C{-}CH_2{-}CH_2{-}CH_3$

(g) $CH_3{-}CH_2{-}CH_2{-}O{-}CH_2{-}CH_3$

(h) $CH_3{-}CH_2{-}CH_2{-}NH_2$

7.12

| | $S_N1$ | $S_N2$ |
|---|---|---|
| (a) | unimolecular | bimolecular |
| (b) | racemization | largely inversion |
| (c) | tert > sec > prim | prim > sec > tert |
| (d) | none | doubles rate (usually) |
| (e) | faster in polar solvents | faster in polar *aprotic* solvents |

7.13 (a) Bromine in carbon tetrachloride will be discolored by cyclohexene.

(b) α-Phenylethyl chloride will give a white precipitate of AgCl when warmed with alcoholic silver nitrate.

(c) 2,4-Dinitrochlorobenzene may be hydrolyzed by boiling with sodium carbonate or sodium hydroxide solution; chlorobenzene will be unreactive under the same conditions.

(d) Benzyl bromide will precipitate AgBr when warmed with a solution of alcoholic silver nitrate. *p*-Bromotoluene will not react under the same conditions.

7.14 $CH_3CH_2\underset{\displaystyle Br}{\underset{|}{C}}HCH_3$ , 2-Bromobutane

7.15 (a) α-Phenylethyl chloride, $C_6H_5-\overset{\displaystyle Cl}{\overset{|}{\underset{\displaystyle H}{\underset{|}{C}}}}-CH_3$

(b) *o*-Methylbenzyl chloride, $o$-$CH_3C_6H_4CH_2Cl$

(c) *p*-Chloroethylbenzene, $p$-$C_2H_5C_6H_4Cl$

7.16 (a) $CH_3$, H, Cl, H (cyclohexane) $\xrightarrow{OH^-}$ $CH_3$, H, H, OH (cyclohexane)

(b) $S_N2$ displacement. The hydroxyl group makes a backside approach. $S_N1$ displacement would have given a mixture of *cis* and *trans* products.

7.17 In aqueous ethanol the three products are *tert*-butyl alcohol, *tert*-butyl ethyl ether and 2-methylpropene (isobutylene). The $S_N1$ products (alcohol and ether) predominate over the E1 product (alkene). The alcohol is obtained in the largest amount because water solvates the intermediate carbonium ion (and the transition state leading to the carbonium ion) somewhat better than ethanol solvates these species. In 80% water and 20% ethanol the yields are 58% alcohol, 29% ether, and 13% alkene.

7.18 An ethanolic solution of potassium hydroxide contains both hydroxide ion and ethoxide ion:

$$CH_3CH_2OH + K^{+-}OH \rightleftharpoons CH_3CH_2O^- K^+ + H_2O$$

Ethoxide ion is a stronger base and better nucleophile than hydroxide ion in ethanolic solution. Therefore, ethoxide competes effectively with hydroxide, even though the equilibrium shown above is displaced to the left (that is, the concentration of $HO^-$ is greater than that of $CH_3CH_2O^-$).

7.19

Br, H, H, H, $CH_2CH_2CH_3$, H → 1-Pentene

Br, H, $CH_2CH_3$, $CH_3$, H, H → *trans*-2-Pentene

Br, $CH_3CH_2$, H, $CH_3$, H, H → *cis*-2-Pentene

In the transition state for E2 elimination (Exercise 7.4) the H-atom and Br-atom must be *anti* (or *trans*) to each other as shown in the Newman projections above. The transition state is stabilized to some extent by the same factors that stabilize the alkene. Therefore, based on this stabilization the 2-alkene should form in preference to the 1-alkene. However, a substantial quantity of 1-pentene is formed because primary H-atoms are less sterically hindered and more easily attacked by base in an E2 elimination. The transition state for *trans*-2-pentene is less hindered than that for *cis*-2-pentene, and the *trans*-isomer is more stable than the *cis*-isomer. Both effects favor the *trans*-alkene over the *cis*-alkene.

7.20 See Exercise 5.17 for conversion of Fischer to Newman structures.

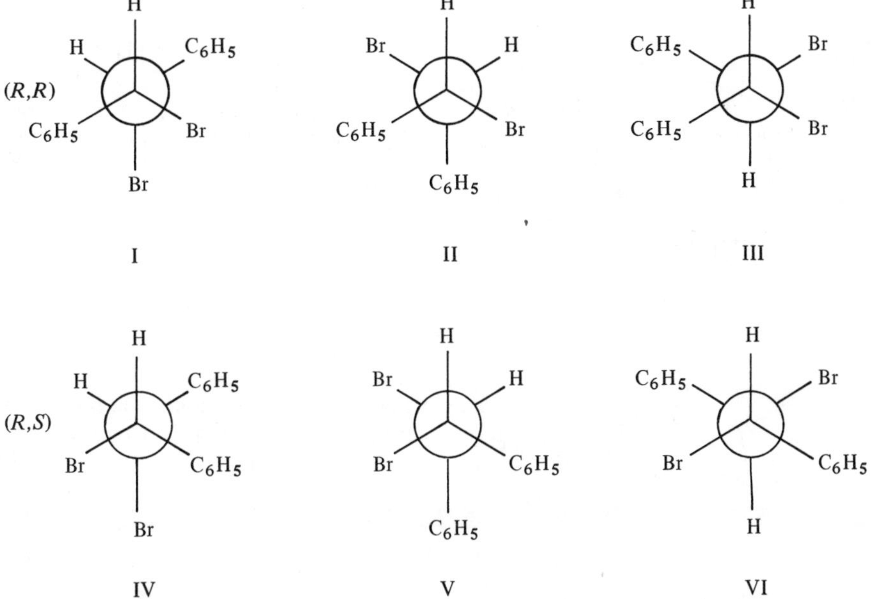


For the (*R, R*)-isomer only conformation I has the proper geometry (*anti* Br– and H-atoms) for E2 elimination, and this conformation will lead to the formation of 1-bromo-*trans*-1,2-diphenylethene. For the (*R, S*)-isomer, both IV and V have the proper geometry (*anti* Br– and H-atoms) for E2 elimination, and both conformations will lead to the formation of 1-bromo-*cis*-1,2-diphenylethene.

7.21 The stability of the anionic intermediate will depend in part on how extensively the negative charge can be dispersed over the entire molecule; that is, the extent of delocalization of the negative charge. In the intermediate from 2,4-dinitro-1-chlorobenzene the negative charge can be delocalized not only into the ring but also into the two nitro groups, giving rise to a highly stabilized intermediate.

7.22 Ion cannot become planar because of ring structure. Thus, ion is destabilized and activation energy for formation of ion is very high.

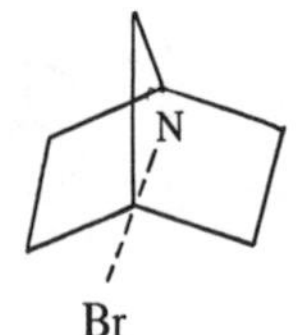

The nucleophile N**:** would have to approach the rear of the C-atom bearing the Br-atom, that is from *inside* the ring—an impossible requirement.

7.23 3.6 2.2 3.3

$Cl-CH_2-CH_2-CH_2-I$

7.24 $C(CH_2-Br)_4$ (Br$-CH_2-$C$-CH_2-$Br with two further $CH_2-$Br groups on the central C)

All protons are chemically equivalent and appear as a single peak at δ 3.3.

## CHAPTER 8

8.1 $CH_3CH_2CH_2CH_2CH_2OH$

1-Pentanol (1°)

$CH_3CH_2CH_2CH(OH)CH_3$

2-Pentanol (2°)

$CH_3CH_2CH(OH)CH_2CH_3$

3-Pentanol (2°)

$CH_3CH(CH_3)CH_2CH_2OH$

3-Methyl-1-butanol (1°)

$CH_3CH(CH_3)CH(OH)CH_3$

3-Methyl-2-butanol (2°)

$CH_3C(CH_3)(OH)CH_2CH_3$

2-Methyl-2-butanol (3°)

$CH_3CH_2CH(CH_3)CH_2OH$

2-Methyl-1-butanol (1°)

$CH_3C(CH_3)_2CH_2OH$

2,2-Dimethyl-1-propanol (1°)

8.2 Order of increasing acidity: *tert*-butyl alcohol < ethyl alcohol < *o*-cresol < phenol < *p*-chlorophenol < *o*-nitrophenol.

The nitro and halogen groups are electron-withdrawing groups. The nitro group stabilizes the anion, and increases acidity. The alcohols show acidic behavior only with electropositive metals–i.e., sodium and potassium.

The methyl group in *o*-cresol is an electron-releasing group, destabilizes the phenoxide ion and decreases acidity.

8.3 (a) $(CH_3)_2C{=}CH{-}CH_3$ 2-Methyl-2-butene

(b) $(CH_3)_3C{-}CH{=}CH_2$ 3,3-Dimethyl-1-butene

8.4 (a) $$CH_3CH{=}CH_2 \xrightarrow{H_2SO_4} CH_3CH(OSO_3H)CH_3 \xrightarrow{H_2O,\ heat} CH_3CH(OH)CH_3$$

(b) $$CH_3CH{=}CH_2 \xrightarrow{HBr,\ H_2O_2} CH_3CH_2CH_2Br \xrightarrow{KOH} CH_3CH_2CH_2OH$$

(c) Isoproyl alcohol (from a) $\xrightarrow{[O]}$ $(CH_3)_2C{=}O$

Acetone

(A)

Isopropyl alcohol + HCl $\xrightarrow{ZnCl_2}$ $CH_3CH(Cl)CH_3$ $\xrightarrow{Mg,\ ether}$ $(CH_3)_2CHMgCl$

(B)

$$\text{(A)} + \text{(B)} \xrightarrow{\text{followed by hydrolysis}} (CH_3)_2C(OH)CH(CH_3)_2$$

2,3-Dimethyl-2-butanol

8.5 H-bonding stabilizes the gauche conformation of ethylene chlorohydrin.

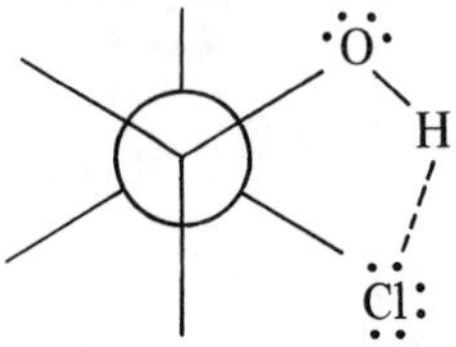

8.7

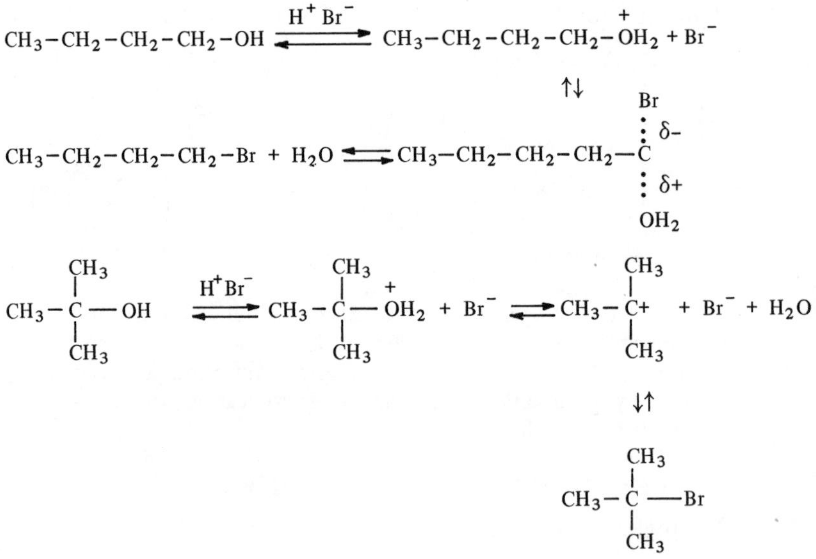

$$CH_3-CH_2-CH_2-CH_2-OH \underset{}{\overset{H^+Br^-}{\rightleftharpoons}} CH_3-CH_2-CH_2-CH_2-\overset{+}{O}H_2 + Br^-$$

$$\updownarrow$$

$$CH_3-CH_2-CH_2-CH_2-Br + H_2O \rightleftharpoons CH_3-CH_2-CH_2-CH_2-\underset{\underset{OH_2}{\vdots\ \delta+}}{\overset{\overset{Br}{\vdots\ \delta-}}{C}}$$

$$CH_3-\underset{CH_3}{\overset{CH_3}{\underset{|}{\overset{|}{C}}}}-OH \overset{H^+Br^-}{\rightleftharpoons} CH_3-\underset{CH_3}{\overset{CH_3}{\underset{|}{\overset{|}{C}}}}-\overset{+}{O}H_2 + Br^- \rightleftharpoons CH_3-\underset{CH_3}{\overset{CH_3}{\underset{|}{\overset{|}{C}}}}+ \quad + Br^- + H_2O$$

$$\updownarrow$$

$$CH_3-\underset{CH_3}{\overset{CH_3}{\underset{|}{\overset{|}{C}}}}-Br$$

8.8 (a) -4 (b) -2 (c) zero (d) +2 (e) +4

8.9

$$3CH_3\underset{OH}{\underset{|}{C}}HCH_3 + Na_2Cr_2O_7 + 4H_2SO_4 \rightarrow 3CH_3\overset{O}{\overset{\|}{C}}CH_3 + Na_2SO_4 + Cr_2(SO_4)_3 + 7H_2O$$

8.10 Most phenolic compounds are acidic enough to form salts in strong basic solution. Solubility in sodium hydroxide solution should distinguish a phenol from an alcohol.

8.11 Ethyl *tert*-butyl ether should be prepared via the Williamson synthesis using potassium *tert*-butoxide as the nucleophile against ethyl bromide.

$$(CH_3)_3CO^- + H-CH_2(CH_3)-Br \rightarrow (CH_3)_3C-O-C_2H_5 + Br^-$$

A tertiary halide, if used as one reactant, would probably undergo elimination to yield an olefin.

8.12 Ethyl alcohol is completely soluble in water. Ethyl ether is soluble in sulfuric acid, and ethyl bromide is soluble in neither water or acid.

8.13 (a) Diisopropyl ether (b) 2-Nitrophenol, *o*-Nitrophenol (c) 2-Butanol, *sec*-Butyl alcohol (d) 2,4-Dichloroanisole (e) 2-Methyl-2-butanol (f) Triphenylcarbinol (g) 1,2-Dihydroxypropane, Propylene glycol (h) 1-Methylcyclohexanol (i) *p*-Cresol, 4-Methylphenol (j) Ethyl phenyl ether

8.14 (a) 1-naphthol (OH on naphthalene) (b) $CH_3-C(OH)(C_6H_5)-CH_3$ (c) $C_6H_5-CH_2OH$

(d) 2,6-di-tert-butylphenol: $(CH_3)_3C-C_6H_3(OH)-C(CH_3)_3$ (e) $C_6H_5-CH(OH)CH_3$

(f) $CH_3-CH=CH-CH_2OH$ (g) $C_6H_5-CH_2-O-CH_2-C_6H_5$

(h) $C_6H_4(OH)NO_2$ (ortho) (i) $C_6H_5-O^-\ Na^+$ (j) $CH_3-C(CH_3)_2-O^-\ K^+$

8.15 (a) $CH_3CH{=}CH_2 + HOCl \rightarrow CH_3CH(OH)CH_2Cl$

(b) $(CH_3)_2CHMgCl$ + Ethylene oxide $\rightarrow (CH_3)_2CH-CH_2CH_2OH$

(c) $(CH_3)_3COH + HCl \rightarrow (CH_3)_3CCl + H_2O$

(d) $C_2H_5OH + H_2SO_4 \xrightarrow{\text{warm}} CH_3CH_2OCH_2CH_3$

(e) Phenol + 3 $Br_2$ → 2,4,6-Tribromophenol + 3 HBr

(f) $C_6H_5CH_2Cl + NaOH \rightarrow C_6H_5CH_2OH + NaCl$

(g) $3\ CH_3CH_2CH_2OH + PI_3 \rightarrow 3\ CH_3CH_2CH_2I + H_3PO_3$

(h) $C_6H_5OCH_3 + HI \xrightarrow{\text{heat}} C_6H_5OH + CH_3I$

(i) $C_2H_5OH + Na \rightarrow C_2H_5ONa + \frac{1}{2}\ H_2$

(j)

$$CH_3CH_2CH_2OH \xrightarrow{\text{oxidize}} CH_3CH_2\overset{H}{\overset{|}{C}}{=}O \longrightarrow CH_3CH_2\overset{O}{\overset{||}{C}}-OH$$

(k) $CH_3CH_2CH{=}CH_2 + (BH_3)_2 \xrightarrow{H_2O_2,\ OH^-} CH_3CH_2CH_2CH_2OH$

(l) Sodium benzene sulfonate + NaOH $\xrightarrow{\text{fusion}} C_6H_5O^-Na^+ \xrightarrow{H^+}$ Phenol

(m) Phenol + $H_2 \xrightarrow{\text{Ni, heat}}$ Cyclohexanol

8.16 $C_2H_5OH + H_2SO_4 \xrightarrow{180°} \underset{[a]}{H_2C{=}CH_2} \xrightarrow{HBr} \underset{[b]}{CH_3CH_2Br}$

$[a] \xrightarrow{Br_2} H_2\underset{Br}{\underset{|}{C}}CH_2Br \xrightarrow[\text{2. } NaNH_2]{\text{1. alcoholic KOH}} \underset{[c]}{C_2H_2}$

(c) + $NaNH_2 \rightarrow Na-C{\equiv}C-H \xrightarrow{C_2H_5Br} \underset{[d]}{H-C{\equiv}C-CH_2CH_3}$

(a) + $HOCl \rightarrow \underset{[e]}{HOCH_2CH_2Cl} \xrightarrow[H_2O]{Na_2CO_3} \underset{[f]}{HOCH_2CH_2OH}$

(e) + $O_2Ag \rightarrow \underset{[g]}{H_2C\overset{O}{—}CH_2}$ (epoxide) $\xrightarrow{H_2O,\ HCl}$ [f]

$CH_3CH_2OH + K_2Cr_2O_7 \xrightarrow{H^+} \underset{[h]}{CH_3-COOH}$

$CH_3CH_2OH + I_2 + NaOH \rightarrow \underset{[i]}{CHI_3} + HCOONa$

(b) + Mg $\xrightarrow{\text{ether}}$ $C_2H_5MgBr$ $\xrightarrow{\text{[g], followed by hydrolysis}}$ $CH_3CH_2CH_2CH_2OH$

[j]

8.17 1-Propanol $\xleftarrow[\text{2. NaOH}]{\text{1. HBr, peroxide}}$ $CH_3CH{=}CH_2$ $\xrightarrow[\text{2. } H_2O_2\text{, NaOH}]{\text{1. } (BH_3)_2}$ 1-Propanol

1-Bromobutane $\xleftarrow[\text{2. HBr, peroxide}]{\text{1. } H_2SO_4\text{, heat}}$ $CH_3CH_2CH_2CH_2OH$ $\xrightarrow{\text{HBr}}$ 1-Bromobutane

8.18

| Solubility in $H_2O$ | Reactivity towards Lucas reagent | Reactivity with sodium metal |
|---|---|---|
| a, f, b, d, c, e | a, e, d, b, f, c | f, c, b, d, a, e |

8.19 (a) 1-Hexyne will discolor $Br_2/CCl_4$ solution, *sec*-butyl alcohol will not.

(b) A small piece of sodium metal in isopropyl alcohol should liberate hydrogen gas.

(c) Ethyl ether is soluble in concentrated sulfuric acid, *n*-pentane is not.

(d) *tert*-Butyl alcohol reacts immediately with Lucas reagent. *n*-Butyl alcohol reacts only when heated.

(e) *sec*-Butyl alcohol will give a positive iodoform reaction.

8.20

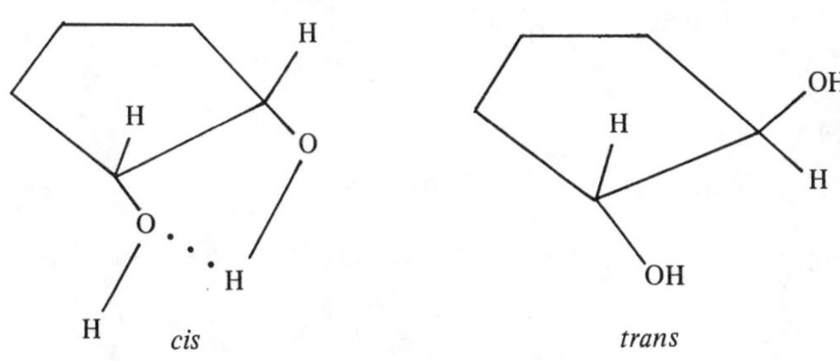


The opportunity for intramolecular hydrogen bonding in the *cis* isomer lessens the extent to which it may H-bond with its neighbors (association). The greater the degree of association in a liquid, the higher its boiling point.

8.21 A symmetrical ether. $R{-}O{-}R + 2\ HI \rightarrow 2\ RI + H_2O$

The atomic weight of iodine = 127

If 0.815 × M.W. of RI = 127, then the M.W. of RI $= \dfrac{127}{0.815} = 156$

156 - 127 = 29 = contributing weight of R

R = $C_2H_5$-, the compound must be diethyl ether.

8.22 $K_f = 1.86°C/mole$

1 mole solute per 1000 g $H_2O$ lowers freezing point 1.86°C

$$\frac{1 \text{ mole}}{1.86°} = \frac{x}{40°}$$

$x$ = 21.5 moles ethylene glycol needed to lower temperature of 1000 g of $H_2O$ 40°C

mol. wt. ethylene glycol = 62 g/mole

$$\frac{62 \text{ g}}{1 \text{ mole}} = \frac{x}{21.5 \text{ moles}}$$

$x$ = 1333 g ethylene glycol (amount needed per 1000 g $H_2O$ to lower temperature 40°C)

density of ethylene glycol = 1.116 g/ml

$$\text{volume} = \frac{1333 \text{ g}}{1.116 \text{ g/ml}} = 1194.7 \text{ ml}$$

volume of solution, if additive, = 1194.7 ml ethylene glycol + 1000 ml $H_2O$ = 2194.7 ml

$$\text{volume of ethylene glycol in quarts} = \frac{1194.7 \text{ ml}}{960 \text{ ml/qt}} = 1.244 \text{ qt}$$

$$\text{volume of solution in quarts} = \frac{2194.7 \text{ ml}}{960 \text{ ml/qt}} = 2.286 \text{ qt}$$

$$\frac{1.244 \text{ qt ethylene glycol}}{2.286 \text{ qt solution}} = \frac{x}{20 \text{ qt}}$$

$x$ = 10.9 qt ethylene glycol needed in 20 qt radiator to lower freezing point to -40°C

8.23

$$CH_3-C(OH)(CH_3)-C(OH)(CH_3)-CH_3 \underset{}{\overset{H^+}{\rightleftharpoons}} CH_3-C(OH)(CH_3)-C(\overset{+}{O}H_2)(CH_3)-CH_3 \overset{-H_2O}{\rightleftharpoons} CH_3-C(OH)(CH_3)-\overset{+}{C}(CH_3)-CH_3$$

$$\downarrow\uparrow$$

$$CH_3-C(=O)-C(CH_3)_2-CH_3 \overset{-H^+}{\rightleftharpoons} \left[ CH_3-C(=\overset{+}{O}H)-C(CH_3)_2-CH_3 \leftrightarrow CH_3-\overset{+}{C}(\ddot{O}H)-C(CH_3)_2-CH_3 \right]$$

8.24 $C_6H_5-CH(OH)-CH_3$ 1-Phenylethanol

8.25 (a) $CH_3-CH=CH_2 \xrightarrow[\text{2. NaOH, } H_2O_2]{\text{1. } B_2H_6} CH_3-CH_2-CH_2-OH$

(b) $CH_3-C(CH_3)_2-CH_2-OH \xrightarrow[\text{2. } -H_2O]{\text{1. } +H^+ (H_2SO_4)} CH_3-C(CH_3)_2-CH_2^+ \rightarrow CH_3-\overset{+}{C}(CH_3)-CH_2-CH_3 \xrightarrow{-H^+} CH_3-C(CH_3)=CH-CH_3$

(c) $CH_3-CH(OH)-C(CH_3)_2-CH_3 \xrightarrow[\text{2. } -H_2O]{\text{1. } +H^+ (HCl)} CH_3-\overset{+}{C}H-C(CH_3)_2-CH_3 \rightarrow CH_3-CH(CH_3)-\overset{+}{C}(CH_3)-CH_3 \xrightarrow{Cl^-} CH_3-CH(CH_3)-C(CH_3)(Cl)-CH_3$

(d) $C_6H_6 \xrightarrow[\text{2. Mg}]{\text{1. } Br_2 + FeBr_3} C_6H_5-MgBr \xrightarrow[\text{2. } H_2O, H^+]{\text{1. } CH_2-CH_2 \text{ (epoxide, O)}} C_6H_5-CH_2-CH_2OH$

(e) $CH_3-CH(CH_3)-CH_2-CH_2-OH \xrightarrow[H_3PO_4]{-H_2O} CH_3-CH(CH_3)-CH=CH_2 \xrightarrow[H^+]{+H_2O} CH_3-CH(CH_3)-CH(OH)-CH_3 \xrightarrow[H_3PO_4]{-H_2O} CH_3-C(CH_3)=CH-CH_3$

(f) $NH_2$ — $MnO_2$, $H_2SO_4$ → O, O — $CH_2{=}CH{-}CH{=}CH_2$, Diels-Alder → O, O

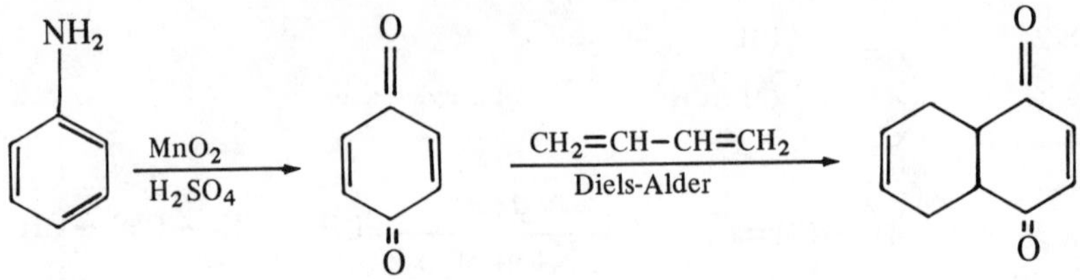

8.26 (a) OH, H, $CH_3$, $CH_3$ — 1. $+H^+$, 2. $-H_2O$ → H, +, $CH_3$, $CH_3$ →

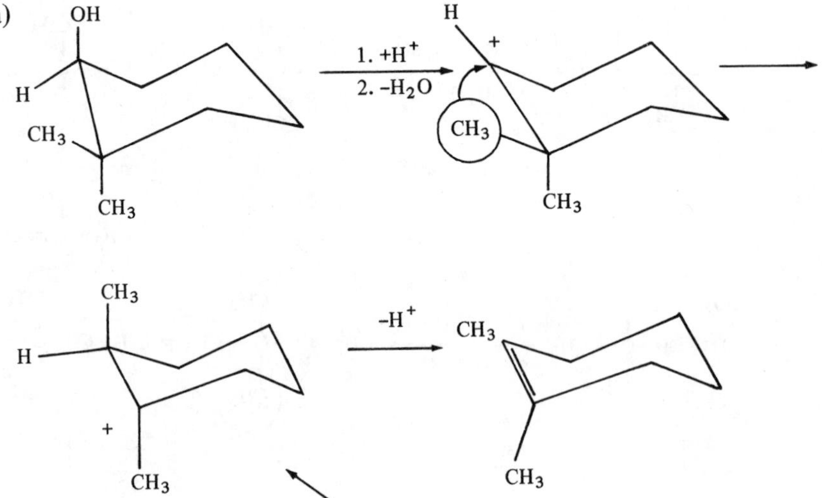

$CH_3$, H, +, $CH_3$ — $-H^+$ → $CH_3$, $CH_3$

(b) $CH_3$, $CH_3$, C, H, OH — 1. $+H^+$, 2. $-H_2O$ → $CH_3$, $CH_3$, C, +, H

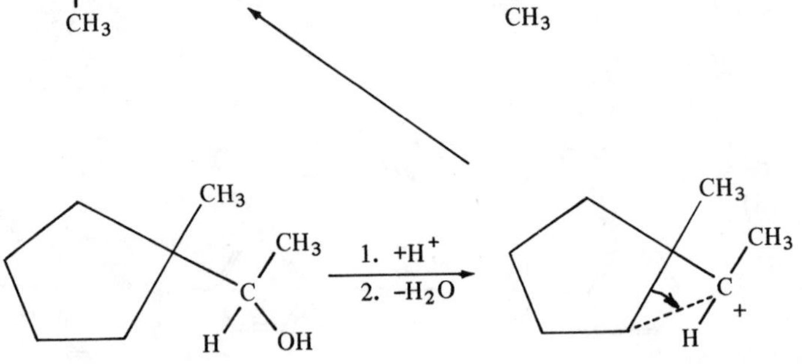

(c) 2 $C_6H_5$-CH=CH$_2$ — $H^+$ → $C_6H_5$-$\overset{+}{C}$H−CH$_3$ + $C_6H_5$-CH=CH$_2$

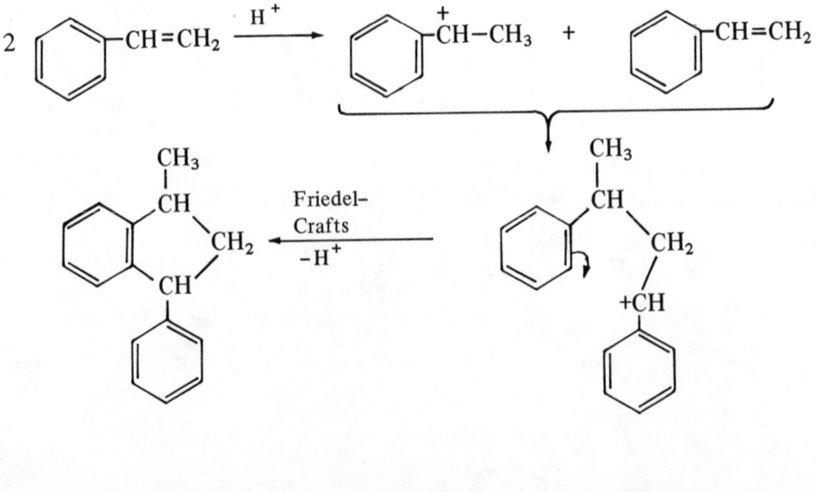

$CH_3$, CH, $CH_2$, CH — Friedel-Crafts, $-H^+$ ← $CH_3$, CH, $CH_2$, +CH

8.27

$$H-C\equiv C-C(OH)(CH_3)-CH_2-CH_3$$

OH 3.33; ≡C–H 2.32; $CH_3$ 1.45; $CH_2$ 1.67; $CH_3$ 0.95

## CHAPTER 9

9.1

$CH_3CH_2CH_2CH_2CHO$ — Pentanal

$CH_3CH(CH_3)CH_2CHO$ — 3-Methylbutanal

$CH_3CH_2CH(CH_3)CHO$ — 2-Methylbutanal

$(CH_3)_3C-CHO$ — 2,2-Dimethylpropanal

$CH_3CH_2CH_2COCH_3$ — 2-Pentanone

$CH_3CH_2COCH_2CH_3$ — 3-Pentanone

$(CH_3)_2CH-COCH_3$ — 3-Methyl-2-butanone

9.2 The strong inductive effect of the trichloromethyl group will make the carbonyl carbon especially susceptible to nucleophilic attack.

$$\underset{\delta-}{Cl_3C}-\underset{\delta+}{C}(=O)CH_3$$

9.3 Resonance structures of acetophenone ($CH_3COC_6H_5$): $C=O$ ↔ $C^+-O^-$ ↔ ring-ortho carbocation ↔ ring-para carbocation ↔ other ortho carbocation.

The reactivity of a carbonyl compound in addition reactions depends in part on the electron-deficient C-atom of the carbonyl group. In acetophenone the interaction of the ring with the carbonyl group delocalizes this positive charge into the ring. As a result nucleophiles are less likely to add to the carbonyl C-atom of aromatic ketones than to that of aliphatic ketones, such as acetone:

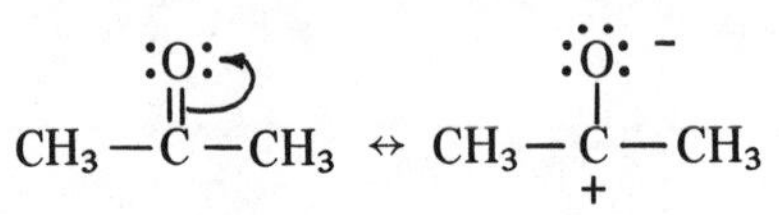

$$CH_3-\overset{\overset{:O:}{\|}}{C}-CH_3 \leftrightarrow CH_3-\underset{+}{\overset{\overset{:\ddot{O}:^-}{|}}{C}}-CH_3$$

(Note: To show the structures affecting the chemical reactivity of acetophenone we could write as many as seven forms: two Kekulé forms with a $\rangle C=O$ group, two Kekulé forms with a $\rangle \overset{+}{C}-O^-$ group, and the three forms on the right above. In other words, we must always write the Kekulé forms for each of the resonance forms of the functional group plus those forms showing interaction of the ring with the functional group.)

9.4 Acetone is a stronger base than acetylene because removal of a proton from the methyl group of acetone gives a resonance-stabilized enolate ion in which the negative charge is delocalized or shared by the α–carbon atom and the oxygen atom. Removal of a proton from acetylene gives an anion in which the charge is localized. The principal structural factor stabilizing the acetylide ion is the triple bond, which in effect keeps the bonding electrons away from the unshared pair of the negative charge.

$$H-C\equiv C-H \rightleftharpoons H-C\equiv C:^- + H^+$$

9.5 The space requirements of the two ethyl groups of diethyl ketone appear to have a shielding effect on the carbonyl group and prevent the formation of a bisulfite addition compound. Cyclopentanone, on the other hand, is a cyclic structure with the carbonyl group open to attack.

9.6 2,3-Dimethyl-2-butanol is a tertiary alcohol which may be obtained through the addition of a Grignard reagent to a ketone.

Step 1. $CH_3CH{=}CH_2 + H_2SO_4 \xrightarrow{\text{hydrolysis, heat}} CH_3\underset{OH}{\underset{|}{C}}HCH_3$

$$CH_3\underset{OH}{\underset{|}{C}}HCH_3 \xrightarrow{\text{oxidize}} \underset{(A)}{CH_3\overset{\overset{O}{\|}}{C}CH_3}$$

Step 2 $CH_3CH{=}CH_2 + HBr \rightarrow (CH_3)_2CHBr \xrightarrow{\text{Mg, ether}} (CH_3)_2CHMgBr$

(B)

$$(A) + (B) \xrightarrow{\text{followed by hydrolysis}} \text{2,3-Dimethyl-2-butanol}$$

9.7

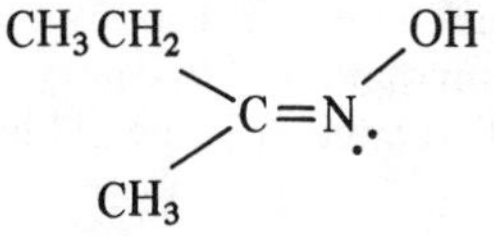

$$\begin{array}{c}CH_3CH_2 \quad\quad OH\\ \diagdown\;\; \diagup\\ C{=}N\!:\\ \diagup\\ CH_3\end{array} \qquad\qquad \begin{array}{c}CH_3CH_2\\ \diagdown\\ C{=}\ddot{N}\\ \diagup \quad \diagdown\\ CH_3 \quad\quad OH\end{array}$$

(a) 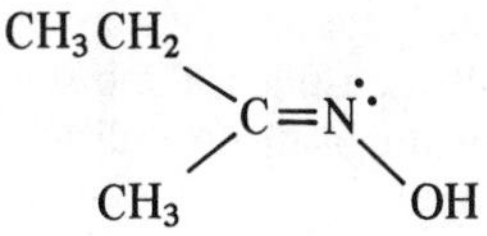 (b)

(a) is *syn* to ethyl, *anti* to methyl (b) is *syn* to methyl, *anti* to ethyl
These isomers are related to *cis* and *trans* isomers of alkenes.

9.8

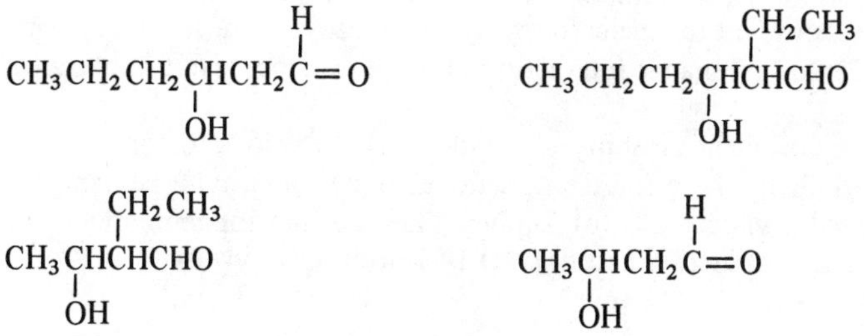

$CH_3CH_2CH_2CH(OH)CH_2C(H){=}O$ $CH_3CH_2CH_2CH(OH)CH(CH_2CH_3)CHO$

$CH_3CH(OH)CH(CH_2CH_3)CHO$ $CH_3CH(OH)CH_2C(H){=}O$

9.9 (1)

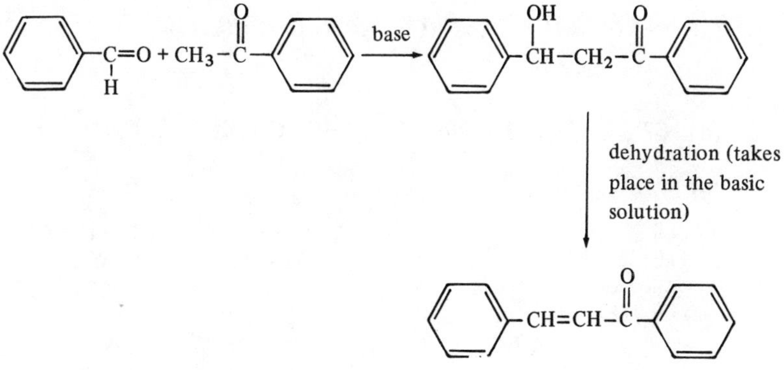

$$C_6H_5{-}C(H){=}O + CH_3{-}C({=}O){-}C_6H_5 \xrightarrow{\text{base}} C_6H_5{-}CH(OH){-}CH_2{-}C({=}O){-}C_6H_5$$

dehydration (takes place in the basic solution)

$$C_6H_5{-}CH{=}CH{-}C({=}O){-}C_6H_5$$

(2)

$$(CH_3)_2C{=}O + CH_3{-}C({=}O){-}CH_3 \xrightarrow[(Ba(OH)_2)]{\text{base}} CH_3{-}C(CH_3)(OH){-}CH_2{-}C({=}O){-}CH_3$$

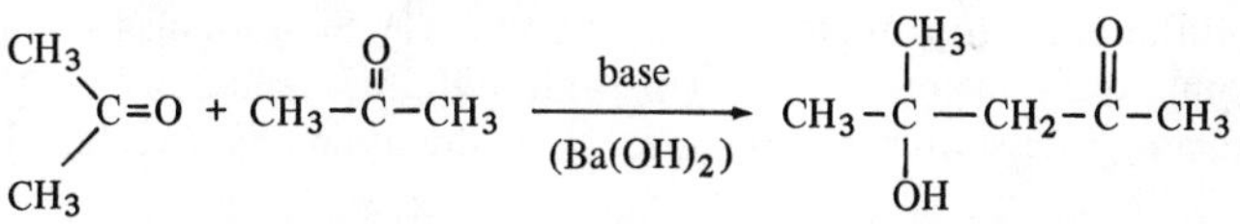

(does not dehydrate readily in base)

(3)

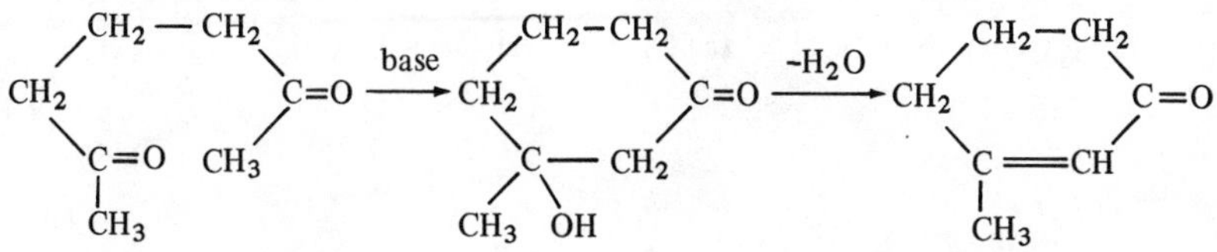

$$CH_3C({=}O)CH_2CH_2CH_2C({=}O)CH_3 \xrightarrow{\text{base}} \text{cyclic: } CH_2{-}CH_2{-}C({=}O){-}CH_2{-}C(CH_3)(OH){-}CH_2 \xrightarrow{-H_2O} \text{cyclic: } CH_2{-}CH_2{-}C({=}O){-}CH{=}C(CH_3){-}CH_2$$

Note that this is an intramolecular aldol condensation leading to the formation of a 6-membered ring followed by a facile dehydration. Reactions of this type are important in the laboratory synthesis of many natural products containing 6-membered rings, such as the steroids.

9.10 Remove cyclohexanone by shaking the mixture with a saturated solution of sodium bisulfite. Collect the bisulfite addition compound by filtration. Treatment of the bisulfite addition compound with a mineral acid will liberate the pure ketone.

9.11

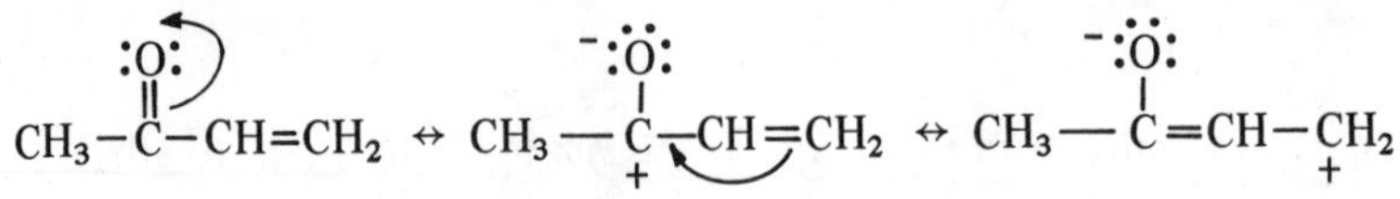

$$CH_3-\overset{\overset{\displaystyle :\!O\!:}{\|}}{C}-CH=CH_2 \leftrightarrow CH_3-\overset{\overset{\displaystyle ^{-}:\!\ddot{O}\!:}{|}}{\underset{+}{C}}-CH=CH_2 \leftrightarrow CH_3-\overset{\overset{\displaystyle ^{-}:\!\ddot{O}\!:}{|}}{C}=CH-\underset{+}{CH_2}$$

It is not necessary (or desirable) to draw the resonance forms of the $-CH{=}CH_2$ separately as we did for the carbonyl group because we have already concluded (Sec. 1.12) that resonance interaction in simple alkenes is not important.

9.12 (a) 2-Methylpropanal (Isobutyraldehyde) (b) Cyclopentanone (c) Methyl phenyl ketone (Acetophenone) (d) 3-Methyl-2-pentanone (Methyl *sec*-butyl ketone) (e) Diphenyl ketone (Benzophenone) (f) 3-Pentanone (Diethyl ketone) (g) 1-(3-Bromophenyl)-2-buten-1-one (h) 3-Methyl-2-cyclohexenone

9.13

(a) $CH_3CH_2CHO$ (b) $(CH_3)_2CH-\overset{\overset{\displaystyle O}{\|}}{C}-CH(CH_3)_2$ (c) $(CH_3)_2CHCHO$

(d) $CH_3(CH_2)_5\overset{\overset{\displaystyle O}{\|}}{C}CH_3$ (e) $C_6H_5-\overset{\overset{\displaystyle O}{\|}}{C}-CH_2Br$ (f) $C_6H_5-CH{=}CH-CHO$

(g) furan ring with CHO at C-2 (O in ring)

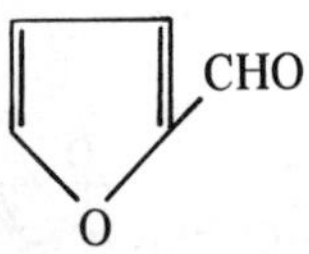

(h) $C_6H_5-CH{=}CH-\overset{\overset{\displaystyle O}{\|}}{C}-CH_3$

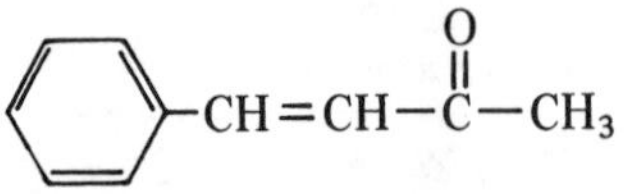

9.14 (a) Acetone or isopropyl alcohol (b) Hydrogen cyanide (c) Acetal (d) 2-Butanone (e) Acetophenone oxime (f) Sodium bisulfite addition compound of *n*-butyraldehyde (g) 4-Methyl-3-pentene-2-one (h) Benzyl alcohol and potassium benzoate (i) Formaldehyde (j) Acetaldehyde

9.15 (a) $CH_3CH_2CH_2OH$ (b) $CH_3CH_2CH(OH)CH_3$ (c) $CH_3CH_2C(CH_3)_2OH$

(d) $(CH_3CH_2)_2C(OH)CH_3$ (e) $C_6H_5-CH(OH)CH_2CH_3$ (f) $C_6H_5-\overset{\overset{\displaystyle OH}{|}}{\underset{\underset{\displaystyle CH_3}{|}}{C}}-CH_2CH_3$

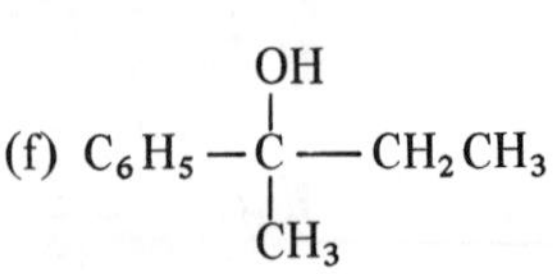

(g)

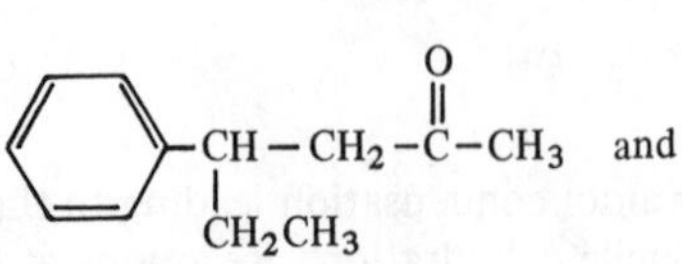

$$C_6H_5-\underset{\underset{\displaystyle CH_2CH_3}{|}}{CH}-CH_2-\overset{\overset{\displaystyle O}{\|}}{C}-CH_3 \quad \text{and} \quad C_6H_5-CH{=}CH-\overset{\overset{\displaystyle OH}{|}}{\underset{\underset{\displaystyle CH_2CH_3}{|}}{C}}-CH_3$$

1,4-Product 1,2-Product

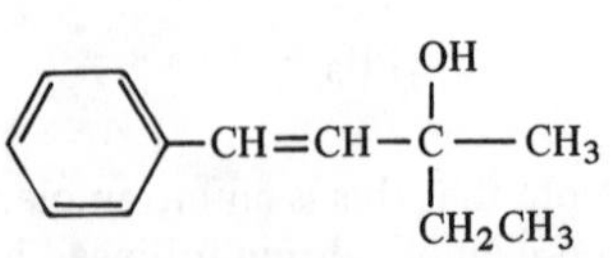

9.16 (a), (b), (e), (f), (g).

9.17

$$C_2H_2 + H_2O \xrightarrow[H_2SO_4]{Hg^{2+}} \underset{(a)}{CH_3CHO} \xrightarrow{I_2 \text{ in KI + NaOH}} \underset{(c)}{CHI_3}$$

$$\underset{(a)}{CH_3CHO} \xrightarrow{[O]} \underset{(d)}{CH_3COOH}$$

$$C_2H_2 \xrightarrow{H_2,\ \text{catalyst}} C_2H_4 \xrightarrow{HBr} \underset{(b)}{C_2H_5Br} \xrightarrow{\text{Mg, ether}} C_2H_5MgBr$$

$$C_2H_5MgBr + (a) \xrightarrow{\text{followed by hydrolysis}} \underset{(e)}{CH_3CH_2CH(OH)CH_3}$$

$$\underset{(e)}{CH_3CH_2CH(OH)CH_3} \xrightarrow{[O]} \underset{(f)}{CH_3CH_2\overset{O}{\overset{\|}{C}}CH_3} \xrightarrow{I_2 \text{ in KI, NaOH}} (c) + CH_3CH_2COONa \xrightarrow{H^+} \underset{(h)}{CH_3CH_2COOH}$$

$$(a) \xrightarrow[\text{warm}]{10\%\ NaOH} \underset{(g)}{CH_3CH{=}CH{-}\overset{H}{\overset{|}{C}}{=}O}$$

9.18 (a) Acetaldehyde (b) 2-Butanone (c) Formaldehyde (d) Acetone
(e) 2-Heptanone

9.19 Phenol is a weak acid and a Grignard reagent can not be prepared in the presence of any acidic compound. The following reaction sequence should lead to the desired product. The *ortho* isomer would form a chelated compound and could be removed by steam distillation.

$$C_6H_5OH + CH_3\overset{O}{\overset{\|}{C}}{-}Cl \rightarrow C_6H_5{-}O{-}\overset{O}{\overset{\|}{C}}{-}CH_3 \xrightarrow{AlCl_3} HO{-}C_6H_4{-}\overset{O}{\overset{\|}{C}}{-}CH_3 \xrightarrow{NaBH_4} HO{-}C_6H_4{-}\overset{OH}{\overset{|}{C}}(H){-}CH_3$$

Fries rearrangement

9.20 (a) Acetaldehyde is easily oxidized by Tollens' reagent. (b) Tollens' reagent will oxidize benzaldehyde. (c) The iodoform reaction would distinguish 2-butanone from 2,2-dimethylpropanal. (d) Cyclopentanone will form an addition compound with saturated sodium bisulfite, 3-pentanone will not. (e) Tollens' reagent will oxidize benzaldehyde, but not acetophenone.

9.21 This student should have tried to classify his unknown as either an aldehyde or a ketone by testing a small sample with Tollens' reagent (reduction would have indicated an aldehyde) or the iodoform reaction which would have been positive with the ketone. He then should have prepared the semicarbazone derivative.

9.22 A = $CH_3CH_2CH(OH)CH_2CH_3$, 3-pentanol

B = $(C_2H_5)_2C{=}O$, 3-pentanone

9.23 $(CH_3)_2C{=}CHCH_2CH_3$, 2-methyl-2-pentene; $CH_3CH_2CH_2C(CH_3){=}CH_2$, 2-methyl-1-pentene; or $CH_3CH(CH_3)C(CH_3){=}CH_2$, 2,3-dimethyl-1-butene

9.24

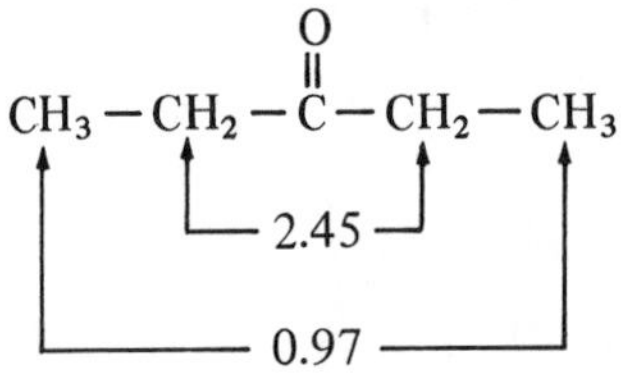

$$CH_3-CH_2-\overset{\overset{\displaystyle O}{\|}}{C}-CH_2-CH_3$$

3-Pentanone, Diethyl ketone

$CH_2$: 2.45; $CH_3$: 0.97

9.25

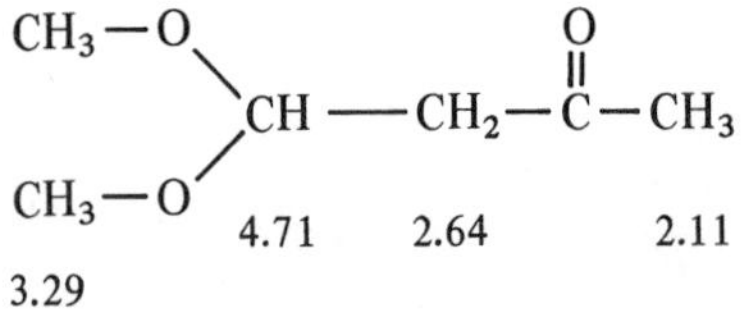

$$(CH_3-O)_2CH-CH_2-\overset{\overset{\displaystyle O}{\|}}{C}-CH_3$$

$CH_3-O$: 3.29; CH: 4.71; $CH_2$: 2.64; $CH_3$: 2.11

9.26

$$CH_3-\underset{\underset{\displaystyle CH_3}{|}}{\overset{\overset{\displaystyle OH}{|}}{C}}-\underset{\underset{\displaystyle CH_3}{|}}{\overset{\overset{\displaystyle OH}{|}}{C}}-CH_3 \xrightarrow{HIO_4} 2\ CH_3-\overset{\overset{\displaystyle O}{\|}}{C}-CH_3$$

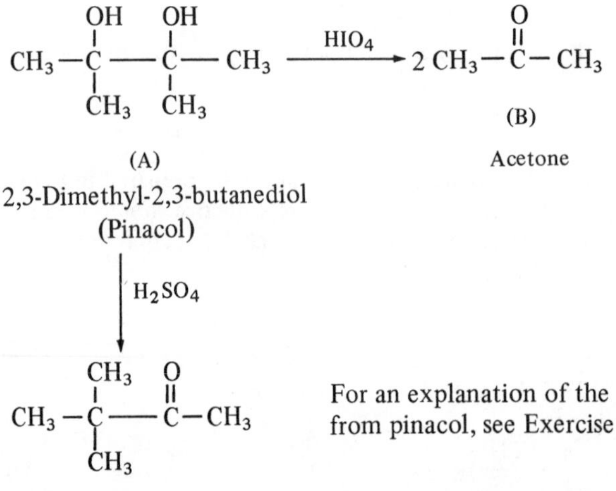

(A)

2,3-Dimethyl-2,3-butanediol

(Pinacol)

(B)

Acetone

$\downarrow H_2SO_4$

$$CH_3-\underset{\underset{\displaystyle CH_3}{|}}{\overset{\overset{\displaystyle CH_3}{|}}{C}}-\overset{\overset{\displaystyle O}{\|}}{C}-CH_3$$

(C)

3,3-Dimethyl-2-butanone

(Pinacolone)

For an explanation of the formation of pinacolone from pinacol, see Exercise 8.23.

9.27 (a)

$$CH_3CH_2CH_2CH{=}CH_2 \xrightarrow[H^+]{H_2O} CH_3CH_2CH_2\overset{OH}{\overset{|}{C}}HCH_3 \xrightarrow[H_2SO_4]{K_2Cr_2O_7} CH_3CH_2CH_2\overset{O}{\overset{\|}{C}}CH_3$$

$$\xrightarrow{CH_3CH_2CH{=}P(C_6H_5)_3} (CH_3CH_2)(H)C{=}C(CH_2CH_2CH_3)(CH_3)$$

(b)

$$CH_3-\overset{O}{\overset{\|}{C}}-CH_3 + CH_3-\overset{P(C_6H_5)_3}{\overset{\|}{C}}-CH_3 \rightarrow (CH_3)_2C{=}C(CH_3)_2 \xrightarrow[H_2O,\ cold]{KMnO_4} CH_3-\underset{CH_3}{\overset{OH}{C}}-\underset{CH_3}{\overset{OH}{C}}-CH_3$$

$$CH_3-\underset{CH_3}{\overset{CH_3}{C}}-\overset{O}{\overset{\|}{C}}-CH_3 \xleftarrow{H_2SO_4} \text{Pinacol}$$

$$(CH_3)_3C-COOH \xleftarrow[2.\ H^+]{1.\ Cl_2 + NaOH} \text{Pinacolone}$$

Pivalic acid — (haloform reaction) — Pinacolone — Pinacol

Note: The pinacol-pinacolone rearrangement is described in Exercise 8.23. Pinacol may also be prepared in the laboratory by treatment of acetone with magnesium amalgam, a reaction called the *pinacol reduction*:

$$2\ \text{Acetone} \xrightarrow{Mg(Hg)} \text{Pinacol}$$

(c)

$$CH_3CH_2CH_2CH{=}CH_2 \xrightarrow[2.\ NaOH + H_2O_2]{1.\ B_2H_6} CH_3CH_2CH_2CH_2CH_2OH \xrightarrow[CH_2Cl_2]{CrO_3(C_5H_5N)_2} CH_3CH_2CH_2CH_2CHO$$

$$CH_3CH_2CH_2CH_2CHO \xrightarrow[NaOH]{2\ CH_2O} CH_3CH_2CH_2\underset{CH_2OH}{\overset{CH_2OH}{C}}-CHO \xrightarrow[NaOH]{CH_2O} CH_3CH_2CH_2\underset{CH_2OH}{\overset{CH_2OH}{C}}-CH_2OH$$

(Cannizzaro) (aldol condensation)

(Note: This sequence illustrates a typical behavior of formaldehyde in the aldol condensation: usually the aldol condensation with this aldehyde is followed by the Cannizzaro reaction to give a molecule with two, three, or four $CH_2OH$ groups.)

(d)

$$HC{\equiv}CH \xrightarrow[2.\ CH_3CH_2CH_2CH_2Br]{1.\ NaNH_2} CH_3CH_2CH_2CH_2C{\equiv}CH \xrightarrow[H_2SO_4,\ HgSO_4]{H_2O} CH_3CH_2CH_2CH_2\overset{O}{\overset{\|}{C}}CH_3$$

(e)

$$C_6H_6 \xrightarrow[\text{AlCl}_3\ \text{Friedel-Crafts}]{CH_3CH_2CH_2CH_2CH_2\overset{O}{\overset{\|}{C}}-Cl} C_6H_5-\overset{O}{\overset{\|}{C}}-CH_2CH_2CH_2CH_2CH_3$$

(Clemmensen reduction) Zn(Hg), HCl

$$\downarrow$$

$$C_6H_5-CH_2CH_2CH_2CH_2CH_2CH_3$$

## CHAPTER 10

10.1 (Strongest) citric acid > lactic acid > ascorbic acid > Barbital (weakest)
At a pH of 1.3 pH units above the acid's p$K_a$ value, the ratio of salt to acid is 95:5. Therefore, the p$K_a$ limit is pH 1.3.

| **Fluid** | **pH** | **p$K_a$ Limit** | **Acids with p$K_a$'s below limit** |
|---|---|---|---|
| beer | 4.5 | 3.2 | citric acid |
| intercellular | 6.1 | 4.8 | citric acid, lactic acid, ascorbic acid |
| cow's milk | 6.6 | 5.3 | citric acid, lactic acid, ascorbic acid |
| intestinal | 8.0 | 6.7 | citric acid, lactic acid, ascorbic acid |

10.2 (d) > (e) > (a) > (b) > (c)

10.3

$$C_6H_5-CH_2CH_3 + 2\ Na_2Cr_2O_7 + 8\ H_2SO_4 \rightarrow C_6H_5-COOH + CO_2 + 10\ H_2O$$

$$+\ 2\ Cr_2(SO_4)_3 + 2\ Na_2SO_4$$

10.4 0.015 1 $\times$ 0.1 equivalent/liter = 0.0015 equivalent
0.228 g = 0.0015 equivalent

$$\frac{0.228 \text{ g}}{0.0015} = 1 \text{ equivalent} = 152 \text{ g} = \text{M.W. of } d,l\text{-mandelic acid}$$

10.5 Addition of aqueous silver nitrate to acetyl bromide would cause silver bromide to precipitate. The bromine of bromoacetic acid is not hydrolyzed this readily.

10.6 The solubilities of acetic anhydride and ethyl acetate in water are nearly the same. However, if a small volume (1 ml) of either compound is shaken with an equal volume of water the acetic anhydride solution will show a strong acid reaction to blue litmus; the ethyl acetate solution will not.

10.7 Solubility in cold dilute (10%) sodium hydroxide solution will distinguish benzoic acid from benzamide.

10.8 $R-\overset{\overset{\large O}{\|}}{C}-Cl + R'_2CuLi \rightarrow R-\overset{\overset{\large O}{\|}}{C}-R'$ (works best with bulky R)

10.10 $$\frac{1.0055 \text{ g/ml} \times 1{,}000 \text{ ml/liter} \times 0.05}{60 \text{ g/mole}} = 0.8378 \text{ mole/liter} = 0.8378\ M$$

10.11

(a) $CH_3(CH_2)_3COOH$ (b) $CH_3(CH_2)_2\underset{\underset{H}{|}}{\overset{\overset{CH_3}{|}}{C}}-\underset{\underset{H}{|}}{\overset{\overset{CH_3}{|}}{C}}-COOH$

(c) $CH_2=\overset{\overset{H}{|}}{C}-CH_2COOH$ (d) $Br-CH_2\overset{\overset{O}{\|}}{C}OC_2H_5$ (e) $CH_3CH_2C(=O)Cl$

(f) 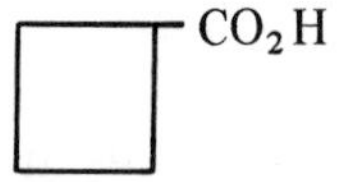 cyclobutane$-CO_2H$ (g) 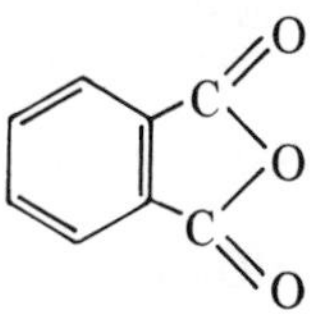 phthalic anhydride (benzene ring fused with C(=O)–O–C(=O))

(h) $CH_3CH_2CH_2C(=O)NH_2$ (i) $(CH_3)_3CCOOH$ (j) 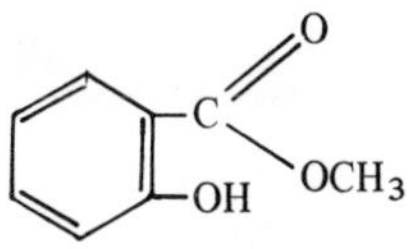 benzene ring bearing $-C(=O)OCH_3$ and $-OH$ on adjacent carbons

(k) $(CH_3-CH_2-\overset{\overset{O}{\|}}{C}-O)_2^-\ Ca^{++}$ (l) $CH_3CH_2CH_2CH_2C\equiv N$

10.12 (a) A = Acetyl chloride; B = Acetamide; C = Acetonitrile

(b) A = Sodium acetate; B = $NH_3$

(c) A = Valeronitrile; B = Valeric acid; C = *n*-Amyl alcohol (1-Pentanol)

(d) A = Ethylmagnesium bromide; B = Propionic acid; C = Propionyl chloride; D = Ethyl propionate; E = *n*-Propyl alcohol; F = Ethyl alcohol.

10.13 (a) 3.54 (b) 4.35 (c) 3.32

10.14 (a) 0.59 (b) $1 \times 10^{-16}$ (c) $1 \times 10^{-17}$ (d) $1.023 \times 10^{-10}$ (e) $1.349 \times 10^{-5}$

10.15 (b) > (a) > (c) > (d).

10.16 (a) Benzamide (b) Propionitrile (Ethyl cyanide) (c) Isopropyl acetate (d) Acetonitrile (e) Bromoform and sodium propionate (f) Sodium acetate and ammonia (g) Acetyl chloride (h) Sodium propionate and ammonia (i) 5-Bromo-2-hydroxybenzoic acid (j) *tert*-Butyl alcohol (k) *p*-Nitrobenzoic acid

10.17

(a) $CaCO_3 \xrightarrow{\text{heat}} CaO + CO_2$; $CaO + 3\,C \rightarrow CaC_2 + CO$

$CaC_2 + H_2O \rightarrow CaO + H{-}C{\equiv}C{-}H$

(b) $H{-}C{\equiv}C{-}H + H_2O \xrightarrow[H_2SO_4]{HgSO_4} CH_3\overset{\overset{H}{|}}{C}{=}O$

(c) $\text{Acetylene} + H_2 \xrightarrow{\text{catalyst}} H_2C{=}CH_2 \xrightarrow{HBr} CH_3CH_2Br$

(d) $CH_3CH_2Br \text{ (from c)} + Mg \xrightarrow{\text{anhy. ether}} C_2H_5MgBr$

$C_2H_5MgBr + \text{product of (b)} \xrightarrow{\text{followed by hydrolysis}} CH_3CH(OH)CH_2CH_3$

(e) $CH_3CHO \text{ (from b)} + I_2 + NaOH \rightarrow CHI_3$

(f) $C_2H_5Br \text{ (from c)} + KCN \rightarrow C_2H_5CN \xrightarrow{\text{hydrolysis}} CH_3CH_2COOH$

(g)

$$H_2C{=}CH_2 \text{ (from c)} \xrightarrow[\text{reductive hydrolysis}]{O_3,} 2\,CH_2O \xrightarrow[\text{heat}]{NaBH_4,} CH_3OH \xrightarrow{HBr} CH_3Br \xrightarrow{Li} CH_3Li$$

$$CH_3COOH \text{ (from K)} + CH_3Li \longrightarrow CH_3{-}\overset{\overset{O}{\|}}{C}{-}CH_3$$

(h) $(CH_3)_2C{=}O$ (from g) + $NaBH_4$ $\xrightarrow{\text{heat}}$ $(CH_3)_2CHOH$

(i) $(CH_3)_2CHOH + HBr \rightarrow CH_3CH(Br)CH_3$

(j) $CH_3CH(OH)CH_3$ (from h) + $H_2SO_4$ $\xrightarrow{\text{heat}}$ $CH_3CH{=}CH_2$

$CH_3CH{=}CH_2 + HBr \xrightarrow{\text{peroxide}} CH_3CH_2CH_2Br$

(k) $CH_3CHO + Cu^{2+} \rightarrow CH_3COOH$

(l) $CH_3COOH$ (from k) + $SOCl_2$ $\rightarrow$ $CH_3COCl$

$+ \longrightarrow (CH_3C(=O))_2O$

$CH_3COOH + NaOH \rightarrow CH_3COONa$

(m)

$(CH_3)_2CHOH$ (from h) + $CH_3COOH$ $\xrightarrow{H^+}$ $CH_3C(=O){-}OCH(CH_3)_2$

(n) $(CH_3)_2CHBr$ (from i) + Mg $\xrightarrow{\text{anhy. ether}}$ $(CH_3)_2CHMgBr$

$(CH_3)_2CHMgBr + CO_2 \xrightarrow{\text{followed by hydrolysis}} (CH_3)_2CH{-}C(=O)OH$

10.18 Dissolve the mixture in ether and extract with cold 10% sodium hydroxide. Separate and warm the alkaline solution on a steam bath to remove dissolved ether. Acidify the alkaline solution with $H_2SO_4$ and distill.

10.19

$$CH_3{-}\overset{O}{\overset{\|}{C}}{-}OH \xrightarrow{LiAlH_4} CH_3CH_2OH \xrightarrow{HBr} CH_3CH_2Br \xrightarrow[2.\ CO_2]{1.\ Mg} CH_3CH_2{-}\overset{O}{\overset{\|}{C}}{-}OH$$

$$CH_3CH_2Br \xrightarrow[2.\ H_2O\ +\ HCl]{1.\ NaCN} CH_3CH_2{-}\overset{O}{\overset{\|}{C}}{-}OH$$

Caution: The route through the nitrile works only with good $S_N2$ halides; whereas the Grignard route is fairly general.

10.20 (a)

Toluene ($C_6H_5CH_3$) $\xrightarrow[FeBr_3]{Br_2}$ p-bromotoluene ($CH_3$ and Br para on the benzene ring)

(b) p-Bromotoluene (ring with $CH_3$ and Br) → 1. Mg, 2. $CO_2$, 3. hydrolysis → p-toluic acid (ring with $CH_3$ and $CO_2H$)

(c) p-Bromotoluene (ring with $CH_3$ and Br) → $Na_2Cr_2O_7$, $H_2SO_4$ → ring with $CO_2H$ and Br → 1. $SOCl_2$, 2. $NH_3$ → ring with $CONH_2$ and Br

(d) p-Bromotoluene (ring with $CH_3$ and Br) → $Br_2$, $h\nu$ → ring with $CH_2Br$ and Br → 1. Mg, 2. $CO_2$, 3. hydrolysis → ring with $CH_2CO_2H$ and Br

(e) Ring with $CH_3$ and $CO_2H$ → $Na_2Cr_2O_7$, $H_2SO_4$ → ring with $CO_2H$ and $CO_2H$

from (b)

(f)

Toluene (ring with $CH_3$) → $Na_2Cr_2O_7$, $H_2SO_4$ → benzoic acid (ring with $CO_2H$)

Toluene (ring with $CH_3$) → $Br_2$, $h\nu$ → ring with $CH_2Br$ → Li → ring with $CH_2Li$

Both products together → care → $C_6H_5$–C(=O)–$CH_2$–$C_6H_5$

10.21 $\dfrac{\text{Wt}}{\text{MW}} = \dfrac{M \times \text{ml}}{1000}$ = number of moles

$$\text{MW} = \frac{\text{Wt} \times 1000}{N \times \text{ml}} = \frac{0.2412 \times 1000}{0.098 \times 12.4}$$

$$= 199$$

Wt = weight of sample in grams
MW = molecular weight or formula weight (FW)
$M$ = molarity or normality ($N$) or formality ($F$)
ml = volume in ml

MW of *o*-bromobenzoic acid = 201.03

MW of 2,4-dibromobenzoic acid = 279.93

Therefore, acid is *o*-bromobenzoic acid.

10.22 (A)

$$CH_3-\underset{CH_3}{\overset{CH_3}{C}}-\underset{H}{\overset{OH}{C}}-CH_3$$

3,3-Dimethyl-2-butanol

(B)

$$CH_3-\underset{CH_3}{\overset{CH_3}{C}}-CO_2Na$$

Sodium 2,2-dimethylpropanoate
Sodium $\alpha,\alpha$-dimethylpropionate

(C)

$$(CH_3)_2C=C(CH_3)_2$$

2,3-Dimethyl-2-butene

(D)

$$CH_3-\underset{Br}{\overset{CH_3}{C}}-\underset{Br}{\overset{CH_3}{C}}-CH_3$$

2,3-Dibromo-2,3-dimethylbutane

Note: See Sec. 8.8.B for explanation of the formation of 2,3-dimethyl-2-butene

(E)

$$CH_3-\overset{O}{\overset{\|}{C}}-CH_3$$

(F)

$$CH_3CO_2Na$$

10.23 A = Acetic acid; B = Isopropyl alcohol. Original compound was isopropyl acetate,

$$CH_3C(=O)OCH(CH_3)_2$$

10.24

$$(CH_3)_2CH-C(=O)-OCH(CH_3)_2$$, Isopropyl isobutyrate.

10.25

A = $C_6H_5-C(=O)-O-C_2H_5$ or $C_6H_5-CH_2-O-C(=O)-CH_3$ or

$C_6H_5-CH_2-C(=O)-O-CH_3$

B = $C_6H_4(COOH)(C_2H_5)$ or $C_6H_4(CH_2CO_2H)(CH_3)$ C = $C_6H_4(CH_3)(CH_2-O-C(=O)-H)$

D = $C_6H_5-CH_2CH_2COOH$ or $C_6H_5-CH(CH_3)COOH$

10.26

$C_6H_5-CH_2-O-C(=O)-CH_3$

$C_6H_5$: 7.29; $CH_2$: 5.05; $CH_3$: 2.00

10.27

$(CH_3)_3C-CH_2-C(=O)-OH$

$CH_3$: 1.08; $CH_2$: 2.20; OH: 11.24

Note: Acids sometimes show ir bands for both the monomer and the dimer.

10.28

$CH_3-CBr(CH_3)-C(=O)-O-CH_2-CH_3$

$CH_3$ (on C–Br): 1.92; $CH_2$: 4.20 (q); $CH_3$ (ethyl): 1.19 (t)

10.29 (a)

$$C_6H_{10}{=}CH_2 \xrightarrow[\text{peroxide}]{HBr} C_6H_{11}-CH_2-Br \xrightarrow[2.\ CO_2]{1.\ Mg} C_6H_{11}-CH_2-CO_2H$$

(b)

$$HO_2C-(CH_2)_5-CO_2H \xrightarrow[HCl]{CH_3CH_2OH} CH_2\begin{cases} CH_2-CH_2-C(=O)-O-CH_2CH_3 \\ CH_2-CH_2-CO_2CH_2CH_3 \end{cases} \xrightarrow{NaOC_2H_5} \text{(cyclohexanone with } CH_3CH_2O_2C \text{ substituent)}$$

Note: This is an intramolecular Claisen condensation and is called the Dieckmann condensation (or cyclization).

(c)

$$CH_3-CH=C(CH_3)_2 \xrightarrow{HBr} CH_3-CH_2-C(CH_3)_2-Br \xrightarrow[2.\ CO_2]{1.\ Mg} CH_3-CH_2-C(CH_3)_2-CO_2H$$

$$CH_3-CH_2-C(CH_3)_2-CO_2H \xrightarrow[2.\ (CH_3)_2CuLi]{1.\ SOCl_2} CH_3-CH_2-C(CH_3)_2-\overset{O}{\overset{\|}{C}}-CH_3$$

$$CH_3-CH_2-C(CH_3)_2-CO_2H \xrightarrow{CH_3Li} CH_3-CH_2-C(CH_3)_2-\overset{O}{\overset{\|}{C}}-CH_3$$

Note: Since the intermediate alkyl bromide is tertiary and will not undergo $S_N2$ reactions, it cannot be converted into a nitrile. Therefore, the Grignard route must be used.

(d)

$$CH_3CH_2CH_2CH_2CO_2H \xrightarrow{LiAlH_4} CH_3CH_2CH_2CH_2CH_2OH \xrightarrow[H_3PO_4]{-H_2O} CH_3CH_2CH_2CH=CH_2$$

$$CH_3CH_2CH_2CH=CH_2 \xrightarrow[\text{(hot)}]{KMnO_4} CH_3CH_2CH_2CO_2H$$

## CHAPTER 11

11.1 Four: *cis–cis, cis–trans, trans–cis, trans-trans.*

11.3 A soap could be made from butter as well as from other animal fats. However, approximately 5% of butter is butyric acid in the form of the glyceryl ester and should hydrolysis of sodium butyrate (a component in the soap) occur, the unpleasant odor of butyric acid would be imparted to the sodium salts of the higher acids.

11.4 (a) Lauryl cerotate (b) Sodium palmitate (c) Glyceryl myristopalmitooleate, (d) Triolein (e) $PGF_{2\beta}$ 16,16-dimethyl

11.5 (a) $CH_3(CH_2)_7-CH=CH-(CH_2)_7COOH$; (b)

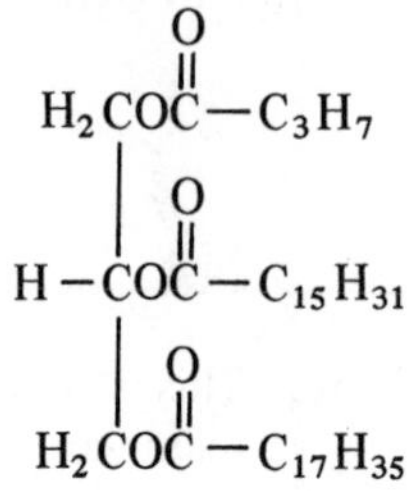

(c) $CH_3(CH_2)_nCO_2^-Na^+$ where $n$ = 12-16

(d)

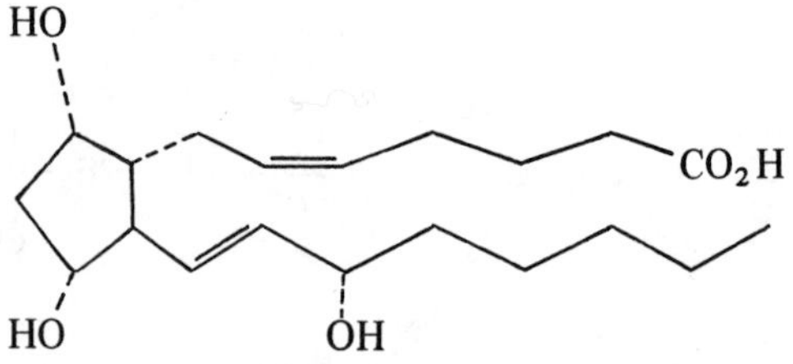

(e) $CH_3(CH_2)_{11}-OSO_3^-\ Na^+$

(f)

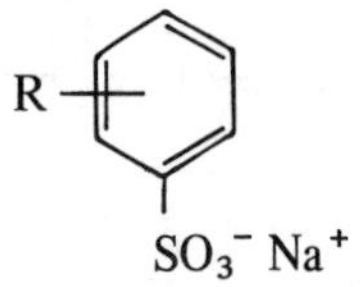

where R $\geq$ 12

(g) $R\overset{+}{N}(CH_3)_3\overset{-}{Cl}$, where R $\geq$ 16

(h)

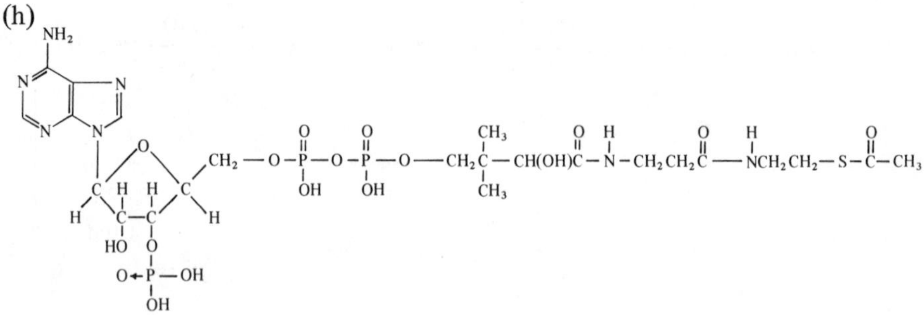

11.6 A formula may be drawn for any one of the five glycerides in this exercise by varying the nature of R, R′, and R″ as indicated.

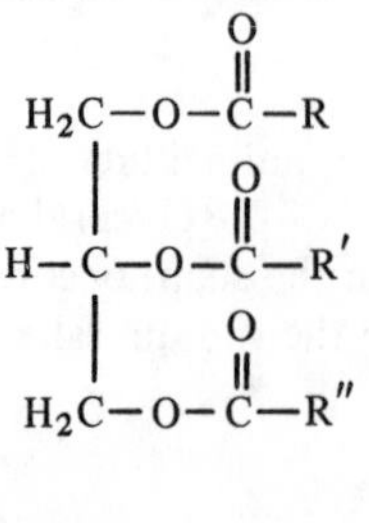

where R = $C_3H_7-$

R′ = $C_{13}H_{27}-$, $C_{15}H_{31}-$, or $C_{17}H_{35}-$

R″ = $C_{17}H_{33}-$, $C_{17}H_{31}-$, or $C_{17}H_{29}-$

Butter and tallow would vary primarily in the amount of stearic acid present. Linseed and safflower oils would vary in the amounts of oleic, linoleic, and linolenic acids esterified, and tung oil would contain one or more eleostearic acid residues.

11.7 This problem may have more than one answer. However, from the percentage value given in Table 11.1 one may assume that a molecule of linseed oil is sufficiently unsaturated to supply four or five double bonds (one linoleic and one linolenic) and to have an iodine value of not less than 170. The molecular weight of the oil would then be given by

$$\text{M.W} = \frac{10 \times 127 \times 100}{170} = 750$$

Its saponification number could then be calculated from

$$\text{Sap. No.} = \frac{168{,}000}{750} = 224$$

11.8 Let us begin this exercise by drawing a general structure usable for all four reactions.

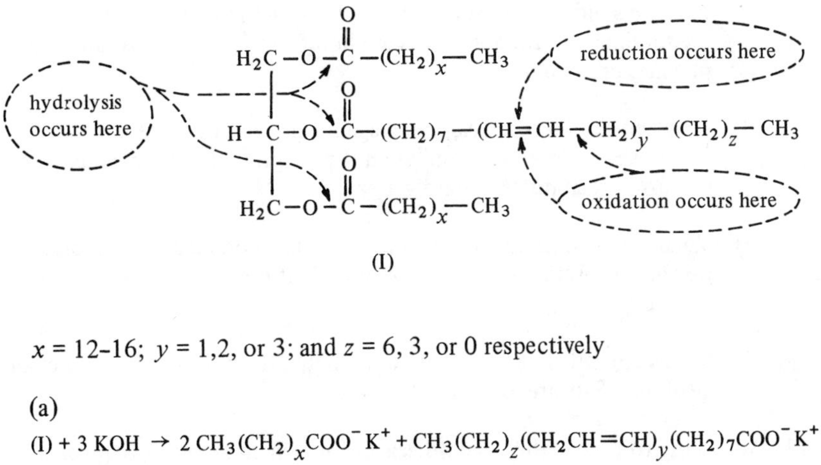

$$\begin{array}{l} \quad\quad\quad\quad\quad\ \ O \\ \quad\quad\quad\quad\quad\ \ \| \\ H_2C-O-C-(CH_2)_x-CH_3 \\ \ \ \ |\quad\quad\quad\ O \\ \ \ \ |\quad\quad\quad\ \| \\ H-C-O-C-(CH_2)_7-(CH{=}CH-CH_2)_y-(CH_2)_z-CH_3 \\ \ \ \ |\quad\quad\quad\ O \\ \ \ \ |\quad\quad\quad\ \| \\ H_2C-O-C-(CH_2)_x-CH_3 \end{array}$$

hydrolysis occurs here (at the ester C–O bonds); reduction occurs here (at the C=C double bond); oxidation occurs here (at the C=C double bond / adjacent CH₂)

(I)

$x$ = 12–16; $y$ = 1,2, or 3; and $z$ = 6, 3, or 0 respectively

(a)

$$\text{(I)} + 3\ KOH \rightarrow 2\ CH_3(CH_2)_xCOO^-K^+ + CH_3(CH_2)_z(CH_2CH{=}CH)_y(CH_2)_7COO^-K^+$$

$$+\ CH_2(OH)-CH(OH)-CH_2(OH)$$

(b)

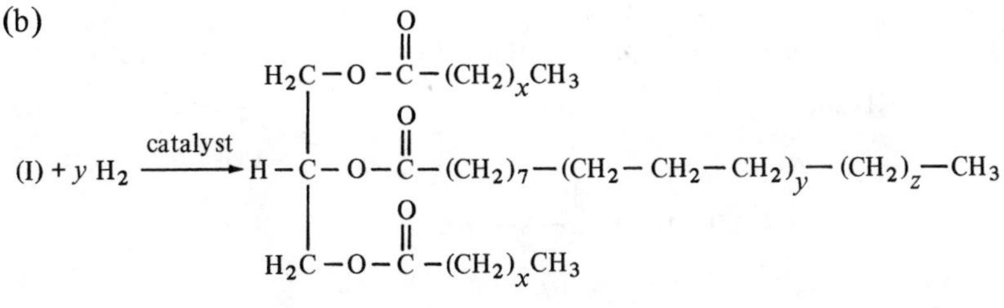

$$\text{(I)} + y\ H_2 \xrightarrow{\text{catalyst}} \begin{array}{l} \quad\quad\quad\quad\quad\ \ O \\ \quad\quad\quad\quad\quad\ \ \| \\ H_2C-O-C-(CH_2)_xCH_3 \\ \ \ \ |\quad\quad\quad\ O \\ \ \ \ |\quad\quad\quad\ \| \\ H-C-O-C-(CH_2)_7-(CH_2-CH_2-CH_2)_y-(CH_2)_z-CH_3 \\ \ \ \ |\quad\quad\quad\ O \\ \ \ \ |\quad\quad\quad\ \| \\ H_2C-O-C-(CH_2)_xCH_3 \end{array}$$

(c) In the drying process only the $—CH{=}CH—CH_2—$ segments in the oil molecule are involved. (See Sec. 11.7.)

(d) The rancidity of butter is due to two reactions: (a) hydrolysis and (b) oxidation. If $x$ = 2,4, or 6, the lower molecular weight carboxylic acids are released on hydrolysis to impart an unpleasant odor to butter. If $y$ = 2 and $z$ = 3, caproic acid may be formed through oxidation of the double bond.

11.9 A grease spot on linen may be removed through the mechanics of soap action. Woolen flannel under the same conditions would probably shrink. Removal of a grease spot from wool by simple solution in an organic solvent would be preferred.

11.10 Most deep well water contains iron, magnesium, and calcium as soluble salts. Such water is commonly referred to as "hard" water. These metallic ions form insoluble salts with the long chain fatty acid anions.

11.11 (a) Saponification is a term specifically applied to the hydrolysis of an ester when the action is carried out in alkaline solution.

(b) Unsaturated oils are used as drying oils. They undergo an oxygen-induced polymerization leading to a hard, dry, tough film.

(c) Fats are solids or semisolids in which the acid portions of the glyceride are predominately saturated. Oils are liquids in which unsaturated acid components predominate.

(d) Soaps and syndets are cleansing agents classified as detergents. Syndets are synthetic detergents and are not precipitated in hard water. Soaps do form precipitates in hard water.

(e) Volatile acids, leading to rancidity, can be produced by hydrolysis reactions (hydrolytic rancidity) and by oxidation reactions (oxidative rancidity).

(f) Waxes are esters of long-chain unbranched fatty acids and long-chain alcohols. Fats are esters of glycerol.

11.12 According to theory the biosynthesis of long chain fatty acids appears to take place in 2-carbon increments via acetyl coenzyme A and malonyl coenzyme A. The building block is the acetic acid unit.

$$\underset{\text{Acetyl CoA}}{CH_3\overset{O}{\overset{\|}{C}}—S—CoA} \xrightarrow{CO_2} \underset{\text{Malonyl CoA}}{HO\overset{O}{\overset{\|}{C}}—CH_2\overset{O}{\overset{\|}{C}}—S—CoA} \xrightarrow{\text{acetyl CoA}} HO\overset{O}{\overset{\|}{C}}CH\overset{O}{\overset{\|}{C}}SCoA \text{ (with } —C{=}O—CH_3 \text{ on CH)}$$

$$\xrightarrow[-(CO_2)]{H_2} \underset{\text{Butyryl CoA}}{CH_3CH_2CH_2\overset{O}{\overset{\|}{C}}—S—CoA}$$

11.13 A saponification number of 200 indicates a relatively low molecular weight and an iodine value of 30 indicates a low degree of unsaturation. A fat with these values is likely to be one with a high percent of the shorter chain acid residues and a low percent of the polyunsaturated fatty acid residues. Palmitic acid would probably represent most of the long chain saturated components and oleic most of the long chain unsaturated components. This fat could be part of the butter mixture.

11.14

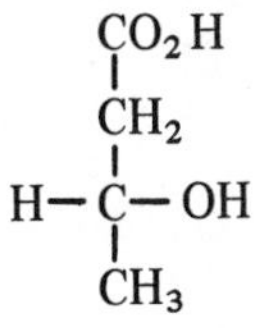

$$\begin{array}{ccc} CO_2H & & CO_2H \\ | & & | \\ CH_2 & & CH_2 \\ | & & | \\ H-C-OH & & OH-C-H \\ | & & | \\ CH_3 & & CH_3 \\ \\ \text{D-} & & \text{L-} \end{array}$$

$\beta$-Hydroxybutyric acid

11.15 HO

$CO_2H$

HO

HO

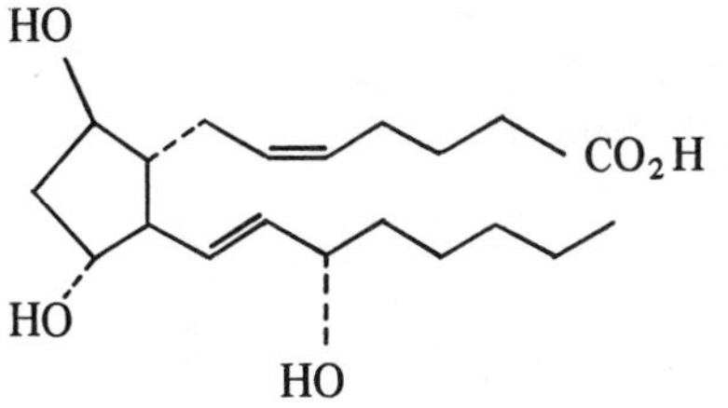

The two OH-groups on the cyclopentane ring are *trans*; in $PGF_{2\alpha}$ they are *cis*.

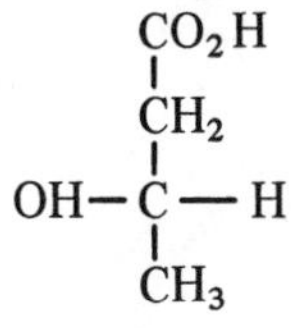

11.16

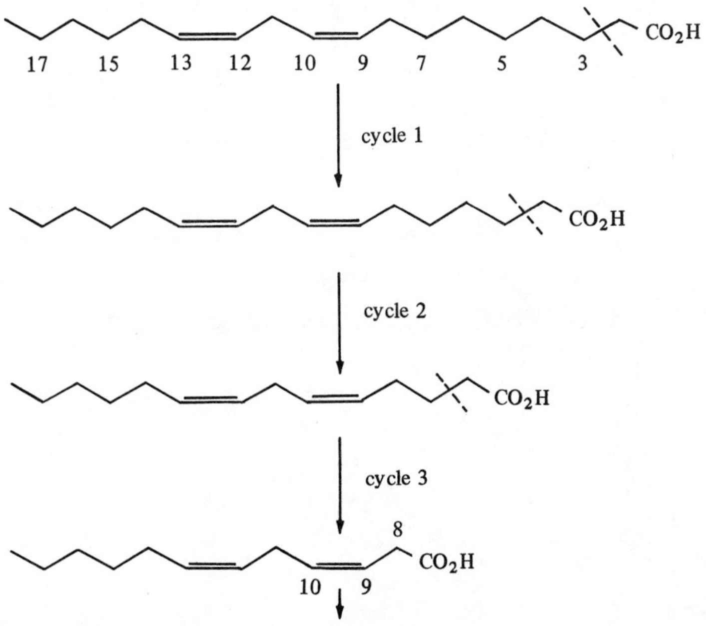

$CO_2H$

17 15 13 12 10 9 7 5 3

cycle 1

$CO_2H$

cycle 2

$CO_2H$

cycle 3

8

$CO_2H$

10 9

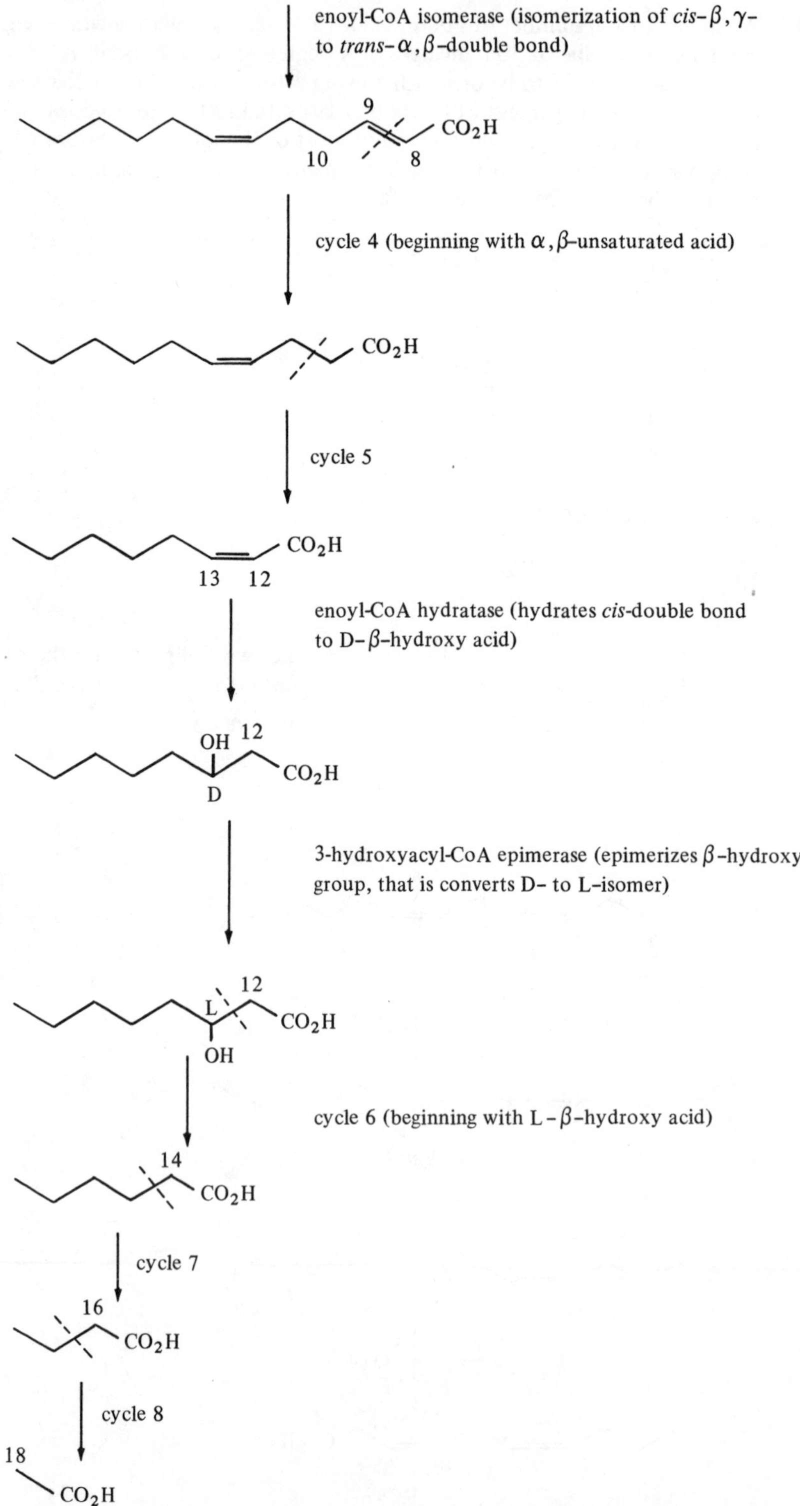
enoyl-CoA isomerase (isomerization of *cis*-β,γ- to *trans*-α,β-double bond)
9
$CO_2H$
10
8
cycle 4 (beginning with α,β-unsaturated acid)
$CO_2H$
cycle 5
$CO_2H$
13
12
enoyl-CoA hydratase (hydrates *cis*-double bond to D-β-hydroxy acid)
OH
12
$CO_2H$
D
3-hydroxyacyl-CoA epimerase (epimerizes β-hydroxy group, that is converts D- to L-isomer)
12
L
$CO_2H$
OH
cycle 6 (beginning with L-β-hydroxy acid)
14
$CO_2H$
cycle 7
16
$CO_2H$
cycle 8
18
$CO_2H$

## CHAPTER 12

12.1 Grignard reagents react with water, alcohols, ammonia, acetylene, and carboxylic acids or any other compound that can provide an acidic hydrogen. Inasmuch as the carboxyl group is part of the halogen compound in this case, a Grignard reagent would not form.

12.2 (a)

$$CH_2(CO_2C_2H_5)_2 \xrightarrow[2.\ CH_3Br]{1.\ NaOC_2H_5} CH_3{-}CH(CO_2C_2H_5)_2 \xrightarrow[2.\ -CO_2]{1.\ H_2O/HCl} CH_3CH_2CO_2H$$

(b)

$$CH_3{-}CH(CO_2C_2H_5)_2 \xrightarrow[2.\ CH_3Br]{1.\ NaOC_2H_5} CH_3{-}C(CH_3)(CO_2C_2H_5)_2 \xrightarrow[2.\ -CO_2]{1.\ H_2O/HCl} (CH_3)_2CH{-}CO_2H$$

(c)

$$CH_2(CO_2C_2H_5)_2 \xrightarrow[2.\ Cl{-}CH_2CO_2C_2H_5]{1.\ NaOC_2H_5} C_2H_5O_2CCH_2{-}CH(CO_2C_2H_5)_2$$

$$\xrightarrow[2.\ -CO_2]{1.\ H_2O/HCl} HO_2C{-}CH_2CH_2{-}CO_2H$$

12.3 In the decarboxylation of malonic acid the enol of acetic acid is formed. In the decarboxylation of acetoacetic acid the enol of acetone is formed. The enol of acetone, a ketone, would be expected to be more stable than the enol of acetic acid, since enols of ketones are more easily formed. Recall that it is necessary to convert acids into acyl halides before they can be halogenated, because acyl halides form enols more readily than acids (Sec. 10.5); thus, acids do not form enols readily. The formation of stable products often affords driving force to a reaction by lowering the activation energy and causing the reaction to go faster.

12.4 Sodioacetoacetic ester (a base) would probably cause a dehydrohalogenation of a tertiary alkyl halide to produce an olefin. Elimination thus would result rather than the intended displacement.

12.5 (a) $Cl{-}\overset{O}{\overset{\|}{C}}{-}(CH)_4{-}\overset{O}{\overset{\|}{C}}{-}Cl$ (b) $CH_3O\overset{O}{\overset{\|}{C}}CH_2CH_2\overset{O}{\overset{\|}{C}}OCH_3$

(c) COOH

COOH

(d) $CH_3CH_2CH(COOH)_2$

(e) $NCCH_2\overset{\overset{O}{\|}}{C}OC_2H_5$

(f) $H_2$ C $H_2C$ $CH_2$ $H_2C$ C O O

(g) $HOOC(CH_2)_2-\overset{\overset{CH_3}{|}}{\underset{\underset{H}{|}}{C}}COOH$

(h) $O-\overset{\overset{O}{\|}}{C}-CH_3$ COOH

12.6 (a) Mandelic acid, $\alpha$-Hydroxyphenylacetic acid

(b) Chloroacetic acid, Chloroethanoic acid

(c) Fumaric acid, *trans*-Butenedioic acid

(d) Acrylic acid, Propenoic acid

(e) $\gamma$-Hydroxyvaleric acid, 4-Hydroxypentanoic acid

(f) Succinic anhydride

(g) $\gamma$-Valerolactone

(h) *cis*-3-Hydroxycyclohexanecarboxylic acid

(i) $\beta$, $\beta$-Dimethyladipic acid, 3,3-Dimethylhexanedioic acid

12.7 (a) $A = HO-CH_2-CH_2-Cl$

Ethylene chlorohydrin

$B = Cl-CH_2-CO_2H$

Chloroacetic acid

$C = NC-CH_2-CO_2^-Na^+$

Sodium cyanoacetate

$D = HO_2C-CH_2-CO_2H$

Malonic acid

(b)

A = $C_6H_5-O^- Na^+$

Sodium phenoxide

B = $C_6H_4(OH)(CO_2^- Na^+)$ (OH and $CO_2^- Na^+$ ortho)

Sodium salicylate

C = $C_6H_4(OH)(CO_2H)$ (OH and $CO_2H$ ortho)

Salicylic acid

D = $C_6H_4(O-\overset{O}{\overset{\|}{C}}-CH_3)(CO_2H)$ (ortho)

Acetylsalicylic acid (Aspirin)

(c)

$$A = CH_3-\underset{OH}{\overset{H}{C}}-CN$$

Acetaldehyde cyanohydrin

$$B = CH_3-\underset{OH}{CH}-CO_2H$$

Lactic acid

C = cyclic dimer: six-membered ring O–C(=O)–CH($CH_3$)–O–C(=O)–CH($CH_3$)– ($H_3C$, H on one ring carbon; H, $CH_3$ on the other; two C=O; two ring O)

Lactide

(d) $$A = CH_3-CH_2-\underset{Br}{CH}-CO_2H$$

α-Bromobutyric acid

$$B = CH_3-CH{=}CH-CO_2H$$

Crotonic acid

$$C = CH_3-\underset{Br}{CH}-CH_2-CO_2H$$

β-Bromobutyric acid

(e) A =

$$C_6H_5-O-CH_2-CH_2-\underset{CO_2C_2H_5}{\overset{CO_2C_2H_5}{CH}}$$

Ethyl β-phenoxyethyl malonate

B = $C_6H_5-OH$

Phenol

$C = Br-CH_2-CH_2-CH_2-CO_2H$

γ-Bromobutyric acid

(f) $A = HO_2C(CH_2)_7-\overset{\overset{\displaystyle O}{\|}}{C}-(CH_2)_7CO_2H$

9-Oxoheptadecanedioic acid

(g)

A = 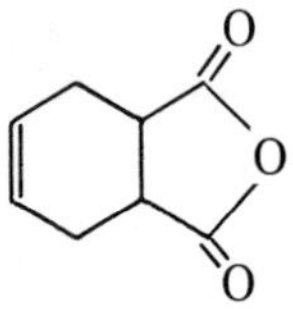

B = 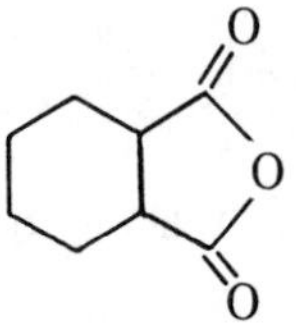

*cis*-1,2,3,6-Tetrahydrophthalic anhydride

Hexahydrophthalic anhydride

12.8 (e) < (a) < (b) < (d) < (c)

12.9

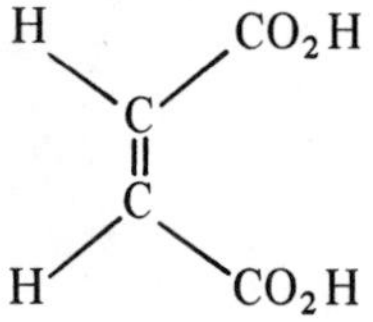

Maleic acid

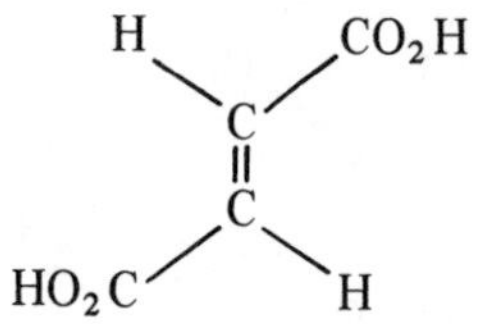

Fumaric acid

The first p$K_a$ for maleic acid is smaller (that is, the acid is stronger) than that for fumaric acid because the ion formed can be stabilized by intramolecular H-bonding. Furthermore, in all dibasic acids the

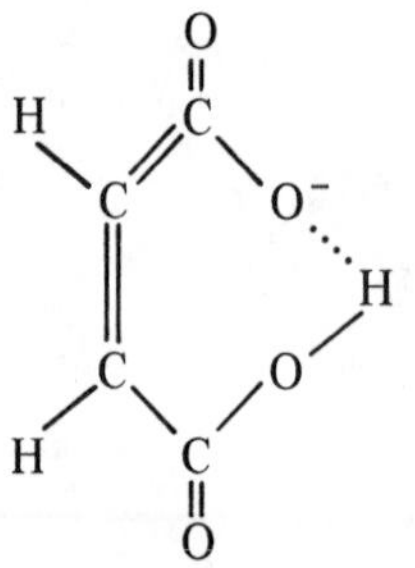

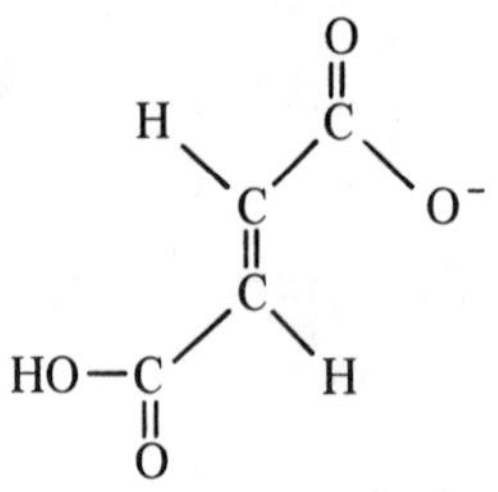

first carboxyl group tends to be a stronger acid than the simple aliphatic acids because the electron-withdrawing effect of the second carboxyl group tends to stabilize the carboxylate anion formed by the ionization of the first carboxyl group. The closer the first carboxyl group is to the second, the greater this electrostatic stabilization will be. In maleic acid the two carboxyl groups are much closer than they are in fumaric acid. In the second ionization step the

H-bond must be broken before maleic acid can form the dianion; thus, the extra stabilization of the H-bond is lost and the loss of the second proton is less favorable than it would be in fumaric acid if all other things were equal. perhaps even more important is the usual acid-weakening effect observed for the second carboxyl group of most dibasic acids caused by the electrostatic repulsion between the two negatively charged centers in the dianion. The two negatively charged carboxylate groups are much closer in the maleate dianion than they are in the fumarate dianion.

$HO_2C$ ⁄\⁄\⁄\ $CO_2^-$    $^-O_2C$ ⁄\⁄\⁄\ $CO_2^-$

Acid strengthening electrostatic effect    Acid weakening electrostatic effect

12.10 (b), (d), and (e) could exist as *cis-trans* isomers; (c), (d), and (e) could exist as optical isomers.

12.11 (a) Aspirin (acetylsalicylic acid) is a substituted aromatic acid, phenyl salicylate is a substituted phenol. Solubility in 10% sodium carbonate solution should distinguish between the two. (b) Acetyl chloride will precipitate silver chloride when treated with alcoholic silver nitrate; chloroacetic acid will not. (c) Maleic acid should decolorize a $Br_2/CCl_4$ solution.

12.12

$$CH_3CH_2OH \xrightarrow{[O]} \underset{(a)}{CH_3COOH} \xrightarrow{C_2H_5OH,\ H^+} \underset{(b)}{CH_3COOC_2H_5}$$

$$(b) + \xrightarrow{C_2H_5O^-\ Na^+\ \text{(from ethyl alcohol + sodium)}} Na^+\ :\overset{-}{C}H_2COOC_2H_5$$

$$Na^+\ :\overset{-}{C}H_2COOC_2H_5 + (b) \rightarrow (c)$$

$$C_2H_5OH \xrightarrow[\text{heat}]{Cu} CH_3CHO \xrightarrow{10\%\ NaOH} \underset{\text{(Aldol)}}{CH_3CH(OH)CH_2CHO}$$

$$\text{Aldol} \xrightarrow{Cu^{2+}} (d) \xrightarrow{\text{heat}} (e)$$

$$(a) + P + Br_2 \rightarrow BrCH_2COOH \xrightarrow{Na^+\ OH^-} BrCH_2COO^-\ Na^+$$

$$BrCH_2COONa \xrightarrow{KCN} NCCH_2COONa \xrightarrow[\text{heat}]{H^+,\ H_2O} (f)$$

$$C_2H_5OH + PBr_3 \rightarrow C_2H_5Br$$

$$(c) \xrightarrow[2.\ C_2H_5Br]{1.\ C_2H_5ONa} CH_3\overset{O}{\overset{\|}{C}}CH(C_2H_5)\overset{O}{\overset{\|}{C}}OC_2H_5 \xrightarrow[\text{heat}]{H^+,\ H_2O} (g) + C_2H_5OH + CO_2$$

$$(e) \xrightarrow{H_2,\ Ni} (h)$$

12.13 $CH_3CH(OH)CH_2COOH$, $\beta$-Hydroxybutyric acid.

(a) $CH_3CH(OH)CH_2COOH + H_2O \rightleftharpoons CH_3CH(OH)CH_2COO^- + H_3O^+$

(b)

$$3\ CH_3CH(OH)CH_2COOH + Na_2Cr_2O_7 + 4\ H_2SO_4 \xrightarrow{heat} 3\ CH_3\overset{\overset{O}{\parallel}}{C}-CH_2COOH + Na_2SO_4 + Cr_2(SO_4)_3 + 7H_2O$$

(blue-green)

(c) $CH_3CH(OH)CH_2COOH \xrightarrow{heat} CH_3-CH=CH-COOH$

(d) $(C) + \text{dilute } KMnO_4 \longrightarrow CH_3-\underset{OH}{CH}-\underset{OH}{CH}-COOH$

(e) $CH_3CH(OH)CH_2COOH + I_2 + NaOH \rightarrow {}^+Na^-OOC-CH_2-COO^-Na^+ + CHI_3$

(yellow solid)

12.14

$$H-C\equiv C-\underset{OH}{\overset{H}{C}}-CO_2H$$

12.15 (a)

$$2\ CH_2(CO_2C_2H_5)_2 \xrightarrow[2.\ ICH_2I]{1.\ NaOC_2H_5} (C_2H_5O_2C)_2CH-CH_2-CH(CO_2C_2H_5)_2$$

$$(C_2H_5O_2C)_2CH-CH_2-CH(CO_2C_2H_5)_2 \xrightarrow[2.\ -CO_2]{1.\ H_2O/HCl} HO_2C-CH_2-CH_2-CH_2-CO_2H$$

or

$$CH_2(CO_2C_2H_5)_2 \xrightarrow[\substack{2.\ CH_2=CHCO_2C_2H_5\ (1,4\text{-addition})^*\\ \text{or}\\ BrCH_2CH_2CO_2C_2H_5\\ 3.\ H_2O/HCl;\ -CO_2}]{1.\ NaOC_2H_5} HO_2C-CH_2-CH_2-CH_2-CO_2H$$

*Carbon anions (carbanions, Grignard reagents, cyanide ion) add 1,4.

(b)

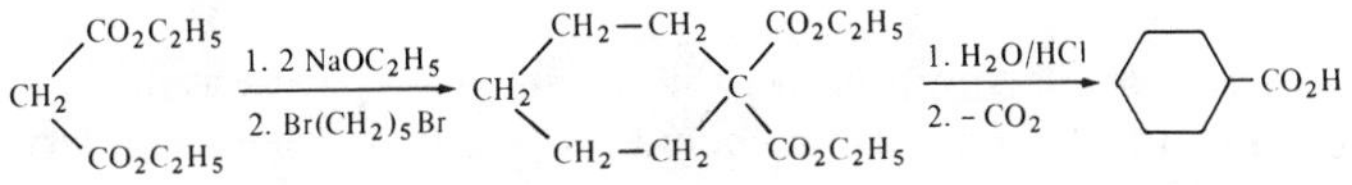

$$CH_2(CO_2C_2H_5)_2 \xrightarrow[2.\ Br(CH_2)_5Br]{1.\ 2\ NaOC_2H_5} \text{(cyclohexane-1,1-}(CO_2C_2H_5)_2\text{: } CH_2(CH_2{-}CH_2)_2C(CO_2C_2H_5)_2) \xrightarrow[2.\ -CO_2]{1.\ H_2O/HCl} \text{cyclohexane}{-}CO_2H$$

(c)

$$CH_2(CO_2C_2H_5)_2 \xrightarrow[2.\ CH_2{=}CHCH_2Br]{1.\ NaOC_2H_5} CH_2{=}CHCH_2CH(CO_2C_2H_5)_2 \xrightarrow[2.\ -CO_2]{1.\ H_2O/HCl} CH_2{=}CHCH_2CH_2CO_2H$$

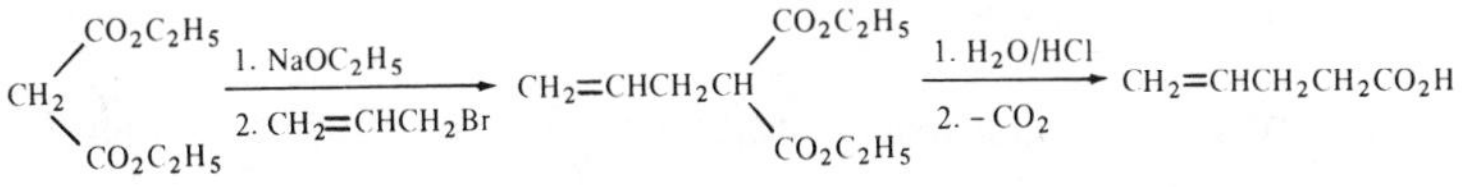

(d)

$$CH_2(COCH_3)(CO_2C_2H_3) \xrightarrow[2.\ C_6H_5CH_2Br]{1.\ NaOC_2H_5} C_6H_5{-}CH_2{-}CH(COCH_3)(CO_2C_2H_5) \xrightarrow[2.\ -CO_2]{1.\ H_2O/HCl} C_6H_5{-}CH_2CH_2\overset{O}{\overset{\|}{C}}CH_3$$

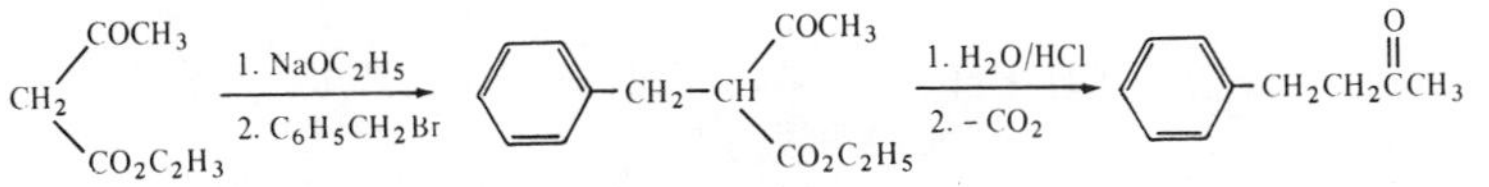

(e)

$$CH_2(COCH_3)(CO_2C_2H_5) \xrightarrow[2.\ CH_2{=}CHCO_2C_2H_5\ (1{,}4\text{-addition})\ \text{or}\ BrCH_2CH_2CO_2C_2H_5]{1.\ NaOC_2H_5} C_2H_5O_2CCH_2CH_2CH(COCH_3)(CO_2C_2H_5) \xrightarrow[2.\ -CO_2]{1.\ H_2O/HCl} HO_2CCH_2CH_2CH_2\overset{O}{\overset{\|}{C}}CH_3$$

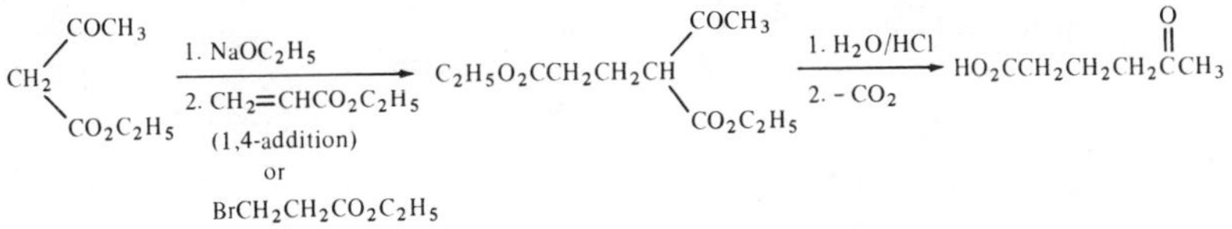

12.16

$$CH_2(CO_2C_2H_5)_2 \xrightarrow{NaOC_2H_5} {}^{-}{:}CH(CO_2C_2H_5)_2 \xrightarrow{CH_2{-}CH_2\ (\text{epoxide},\ {:}\ddot{O}{:})} (C_2H_5O_2C)CH(CH_2CH_2\ddot{O}{:}^{-})C({=}O)OC_2H_5$$

$\downarrow$ addition

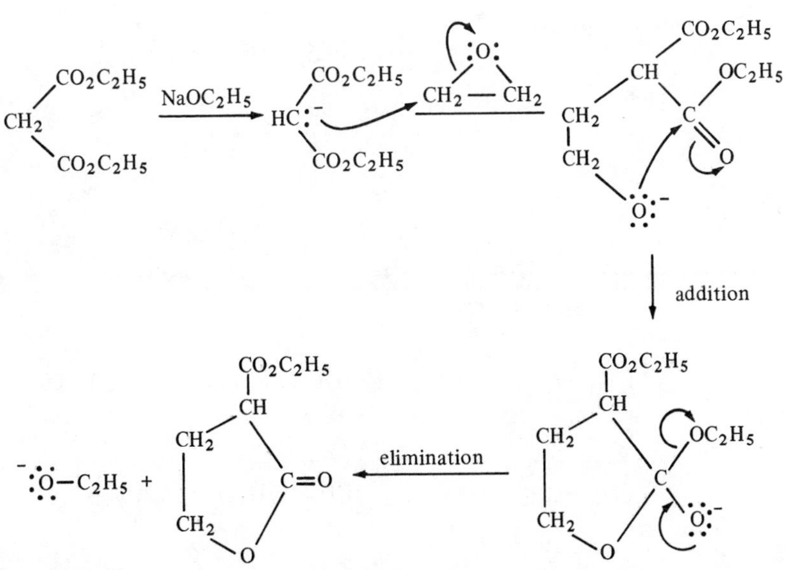

$$\text{tetrahedral intermediate: } (C_2H_5O_2C)CH{-}CH_2{-}CH_2{-}O{-}C(OC_2H_5)({:}\ddot{O}{:}^{-}) \xrightarrow{\text{elimination}} {}^{-}{:}\ddot{O}{-}C_2H_5 + \text{lactone } (C_2H_5O_2C)CH{-}CH_2{-}CH_2{-}O{-}C{=}O$$

## CHAPTER 13

13.1 Four alcohols are permitted for the formula $C_4H_{10}O$ or as many alcohols as there are butyl groups. These include two primary, one secondary, and one tertiary alcohol.

In addition to the *four* primary amines afforded by the four different butyl groups, there are also *three* secondary amines, and *one* tertiary amine of formula $C_4H_{11}N$.

13.2 (a) Allyl chloride is converted to allyl cyanide and the cyanide then reduced.

$$CH_2{=}CH{-}CH_2Cl \xrightarrow{KCN} CH_2{=}CH{-}CH_2CN \quad (S_N1)$$

$$CH_2{=}CH{-}CH_2CN \xrightarrow[\text{ether}]{LiAlH_4} CH_2{=}CH{-}CH_2CH_2NH_2$$

4-Amino-1-butene
(allylcarbinylamine)

(b) Pure *n*-butylamine can be obtained from *n*-butyl bromide via Gabriel's potassium phthalimide synthesis.

$$C_6H_4(CO)_2N^-K^+ + CH_3CH_2CH_2CH_2Br \xrightarrow{(S_N2)} C_6H_4(CO)_2N{-}CH_2CH_2CH_2CH_3 + KBr$$

$$C_6H_4(CO)_2N{-}CH_2CH_2CH_2CH_3 \xrightarrow{Na^+OH^-} C_6H_4(COO^-Na^+)_2 + CH_3CH_2CH_2CH_2NH_2$$

Sodium phthalate

*n*-Butylamine

13.3 $H\ddot{\underset{..}{O}}{:}^- \rightarrow H{-}CH_2{-}CH_2{-}\overset{+}{N}(CH_3)_3 \longrightarrow CH_2{=}CH_2 + H_2O + (CH_3)_3N{:}$

Leaving group = $(CH_3)_3N{:}$ (trimethylamine), a small, neutral molecule

$H\ddot{\underset{..}{O}}{:}^- \rightarrow H{-}CH_2{-}CH_2{-}\ddot{N}(CH_3)_2 \longrightarrow CH_2{=}CH_2 + H_2O + (CH_3)_2\ddot{N}{:}^-$

Leaving group = $(CH_3)_2\ddot{N}{:}^-$ (dimethylamide ion)

Good leaving groups are small, neutral molecules ($H_2O$, $CH_3OH$, $(CH_3)_3N$, $N_2$, etc.) or stable anions ($Cl^-$, $Br^-$, $HSO_4^-$, etc.) that are *weak* bases (that is, anions derived from strong acids such as HCl, HBr, $H_2SO_4$, etc.). Anions ($OH^-$, $CH_3O^-$, $(CH_3)_2\ddot{N}\!:^-$, acetate, etc.) that are strong bases (that is, anions derived from weak acids such as $H_2O$, $CH_3OH$, $(CH_3)_3N\!:$ , acetic acid, etc.) are poor leaving groups.

13.4 The reaction of a primary amine with nitrous acid results in the formation of a diazonium ion which decomposes to yield nitrogen and a carbonium ion. The latter, if alkyl, may either be solvated to yield an alcohol or undergo elimination (El) to produce an olefin. If the cation intermediate is that formed from the decomposition of a benzenediazonium ion, then the elimination route is not possible and only solvation (or combination with other anions) is possible.

13.5 Aniline is an organic base and in a strong acid solution forms the anilinium ion, $C_6H_5\overset{+}{N}H_3$. A positively charged group on the benzene ring would be a very strong *meta*-director.

13.6 The electron-withdrawing nitro groups in positions *ortho* and *para* to the diazonium group makes the latter a much stronger electrophile than it would be if unsubstituted.

13.7 (a) *tert*-butylamine (1° amine) (b) trimethylamine (3° amine) (c) 1-naphthylamine, $\alpha$-naphthylamine (1° amine) (d) 2-(N,N-dimethylamino) ethanol (3° amine) (e) N,N-dimethylaniline (3° amine) (f) cyclohexylamine, aminocyclohexane (1° amine) (g) 1,5-diaminopentane (1° amine) (h) phenylhydrazine (i) phenyl isocyanate (j) 2-butanethiol (k) diallyl sulfide

13.8 (a) $C_6H_5-NH_2$ (b) $C_6H_5-\overset{+}{N_2}\ {}^-OSO_3H$

(c) $H_2N-C_6H_4-SO_3H$ (*para*) (d) $(CH_3)_4N^+\ Cl^-$

(e) $C_6H_5-\overset{H}{\overset{|}{N}}-\overset{O}{\overset{\|}{C}}-CH_3$

(f)

$$H_2N-C_6H_4-\overset{O}{\overset{\|}{C}}-O-CH_2CH_2-\overset{+}{N}H(C_2H_5)_2\quad Cl^-$$

(g) 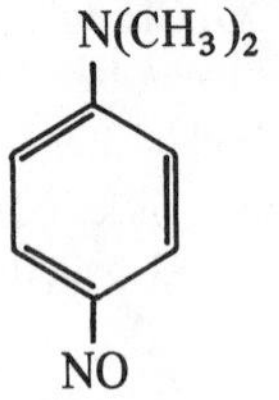 $N(CH_3)_2$ / NO (h) 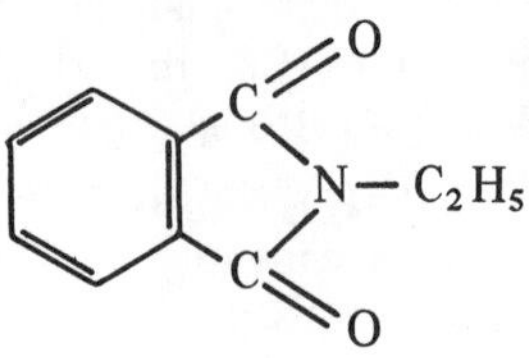 O, C, N—$C_2H_5$, C, O

(i) 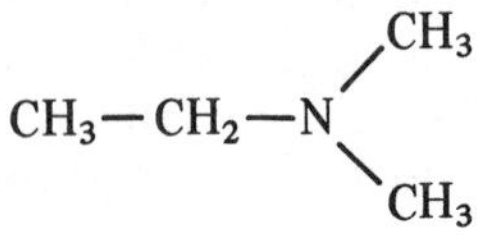 $CH_3—CH_2—N(CH_3)_2$

13.9 (c) > (d) > (b) > (a) > (e)

13.10

<table>
<tr><td colspan="3">Aniline, benzoic acid, acetophenone, bromobenzene. — Dissolve in ether, extract with 10% HCl</td></tr>
<tr><td>Aqueous layer: $C_6H_5\overset{+}{N}H_3Cl^-$<br>Neutralize with NaOH</td><td colspan="2">Ether layer: Benzoic acid, bromobenzene, acetophenone. Extract with 10% NaOH.</td></tr>
<tr><td rowspan="3">$C_6H_5—NH_2$</td><td>Aqueous layer:<br>$C_6H_5COO^-Na^+$</td><td>Ether layer: Bromobenzene, acetophenone.</td></tr>
<tr><td>Neutralize with HCl.</td><td rowspan="2">Remove ether on steam bath, fractionally distill. Bromobenzene, b.p., 156°; Acetophenone, b.p., 202°.</td></tr>
<tr><td>$C_6H_5COOH$</td></tr>
</table>

13.11 (a) Oxime formation followed by reduction. (b) Nitrile formation followed by reduction. (c) Hofmann hypohalite degradation. (d) Hofmann elimination. (e) Reduction with tin and hydrochloric acid. (f) Treatment of aniline with either acetyl chloride or acetic anhydride. (g) Solution of aniline hydrochloride treated with cold sodium nitrite. (h) Treatment of cold diazonium salt solution with cuprous cyanide. (i) Nitration followed by reduction. (j) Diazotization of aniline carried out in an excess of aniline. (k) Formation of the Grignard reagent, followed by treatment with sulfur and acid hydrolysis (Note: The isothiouronium salt route cannot be used because *tert*-butyl bromide is a poor $S_N2$ halide).

13.12 (a) Hinsberg test (b) solubility in water (c) iodoform reaction (d) reaction with cold nitrous acid (e) If smell alone would not suffice (1-butanethiol has a delicate aroma reminiscent of skunk), the solubility of 1-butanethiol in dilute base would distinguish it from 1-butanol.

13.13 *sec*-Butylamine

13.14 (a)

$C_6H_5-\ddot{O}:^- Na^+$ Sodium phenoxide

(b)

$Cl-C_6H_4-N_2^+ Cl^-$ *p*-Chlorobenzenediazonium chloride

(c)

$HO-C_6H_4-N{=}N-C_6H_4-Cl$ 4-Chloro-4′-hydroxyazobenzene

(d)

$O_2N-C_6H_3(NO_2)-NH_2$ 2,4-Dinitroaniline

(e)

$C_6H_5-CH(CH_3)-NH_2$ α-Phenylethylamine

13.15

$$CH_3-C_6H_4-CO_2H \xrightarrow[2.\ NH_3]{1.\ SOCl_2} CH_3-C_6H_4-CONH_2 \xrightarrow[2.\ NaOH]{1.\ Br_2} CH_3-C_6H_4-NH_2$$

13.16

$$\text{(cyclobutyl)}-CH_2-NH_2 \xrightarrow[H^+]{HNO_2} \text{(cyclobutyl)}-CH_2-N_2^+ \longrightarrow \text{(cyclobutyl)}-CH_2^+$$

$$\longrightarrow \text{(cyclopentyl cation)}^+ \xrightarrow{H_2O} \text{(cyclopentyl)}-OH$$

13.17

$$pH = pK_a + \log \frac{[A^-]}{[HA]} = pK_a + \log \frac{[CH_3CH_2NH_2]}{[CH_3CH_2NH_3^+]}$$

$pK_a$ of $CH_3CH_2NH_3^+$ = 10.63

$$\text{Therefore, pH} = 10.63 + \log \frac{[CH_3CH_2NH_2]}{[CH_3CH_2NH_3^+]}$$

| pH | $[CH_3CH_2NH_2] / [CH_3CH_2NH_3^+]$ |
|---|---|
| 6 | $2.34 \times 10^{-5}$ |
| 8 | $2.34 \times 10^{-3}$ |
| 10 | 0.234 |
| 12 | 23.4 |

Note: At pH = 10.63 the ratio would be approximately 1.00; thus, even in fairl basic solution there is a reasonable amount of the ammonium ion present.

13.18

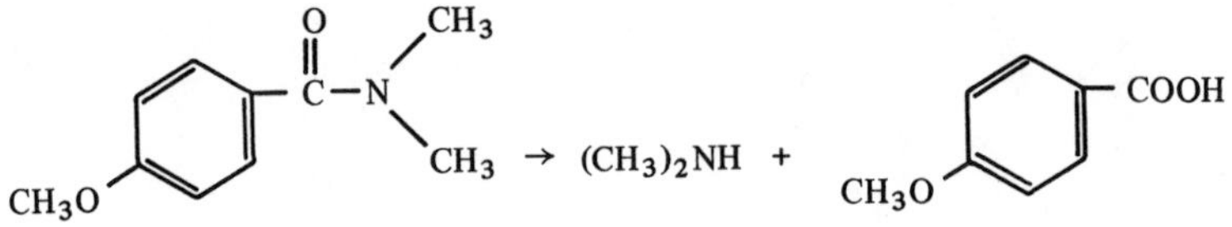

(A) p-$CH_3O$–$C_6H_4$–C(=O)–N($CH_3$)$_2$ → $(CH_3)_2NH$ + p-$CH_3O$–$C_6H_4$–COOH

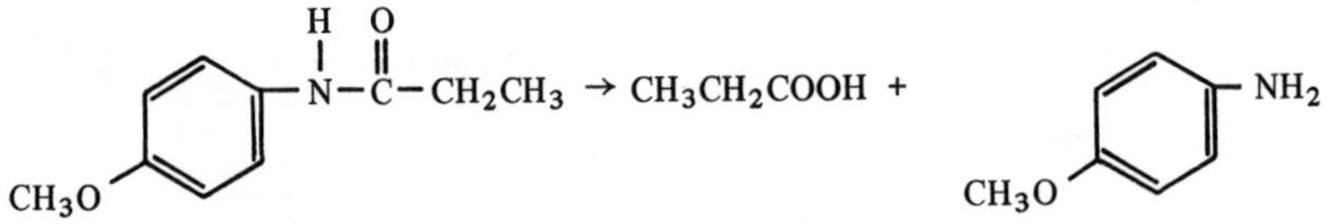

(B) p-$CH_3O$–$C_6H_4$–NH–C(=O)–$CH_2CH_3$ → $CH_3CH_2COOH$ + p-$CH_3O$–$C_6H_4$–$NH_2$

(C) 1,2-di-$C_2H_5$-4-$NO_2$-benzene ($C_2H_5$, $C_2H_5$, $NO_2$); 4-$NO_2$-phthalic acid ($NO_2$, COOH, COOH)

13.19 Benzanilide, $C_6H_5$–C(=O)–NH–$C_6H_5$

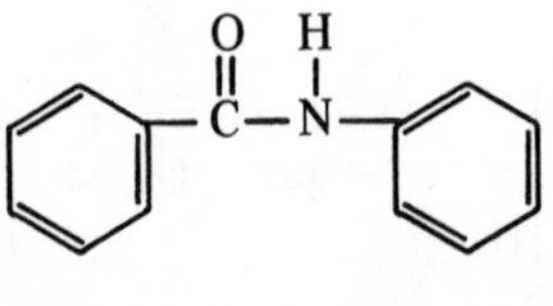

13.20 Pyramidal

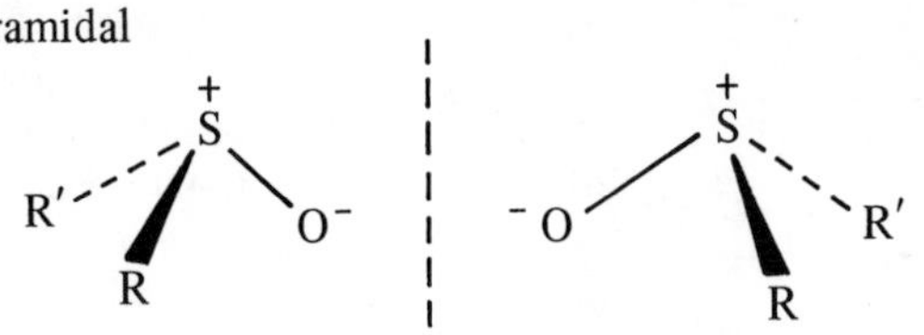

Enantiomeric sulfoxides would not be possible if the molecule were planar.

13.21

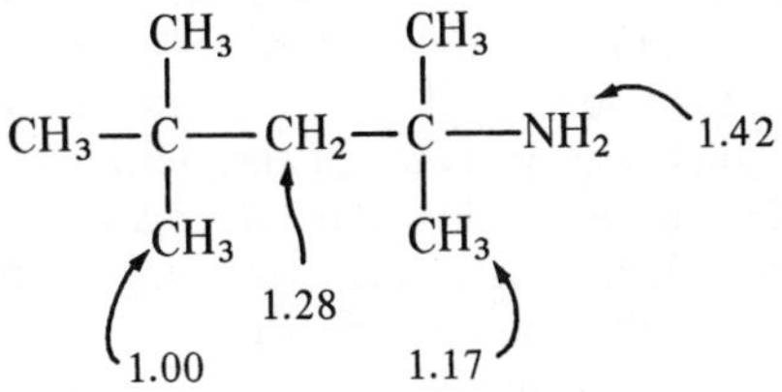

1.00, 1.28, 1.17, 1.42

13.22

2.88, 0.67, 1.00

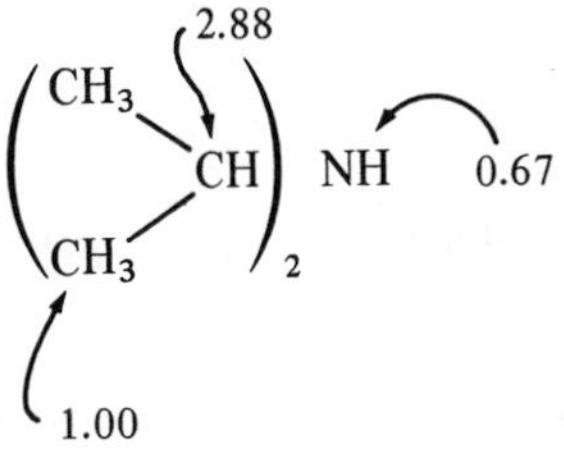

13.23 $C_{63.01/12.01}H_{12.24/1.008}N_{24.65/14.01} = C_{5.25}H_{12.14}N_{1.76}$

$= C_{2.98}H_{6.90}N_{1.00}$

$= C_3H_7N$

Since compound C reacted with two moles of R-X, two nitrogens must be present. Therefore, C's molecular formula = $C_6H_{14}N_2$.

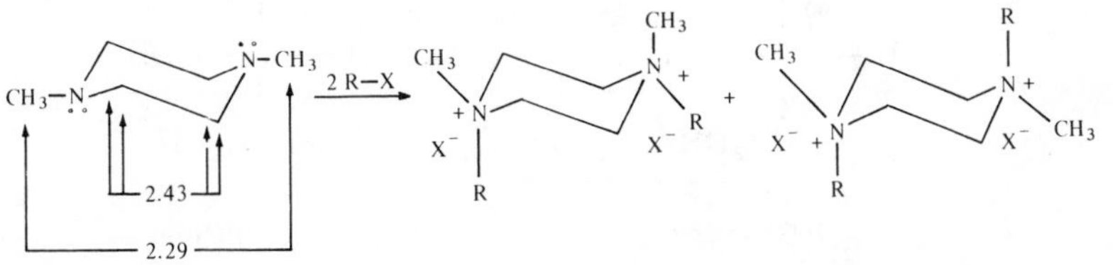

2 R–X

2.43

2.29

13.24 $(^{*}CH_3)(CH_3)N{-}CH{=}O \xrightarrow[\text{C—N bond}]{\text{rotate about}} (CH_3)(^{*}CH_3)N{-}CH{=}O$

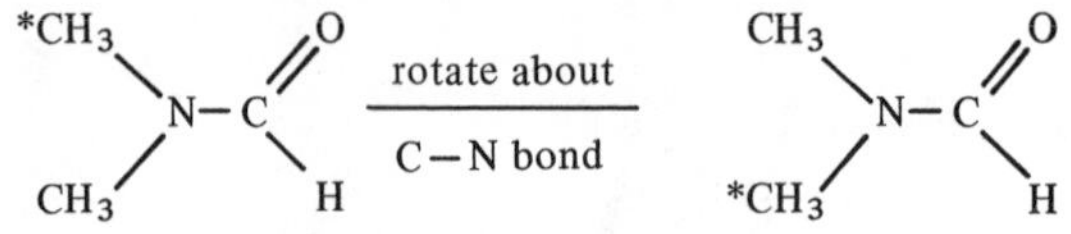

$CH_3$ and $^{*}CH_3$ will not be in same environment unless there is rotation about C—N bond.

$$\left[(CH_3)_2\ddot{N}{-}CH{=}\ddot{O}: \;\leftrightarrow\; (CH_3)_2\overset{+}{N}{=}CH{-}\ddot{O}:^{-}\right]$$

Because of the resonance interactions of the amide group, rotation around the C—N bond is somewhat restricted (that is, rotation is slow on the nmr time scale at room temperature). Thus, the two methyl groups are not in exactly the same environment on the nmr time scale and are not chemically equivalent. In different solvents the chemical shifts will be different. When a solution of the amide is heated, the rotation rate increases, finally reaching the point at which rotation is so rapid that both methyl groups *appear* to be in the same environment to the nmr instrument. This is similar to the rapid exchange effects discussed in the text.

## CHAPTER 14

14.1 Galactose, on oxidation with nitric acid, would have given an optically inactive meso form.

14.2 Fructose and glucose give the same osazone but are not epimers. Osazone formation involves carbon atoms 1 and 2 of fructose, glucose, and mannose. Inasmuch as the configurations of carbons other than 1 and 2 of all three sugars are the same, all three will yield identical osazones. Glucose and mannose differ only in the configuration of carbon 2 (epimers) and therefore give the same osazone.

14.3

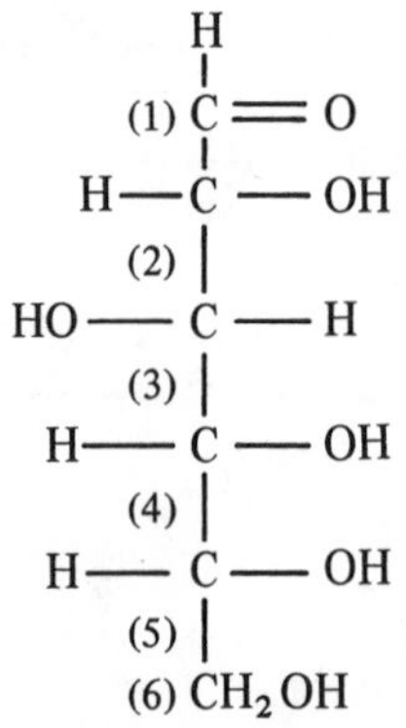

$$\begin{array}{ccc} & H & \\ & | & \\ (1) & C{=}O & \\ H{-} & C & {-}OH \\ (2) & | & \\ HO{-} & C & {-}H \\ (3) & | & \\ H{-} & C & {-}OH \\ (4) & | & \\ H{-} & C & {-}OH \\ (5) & | & \\ (6) & CH_2OH & \end{array}$$

D-Glucose

$$\begin{array}{ccc} (1) & CH_2OH & \\ & | & \\ (2) & C{=}O & \\ & | & \\ HO{-} & C & {-}H \\ (3) & | & \\ H{-} & C & {-}OH \\ (4) & | & \\ H{-} & C & {-}OH \\ (5) & | & \\ (6) & CH_2OH & \end{array}$$

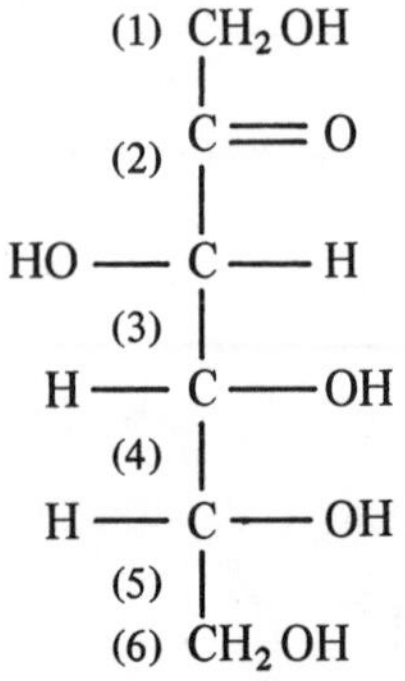

D-Fructose

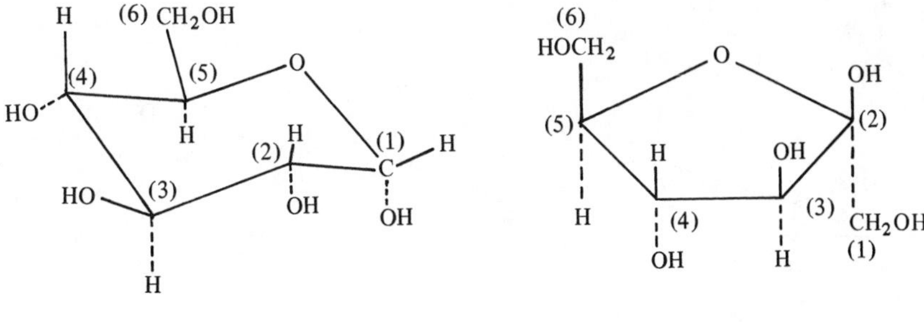


α-D-Glucose

β-D-Fructofuranose

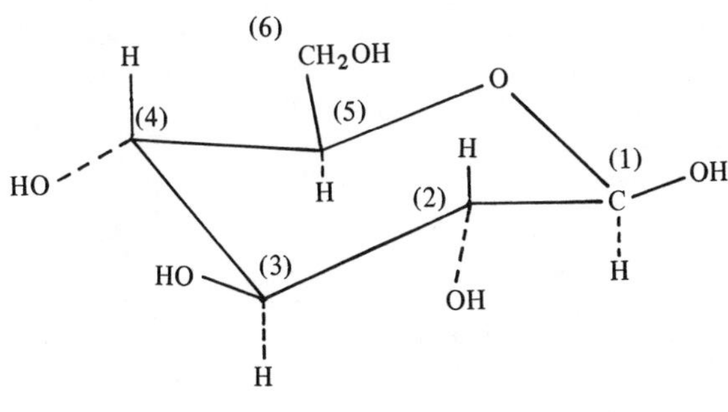


β-D-Glucose

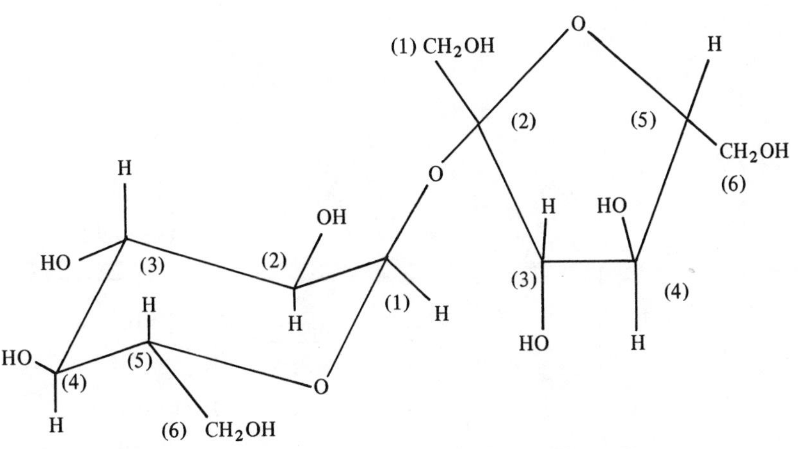


α-D-Glucose unit

β-D-Fructose unit

Sucrose

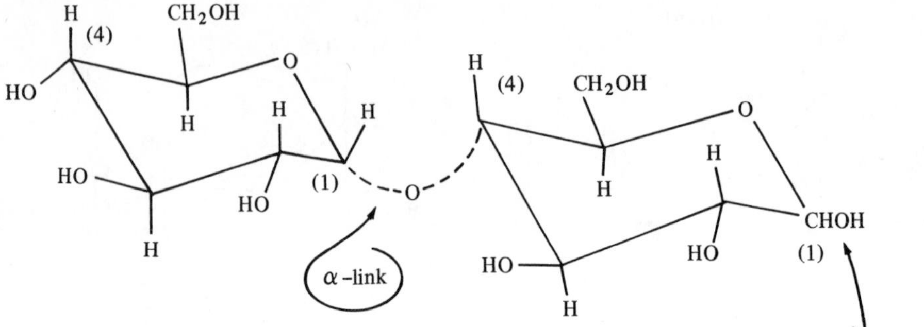

H (4) CH$_2$OH O HO H H H HO (1) O H (4) CH$_2$OH O HO H H H HO CHOH (1) HO H α-link HO H

α-D-Glucose unit

D-Glucose unit

Maltose

(Potential aldehyde group in hemiacetal structure, may be α or β)

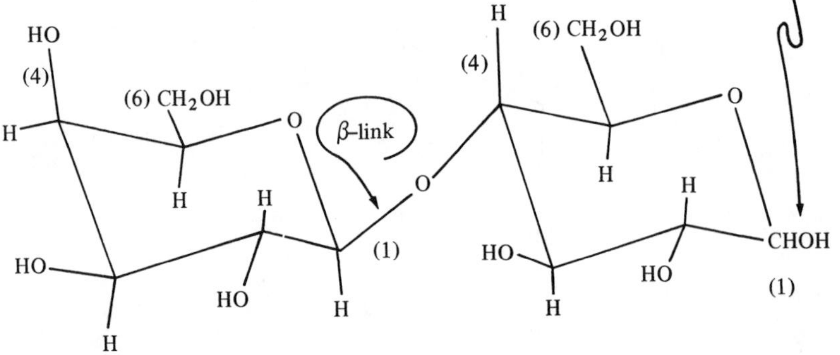

HO (4) H (6) CH$_2$OH O β-link H H O (1) HO HO H H H (4) (6) CH$_2$OH O H H CHOH HO HO (1) H

β-D-Galactose unit

D-Glucose unit

Lactose

14.4 (a)

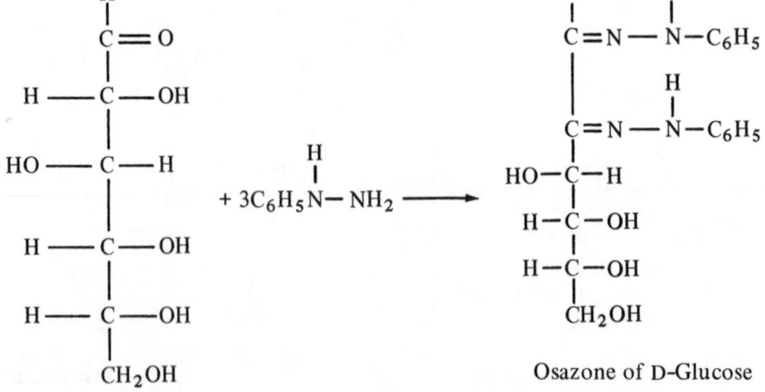

$$\begin{array}{c} H \\ | \\ C{=}O \\ | \\ H{-}C{-}OH \\ | \\ HO{-}C{-}H \\ | \\ H{-}C{-}OH \\ | \\ H{-}C{-}OH \\ | \\ CH_2OH \end{array} + 3C_6H_5\overset{H}{\overset{|}{N}}{-}NH_2 \longrightarrow \begin{array}{l} \;\;H \qquad\;\; H \\ \;\;| \qquad\;\;\; | \\ C{=}N{-}N{-}C_6H_5 \\ | \qquad\quad\; H \\ | \qquad\quad\;\, | \\ C{=}N{-}N{-}C_6H_5 \\ | \\ HO{-}C{-}H \\ H{-}C{-}OH \\ H{-}C{-}OH \\ CH_2OH \end{array}$$

Osazone of D-Glucose

$+ C_6H_5NH_2 + NH_3 + 2\,H_2O$

(b) A simple aldehyde when treated with an alkaline solution of cupric ion yields an acid. α-Hydroxy aldehydes and α-hydroxy ketones in alkaline solutions give rearranged and complex oxidation products. The diagnostic value of Fehling's reaction lies in the color change from the blue of the cupric ion complex to the uncomplexed red cuprous oxide.

(c) H–C=O (open-chain glucose, terminating in $CH_2OH$) $+ 2\,CH_3OH \xrightarrow[HCl]{anhydrous}$ H–C–$OCH_3$ (ring through O, terminating in $CH_2OH$)

α-Methylglucoside

$+$ $CH_3O$–C–H (ring through O, terminating in $CH_2OH$) $+ 2H_2O$

β-Methylglucoside

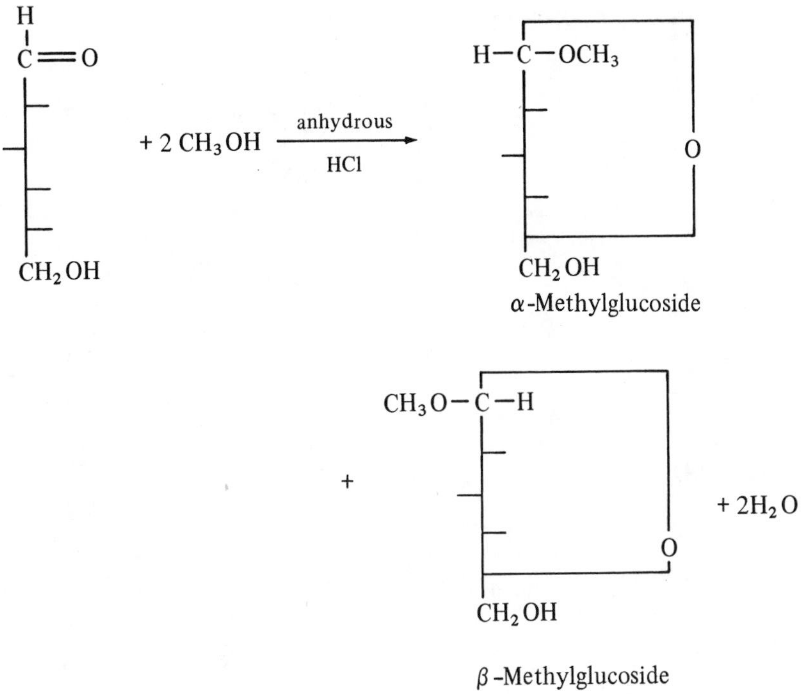

(d) $C_6H_{12}O_6 + 2\,HNO_3 \rightarrow HOOC{-}(CHOH)_4{-}COOH + 2\,NO + 2\,H_2O$

(e) $C_6H_{12}O_6 + 5\,(CH_3CO)_2O \rightarrow CH_3\overset{\overset{O}{\|}}{C}{-}O{-}CH_2(CHO\overset{\overset{O}{\|}}{C}CH_3)_4{-}CHO + 5\,CH_3COOH$

(f) $C_6H_{12}O_6 \xrightarrow{zymase} 2\,C_2H_5OH + 2\,CO_2$

(g) $C_6H_{12}O_6 + Br_2 + H_2O \rightarrow HOCH_2(CHOH)_4COOH + 2\,HBr$

(h) $C_6H_{12}O_6 + H_2NOH \rightarrow HOCH_2(CHOH)_4CH{=}NOH + H_2O$

14.5 The anomeric carbon atoms of glucose and of fructose are involved in the glycosidic link to form sucrose. Maltose and lactose each have an anomeric carbon atom which is not involved in the glycosidic link. (See structures in the answer to Exercise 14.3.)

14.6 Simple ketones do not reduce Fehling's solution. However, in alkaline solution aldoses and ketoses may be interconverted and also decomposed into products that do have reducing properties. Since Fehling's reagent is an alkaline solution of a cupric ion complex, these changes take place to reduce the cupric ion to cuprous oxide.

14.7 (a) See Exercise 14.4 (c)

(b) See Exercise 14.3

(c)

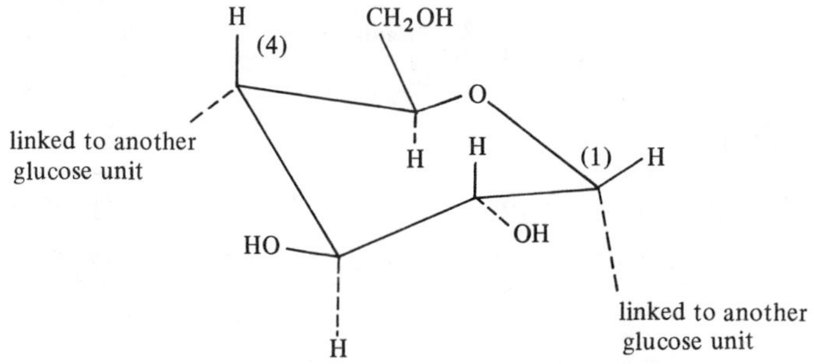

H (4) $CH_2OH$ O
linked to another glucose unit
H H (1) H
HO OH
H
linked to another glucose unit

(d)

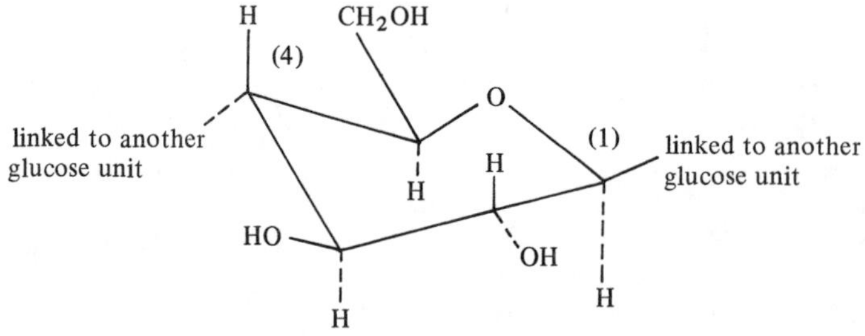

H (4) $CH_2OH$ O
linked to another glucose unit
(1) linked to another glucose unit
H H
HO OH
H H

14.8 (a)

$$ROH + (CH_3\overset{O}{\overset{\|}{C}})_2O \longrightarrow R{-}O{-}\overset{O}{\overset{\|}{C}}{-}CH_3 + CH_3{-}\overset{O}{\overset{\|}{C}}{-}OH$$

(b) $ROH + HNO_3 \longrightarrow R{-}O{-}NO_2 + H_2O$

(c)

$$ROH + NaOH + CS_2 \longrightarrow R{-}O{-}\overset{S}{\overset{\|}{C}}{-}S^- Na^+ + H_2O$$

(d)

$$R{-}O{-}\overset{S}{\overset{\|}{C}}{-}S^- Na^+ \text{ (from c)} + CH_3CH_2Cl \rightarrow R{-}O{-}CH_2CH_3 + CS_2 + NaCl$$

14.9 A (a or d); B (d); C (c or d); D (b, c, or d).

14.10 Possible structures for (A) are those of D-glucose or D-mannose; a possible structure for (B) is that of D-arabinose.

14.11 α-D-Glucopyranosyl-β-D-fructofuranoside is another name for sucrose (table sugar).

14.12 (a) Cellulose nitrate (b) Glycogen (c) Inulin (d) Lactose (e) Rayon (f) Cellulose

14.13

CHO
H—C—OH
H—C—OH
$CH_2OH$

D-Erythrose

14.14 $[\alpha]_{\text{equilibrium}} = n \times [\alpha]_\alpha + (1-n)\,[\alpha]_\beta$

where
$n$ = fraction of the α-anomer
$1-n$ = fraction of the β-anomer
$+14.5 = n(+29.3) + (1-n)(-16.3)$
$n = 0.675$ (α-anomer)
$1-n = 0.325$ (β-anomer)

In solution, 67.5% of the D-mannose is in the α-form and 32.5% in the β-form.

14.15 (a)

α-D-gulose

(b)

α-L-Gulose

(c)

β-D-Gulose

(d)

α-D-Idose

(e)

4-O-α-D-Gulopyranosyl-α-D-glucopyranose

(could also draw structures using β-forms or mixture of α- and β-forms)

14.16

| CHO | CHO | CHO |
|---|---|---|
| HO—H | HO—H | H—OH |
| HO—H | HO—H | HO—H |
| H—OH | HO—H | HO—H |
| $CH_2OH$ | H—OH | H—OH |
| | $CH_2OH$ | $CH_2OH$ |
| D-Lyxose | D-Talose | D-Galactose |

14.17

| CHO | CHO |
|---|---|
| H—OH | HO—H |
| H—OH | H—OH |
| HO—H | HO—H |
| H—OH | H—OH |
| H—OH | H—OH |
| $CH_2OH$ | $CH_2OH$ |
| (A) | (B) |

14.18

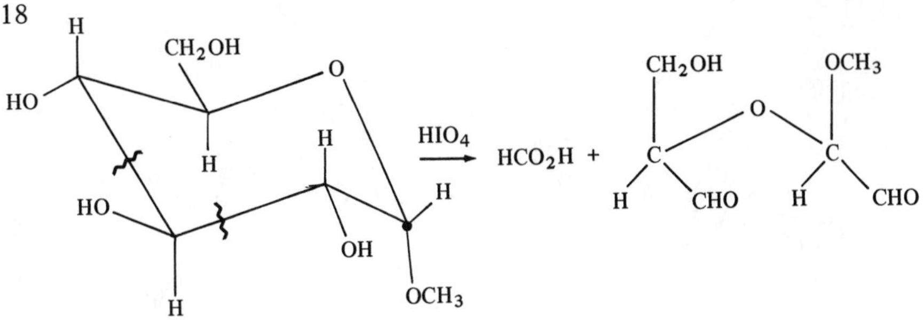

Methyl α-D-glucopyranoside

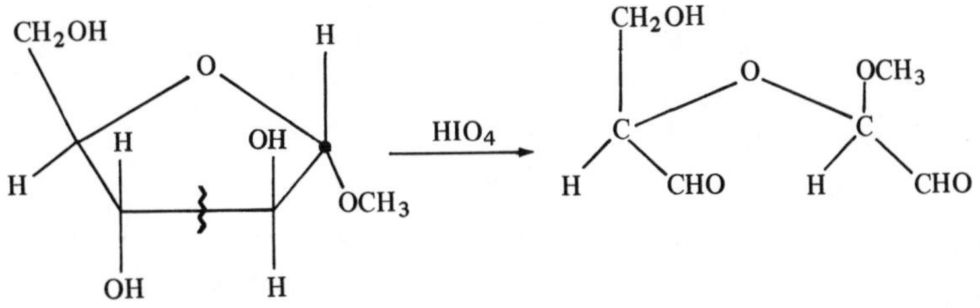

Methyl α-D-arabinofuranoside

14.19

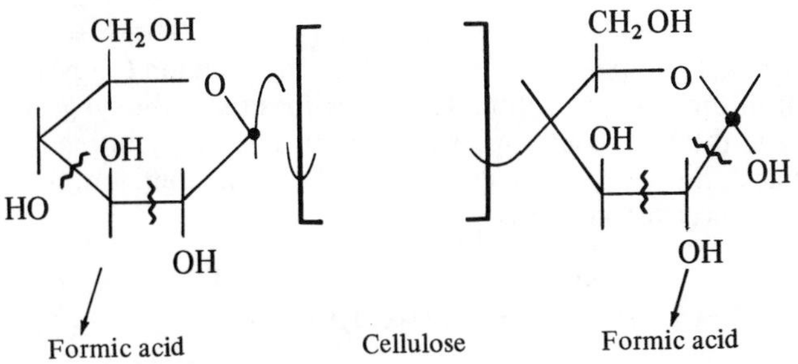

Formic acid Cellulose Formic acid

From each molecule of cellulose we expect two molecules of formic acid.

Therefore, moles of cellulose $= \dfrac{1.00 \times 10^{-6}}{2}$

$= 5.00 \times 10^{-7}$

MW of cellulose $= \dfrac{0.243}{5.00 \times 10^{-7}}$

$= 4.86 \times 10^{5}$

MW of glucose unit $= 162$

Number of glucose units $= \dfrac{4.86 \times 10^{5}}{162}$

$= 3000$

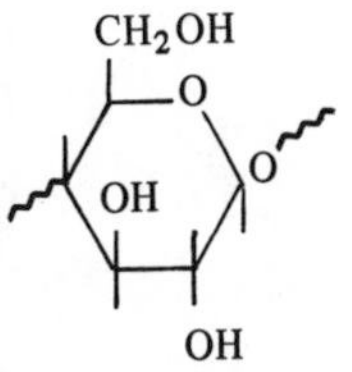

Glucose unit in cellulose

14.20

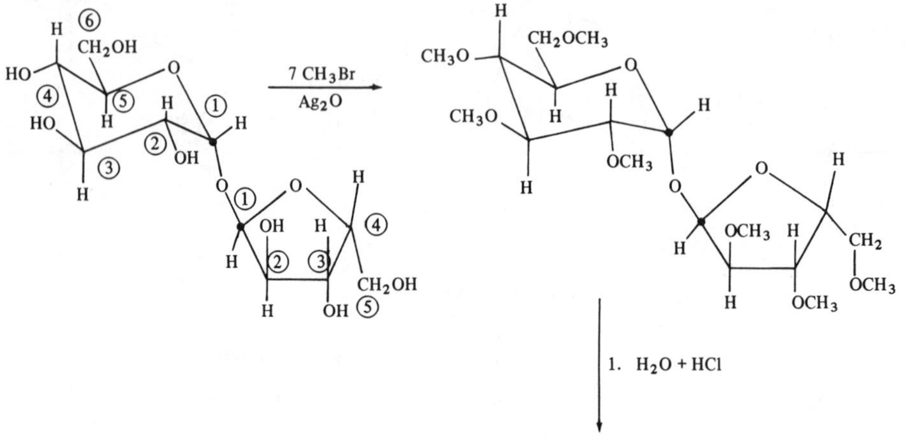

$\xrightarrow[Ag_2O]{7\ CH_3Br}$

1. $H_2O + HCl$

The unmethylated oxygens (the OH groups) show where the rings were closed in the disaccharide.

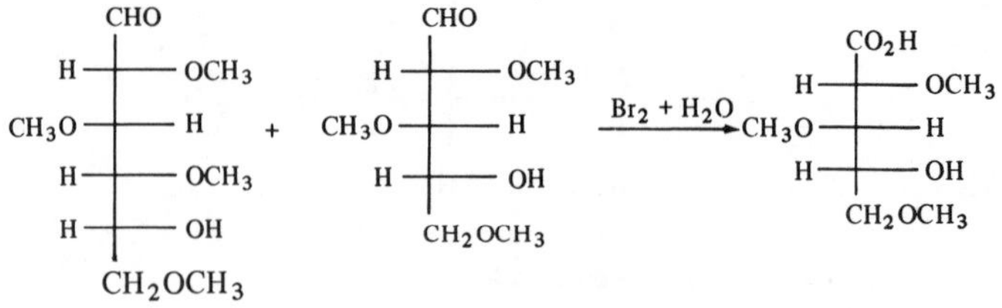

$\xrightarrow{Br_2 + H_2O}$

## CHAPTER 15

15.1 The carboxyl group ($-COOH$) and the ammonium ion ($-\overset{+}{N}H_3$) both are capable of providing a proton. In such cations the carboxyl group is more acidic and the first to give up its proton. In a neutral solution of a simple amino acid, however, the acidic group is the ammonium ion, the basic group is the carboxylate ion.

15.2

$$CH_2{=}CH-COOH \xrightarrow{HBr} BrCH_2CH_2COOH$$

$$BrCH_2CH_2COOH \xrightarrow{NaOH} BrCH_2CH_2COO^-\ Na^+$$

$$BrCH_2CH_2COO^-\ Na^+ \xrightarrow{NaCN} NC-CH_2CH_2COO^-\ Na^+$$

$$NC-CH_2CH_2COO^-\ Na^+ \xrightarrow[H_2O]{HCl,} HOOC-CH_2CH_2-COOH$$

$$HOOC-CH_2CH_2-COOH \xrightarrow{Br_2,\ P} HOOC-\underset{\displaystyle Br}{CH}CH_2-COOH$$

$$HOOC-\underset{\displaystyle Br}{CH}CH_2-COOH \xrightarrow{\text{excess } NH_3} HOOC-\underset{\displaystyle NH_2}{CH}CH_2-COOH$$

15.3

$$\text{Dansyl chloride (5-(CH_3)_2N-naphthalene-1-SO_2Cl)} + \text{Peptide} \xrightarrow[\text{hydrolysis}]{\text{followed by}} \text{5-(CH_3)_2N-naphthalene-1-}SO_2\text{-NH-CH(R)-}CO_2H + \text{Other } \alpha\text{-amino acids and small peptides}$$

15.4 Unsubstituted bromobenzene does not undergo a nucleophilic displacement at the ring carbon bonded to bromine. 2,4-Dinitrofluorobenzene, on the other hand, has the strong inductive effect of the fluorine atom reenforced by the electron-withdrawing power of nitro groups in the *ortho* and *para* positions. These effects make the ring carbon at fluorine especially attractive to a nucleophile–in this case the α-amino group.

15.5 The eyes of newborn infants are treated with a silver nitrate solution (and in present day medical practice also with certain of the antibiotics) as a prophylaxis against gonococcus and the syphilis spirochete.

15.6

$$CH_3{-}S{-}CH_2{-}CH_2{-}CH(NH_2){-}COOH$$

Methionine

$$HS{-}CH_2{-}CH(NH_2){-}COOH$$

Cysteine (can form disulfide bonds)

15.7

$$C_6H_5{-}CH_2{-}CH(NH_2){-}COOH$$

Phenylalanine

$$\text{(indol-3-yl)}{-}CH_2{-}CH(NH_2){-}COOH$$

Tryptophan

$$HO{-}C_6H_4{-}CH_2{-}CH(NH_2){-}COOH$$

Tyrosine

15.8 The synthesis of D, L-alanine may begin with *n*-propyl alcohol. The first step would require oxidation of the alcohol to propionic acid. The next step is the conversion of the latter to D, L-$\alpha$-bromopropionic acid via the Hell-Volhard-Zelinsky reaction. Reaction of the $\alpha$-bromo acid with N-potassium phthalimide, followed by hydrolysis should yield D, L-alanine.

15.9

(a) → $CH_3CH(OH)COOH + N_2 + NaCl + H_2O$

(b) → $C_6H_5CONHCH_2COOH + HCl$

(c) →

$$HOOC-\underset{NH_2}{\overset{H}{C}}-CH_2-S-S-CH_2-\underset{NH_2}{\overset{H}{C}}-COOH + 2\ HI$$

(d) → $CH_3CH(NH_2)COO^- + Na^+ + H_2O$

(e) → $C_6H_5CH_2OCONHCH(CH_3)COOH + HCl$

(f) $SOCl_2$

→ $C_6H_5CH_2OCONHCH(CH_3)C(=O)Cl + HCl + SO_2$

(g) $CH_3CH(NH_2)COOH$

→ $C_6H_5CH_2OCONHCH(CH_3)CONHCH(CH_3)COOH + HCl$

15.10 Gly-Ala-Phe; Phe-Gly-Ala; Phe-Ala-Gly; Ala-Gly-Phe; Gly-Phe-Ala; Ala-Phe-Gly.

15.11 If each molecule of hemoglobin contains four heme structures, each mole of hemoglobin must contain (4)(55.85)g of iron. Since we know that a kilogram of hemoglobin contains 3.33 g of iron, we can use a simple proportion to determine the molecular weight of the protein.

$$\frac{3.33 \text{ g Fe}}{1000 \text{ g hemoglobin}} = \frac{4(55.85) \text{ g Fe}}{x \text{ g hemoglobin}}$$

$x = 67{,}100$ molecular weight of the protein

15.12 One gram-molecular volume of nitrogen gas is liberated by each amino group in an amino acid. Our first step in this problem is to convert the volume of nitrogen collected to that at standard conditions.

$$2.50 \text{ ml} \times \frac{273}{298} \times \frac{740}{760} = 2.23 \text{ ml } N_2 \text{ at S.T.P.}$$

2.23 ml $N_2$ = 0.0001 mole $N_2$

8.74 mg = 0.00874 g = 0.0001 mole of amino acid having only one amino group

In this case the molecular weight of the unknown amino acid is 87.4 g. However, an isoelectric value of 10.8 suggests the presence of more than one amino group. The unknown amino acid appears to be arginine, molecular weight 174.

15.13

```
                 Pro-Gly
                     Gly—Phe—Ser
                              Ser—Pro—Phe
   Arg—Pro          Phe—Ser—Pro
                                   Pro—Phe—Arg
                                            Arg—Pro
-------------------------------------------------------
   Arg—Pro-Gly—Phe—Ser—Pro—Phe—Arg—Pro
```

15.14 (a) 6.07; (b) 6.02; (c) 5.66; (d) 5.68

15.15 Phenolphthalein changes color within a pH range of 8.6–10.0. At this alkalinity it is probable the glycine molecule would exist as an alkali metal salt and the acid anion in this case would migrate to the anode.

15.16

| | $pK_1$ | $pK_2$ | $pK_3$ | pH | Fraction in form shown | | | | | | Net charge |
|---|---|---|---|---|---|---|---|---|---|---|---|
| | | | | | $NH_3^+$ | COOH | $CO_2^-$ | $NH_2$ | 2nd COOH | 2nd $CO_2^-$ | |
| alanine | 2.29 | 9.74 | | 1.0 | 1 | 1 | 0 | 0 | - | - | +1 |
| | | | | 2.2 | 1 | ½ | ½ | 0 | - | - | +½ |
| | | | | 4.0 | ~1 | ~0 | ~1 | 0 | - | - | 0 |
| | | | | 9.8 | ½ | 0 | 1 | ½ | - | - | -½ |
| aspartic acid | 2.10 | 3.86 | 9.82 | 1.0 | 1 | 1 | 0 | 0 | 1 | 0 | +1 |
| | | | | 2.2 | 1 | ½ | ½ | 0 | 1 | 0 | +½ |
| | | | | 4.0 | ~1 | ~0 | ~1 | 0 | ½ | ½ | -½ |
| | | | | 9.8 | ½ | 0 | 1 | ½ | 0 | 1 | -1½ |
| asparagine | 2.02 | 8.80 | | 1.0 | 1 | 1 | 0 | 0 | - | - | +1 |
| | | | | 2.2 | 1 | ½ | ½ | 0 | - | - | +½ |
| | | | | 4.0 | ~1 | ~0 | ~1 | 0 | - | - | 0 |
| | | | | 9.8 | <½ | 0 | 1 | >½ | - | - | >-½ |

15.17 (a) pH 3 $\overset{+}{H_3N}-CH_2CH_2CH_2CH_2\underset{\underset{NH_3^+}{|}}{CH}-COOH + \overset{+}{H_3N}-CH_2CH_2CH_2CH_2\underset{\underset{NH_3^+}{|}}{CH}-COO^-$

pH 10 $\overset{+}{H_3N}-CH_2CH_2CH_2CH_2\underset{\underset{NH_2^+}{|}}{CH}-COO^- + \overset{+}{NH_2}-CH_2CH_2CH_2CH_2-\underset{\underset{NH_2}{|}}{CH}-COO^-$

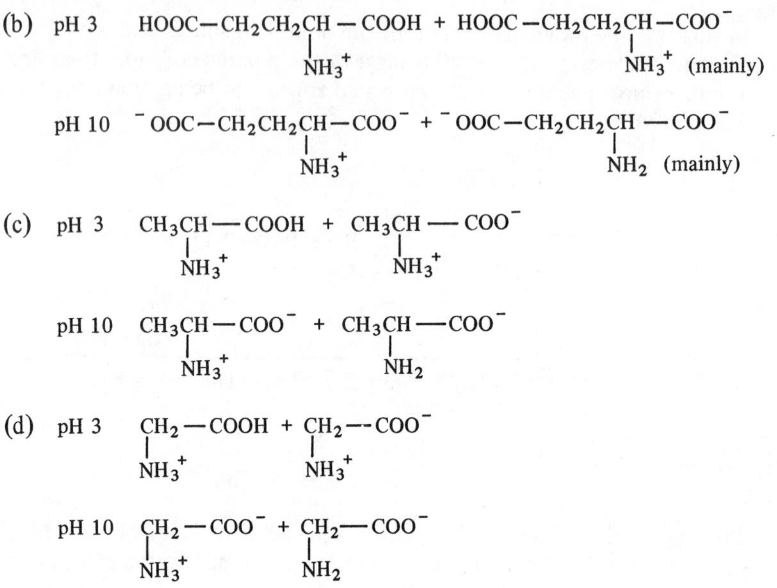

(b) pH 3 $HOOC{-}CH_2CH_2CH(NH_3^+){-}COOH + HOOC{-}CH_2CH_2CH(NH_3^+){-}COO^-$ (mainly)

pH 10 $^-OOC{-}CH_2CH_2CH(NH_3^+){-}COO^- + {}^-OOC{-}CH_2CH_2CH(NH_2){-}COO^-$ (mainly)

(c) pH 3 $CH_3CH(NH_3^+){-}COOH + CH_3CH(NH_3^+){-}COO^-$

pH 10 $CH_3CH(NH_3^+){-}COO^- + CH_3CH(NH_2){-}COO^-$

(d) pH 3 $CH_2(NH_3^+){-}COOH + CH_2(NH_3^+){-}COO^-$

pH 10 $CH_2(NH_3^+){-}COO^- + CH_2(NH_2){-}COO^-$

15.18

$$H_2\ddot{N}{-}C(=NH_2^+){-}\ddot{N}H{-}CH_2CH_2CH_2CH(\ddot{N}H_2){-}COO^- \leftrightarrow H_2N{=}\overset{+}{C}(\ddot{N}H_2){-}\ddot{N}H{-}CH_2CH_2CH_2CH(\ddot{N}H_2){-}COO^-$$

$$\updownarrow$$

$$H_2\ddot{N}{-}C(\ddot{N}H_2){=}\overset{+}{N}H{-}CH_2CH_2CH_2CH(\ddot{N}H_2){-}COO^-$$

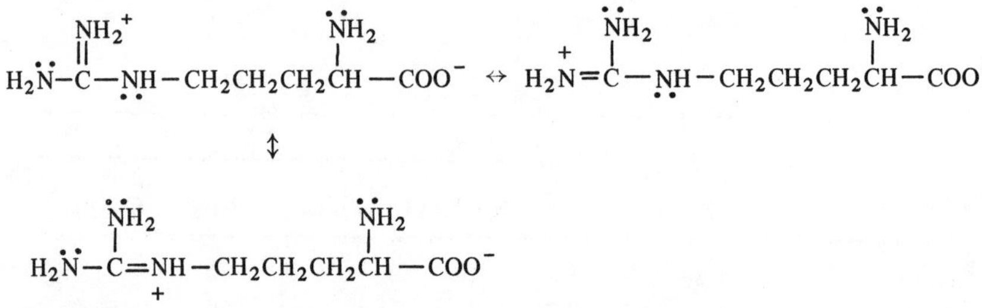

Resonance interaction in the guanadino ion stabilizes that ion greatly. Note that two of the three contributing forms are equivalent and the third form is not greatly different. The result is that the guanadino group is a very basic group, much more so than a simple amino group ($-NH_2$), and accepts a proton to form the cation more readily than the amino group.

15.19 (a)

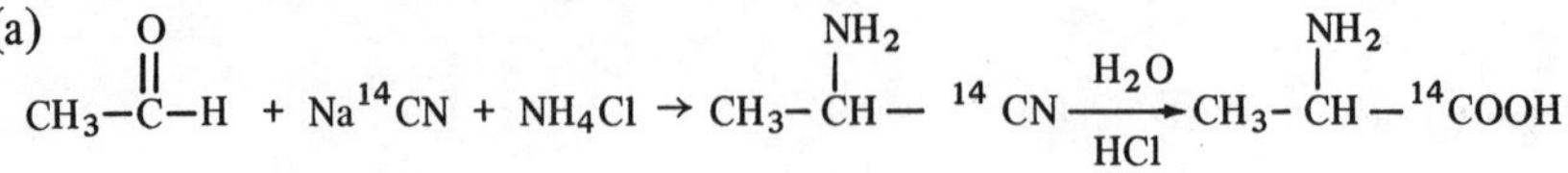

$$CH_3{-}\overset{O}{\overset{\|}{C}}{-}H + Na^{14}CN + NH_4Cl \rightarrow CH_3{-}CH(NH_2){-}{}^{14}CN \xrightarrow[HCl]{H_2O} CH_3{-}CH(NH_2){-}{}^{14}COOH$$

(b) $Ba^{14}CO_2 + H_2SO_4 \rightarrow {}^{14}CO_2$

$$CH_3MgBr + {}^{14}CO_2 \rightarrow CH_3{-}{}^{14}COOH \xrightarrow{LiAlH_4} CH_3{-}{}^{14}CH_2OH$$

$$CH_3{-}{}^{14}CH_2OH \xrightarrow{CrO_3(C_5H_5N)_2} CH_3{-}{}^{14}C(=O)H$$

$$CH_3{-}{}^{14}C(=O)H \xrightarrow[2.\ H_2O + HCl]{1.\ NaCN + NH_4Cl} CH_3{-}{}^{14}CH(NH_2){-}COOH$$

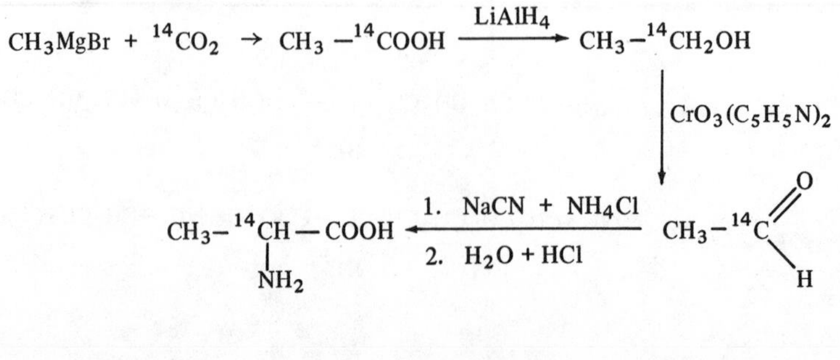

(c)

$CH_2{=}CH{-}COOH \xrightarrow[\text{1,4-addition}]{HBr} BrCH_2CH_2COOH \xrightarrow{Na^{14}CN} N^{14}CCH_2CH_2COOH$

$CH_2{=}CH{-}COOH$ → (1. $Na^{14}CN$(1,4-addition); 2. $H^+$) → $N^{14}CCH_2CH_2COOH$

$N^{14}CCH_2CH_2COOH \xrightarrow{Br_2 + P} N^{14}CCH_2CH(Br){-}COOH \xrightarrow{NH_3} N^{14}CCH_2CH(NH_2){-}COOH$

$N^{14}CCH_2CH(NH_2){-}COOH \xrightarrow[HCl]{H_2O} HOO^{14}C{-}CH_2CH(NH_2){-}COOH$

(d)

$C_6H_5{-}MgBr + {}^{14}CO_2 \xrightarrow{\text{then HX, } H_2O} C_6H_5{-}{}^{14}COOH \xrightarrow{LiAlH_4} C_6H_5{-}{}^{14}CH_2{-}OH$

$C_6H_5{-}{}^{14}CH_2{-}OH \xrightarrow{HBr} C_6H_5{-}{}^{14}CH_2Br$

$C_6H_5{-}{}^{14}CH_2Br$ → (1. $NaOC_2H_5$; 2. malonic ester; 3. $H_2O + HCl$) → $C_6H_5{-}{}^{14}CH_2CH_2COOH$

$C_6H_5{-}{}^{14}CH_2CH_2COOH$ → (1. $Br_2 + P$; 2. $NH_3$) → $C_6H_5{-}{}^{14}CH_2CH(NH_2){-}COOH$

15.20 (a) Ala — Leu — Cy — Gly — Ser — Leu — His — Cy — Phe, with the two Cy residues joined by S — S (Cy of Ala–Leu–Cy bonded through S—S to Cy bearing Phe; His–Cy linked; Ser–Leu–His closing the ring)

(b) 3-phenyl-2-thiohydantoin of alanine: ring N(C₆H₅) — C(=S) — NH — C(H)(CH₃) — C(=O) — N

$C_6H_5{-}N{-}C(=S){-}NH{-}C(CH_3)(H){-}C(=O)$ (five-membered ring closing at N)

(c)

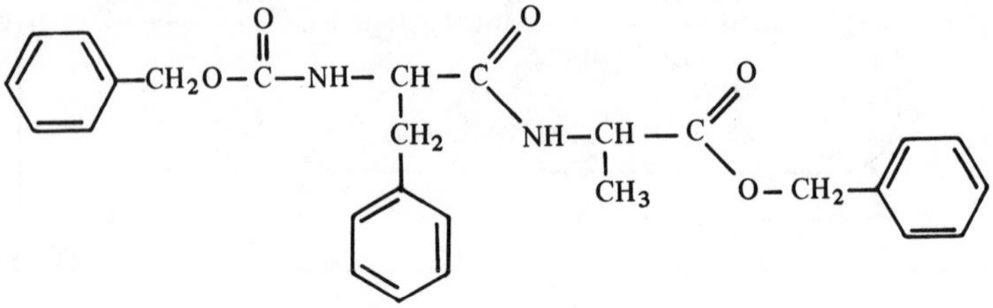

## CHAPTER 16

16.1 (a)

Pyrimidine

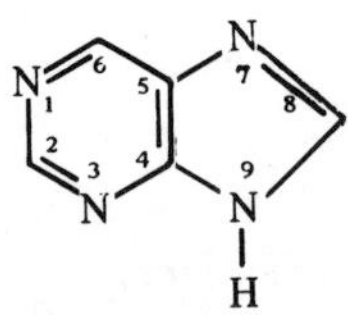

Purine

(b) A sugar molecule + heterocyclic base → a nucleoside
A nucleoside + phosphoric acid → a nucleotide

(c)

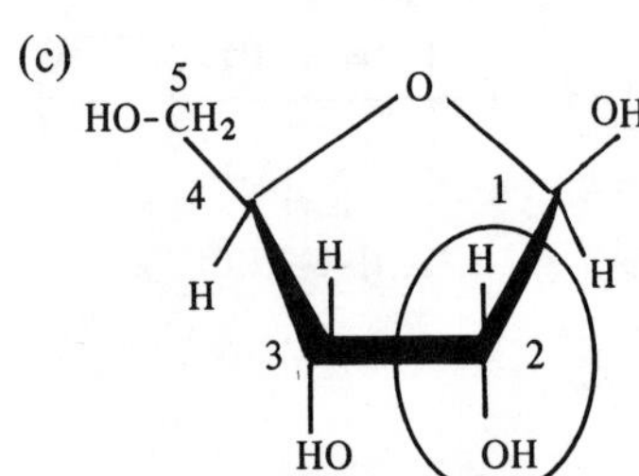

Ribose
($\beta$-D-Ribofuranose)

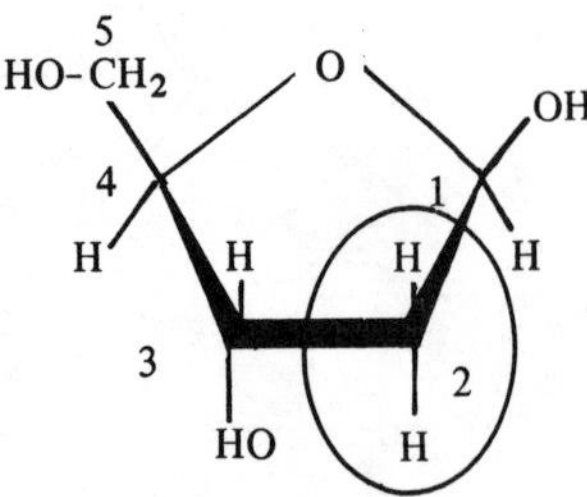

Deoxyribose
(2-Deoxy-$\beta$-D-ribofuranose)

(d) Three bases in a specific sequence in *m*-RNA is called a *codon* for a single amino acid. The corresponding complementary sequence of three bases for the same amino acid in *t*-RNA is the *anticondon.*

16.2 (a)

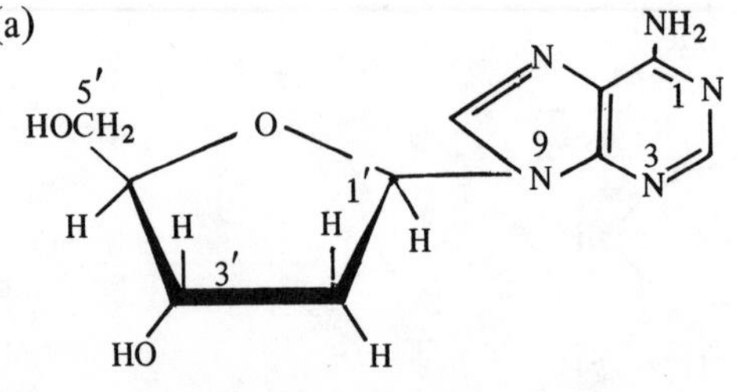

(b)

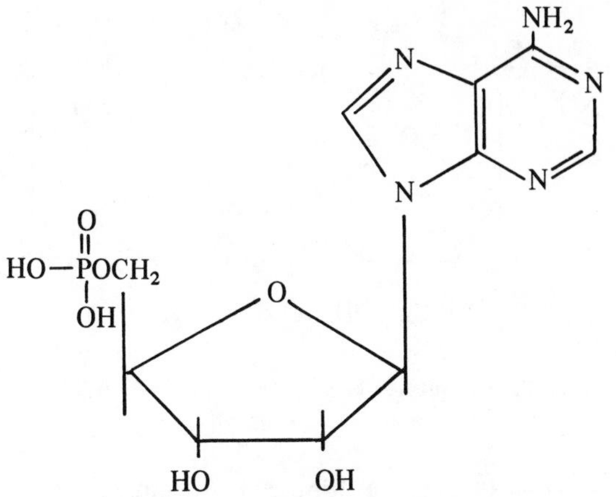

NH$_2$
N
N
N
N
O
HO—POCH$_2$
OH
O
HO OH

(c)

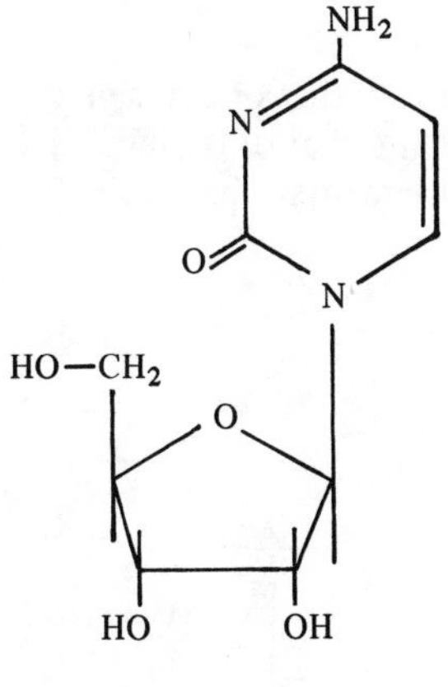

NH$_2$
N
O
N
HO—CH$_2$
O
HO OH

(d)

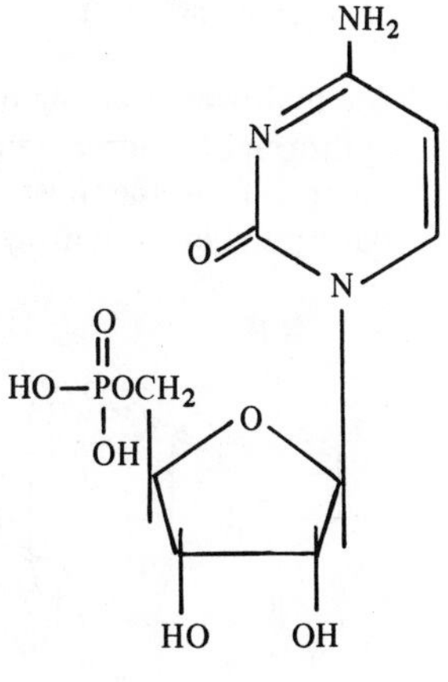

NH$_2$
N
O
N
O
HO—POCH$_2$
OH
O
HO OH

(e)

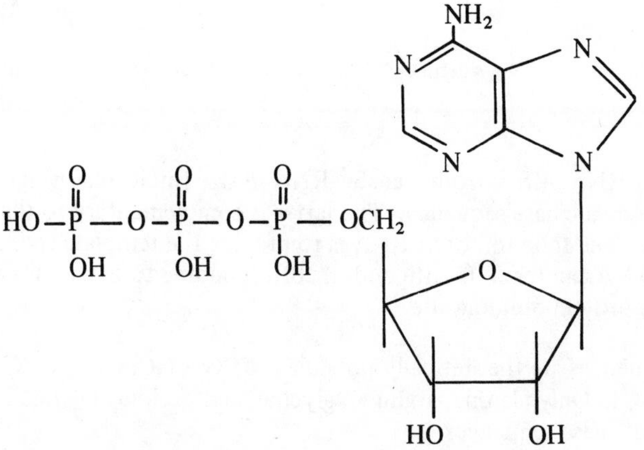

NH$_2$
N
N
N
N
O O O
HO—P—O—P—O—P—OCH$_2$
OH OH OH
O
HO OH

(f)

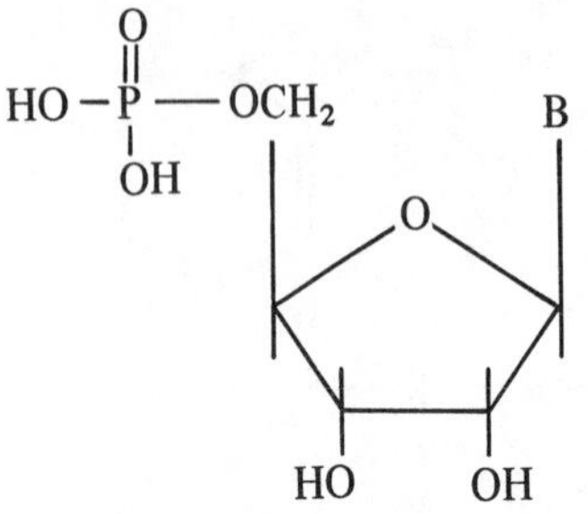

$B$ = A, G, C, or U but not T

(g)

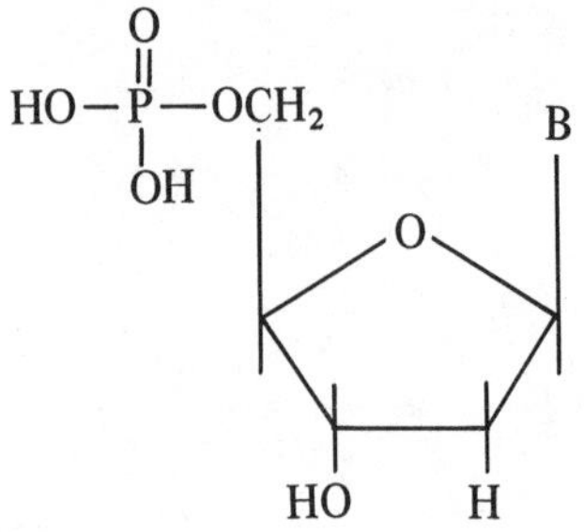

$B$ = A, G, C, or T but not U

16.3

$$ATP \xrightarrow{\text{hydrolysis}} 3\ H_3PO_4 + \text{Ribose} + \text{Adenine}$$

16.4 Cytosine appears opposite quanine in the complementary DNA strand, but thymine in the same strand with cytosine appears opposite adenine. The number of cytosine structures, therefore, need not be equal to the number of thymine structures.

16.5 The hydrogen bonding potential of adenine is 2; that of cytosine is 3. The hydrogen bonding potential of guanine is 3 and that of thymine is 2. In order to form the most stable pairing, complementary base pairs form the same number of hydrogen bonds.

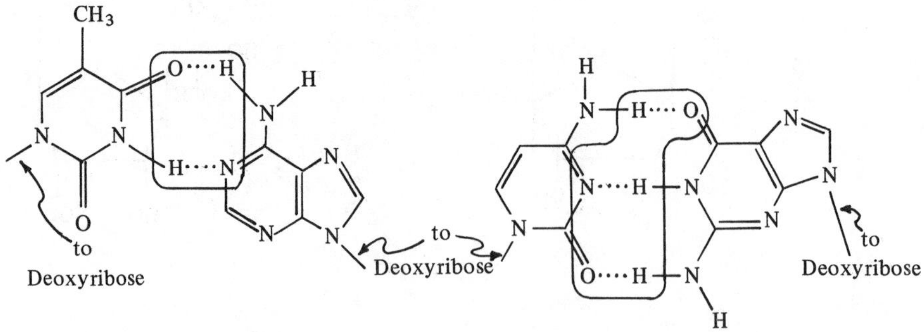

Thymine Adenine Cytosine Guanine

16.6 Messenger RNA differs from transfer RNA in size (molecular weight) and in the heterocyclic base sequence. The latter is complementary to that found in *t*-RNA. The function of *m*-RNA is to provide the template (codon) from which *t*-RNA can form its anticondon corresponding to the amino acid called for at the protein-building site.

16.7 Eight sequences are theoretically possible. (CCG, CGC, GCC, GGC, GCG, CGG, CCC, GGG). Only alanine, arginine, glycine, and proline are coded for by one or more of these sequences.

16.8

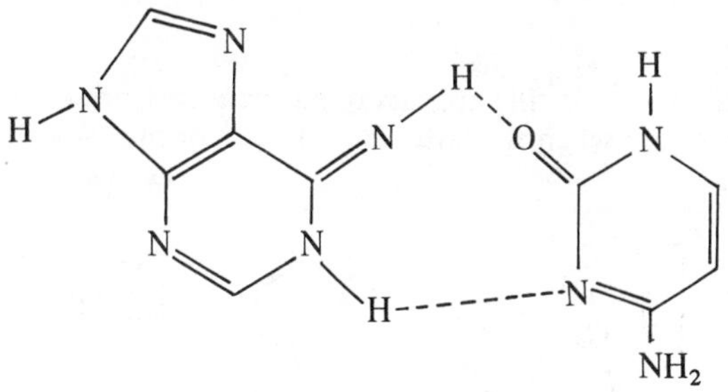

Adenine (imino form) — Cytosine

16.9 C—A*—G produces G—C—C as the resulting codon in *m*-RNA.
G—C—C produces C—G—G as the resulting anticodon in *t*-RNA.
Arginine would erroneously be incorporated into the polypeptide chain in place of glutamine.

16.10 (a) *5′ end* *3′ end* *m*-RNA

GGU GCA AAG UCC UGA CAC AUA

*Amino end* *Carboxyl end*

Gly — Ala — Lys — Ser — Trp — His — Ile

(b) Cys — Val — Ile — Leu — Phe — Pro — Asn

(c) Lys — Pro — Phe — Phe — Pro — Arg — Glu

16.11 DNA: (*5′ end*) AAA–CAC–TTC–TCG–ACG–ATC–GGC–GGC–TAC (*3′ end*)

(a) (*3′ end*) TTT–GTG–AAG–AGC–TGC–TAG–CCG–CCG–ATG (*5′ end*)

(b) (*3′ end*) UUU–GUG–AAG–AGC–UGC–UAG–CCG–CCG–AUG (*5′ end*)

(c) *Carboxyl end* Phe — Val — Glu — Arg — Arg — Asp — Ala — Ala — Val *Amino end*

16.12 Sequence of amino acids:

*Carboxyl end* *Amino end*

Gly — Leu — Pro — Cys — Asn — Gln — Ile — Tyr — Cys Protein

*5′ end* *3′ end*

UCC UAG UGG GCA GUU UUG UAU GUA GCA *t*-RNA

*3′ end* *5′ end*

AGG — AUC — ACC — CGU — CAA — AAC — AUA — CAU — CGU *m*-RNA

## CHAPTER 17

17.1 The nitrogen atom in the pyridine ring appears to have an electron-withdrawing effect upon the methyl group when the latter is in the $\alpha$-, or the $\gamma$-position.

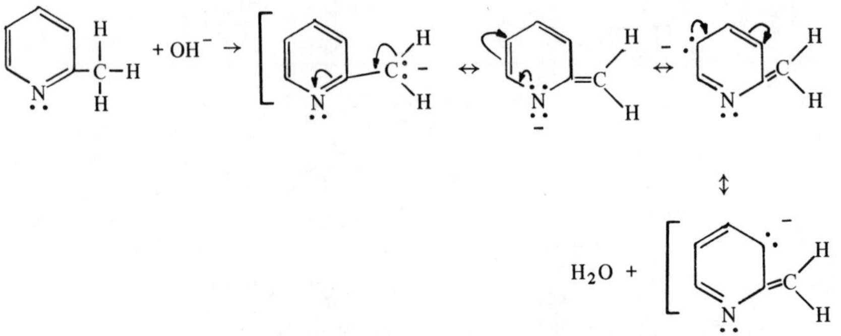

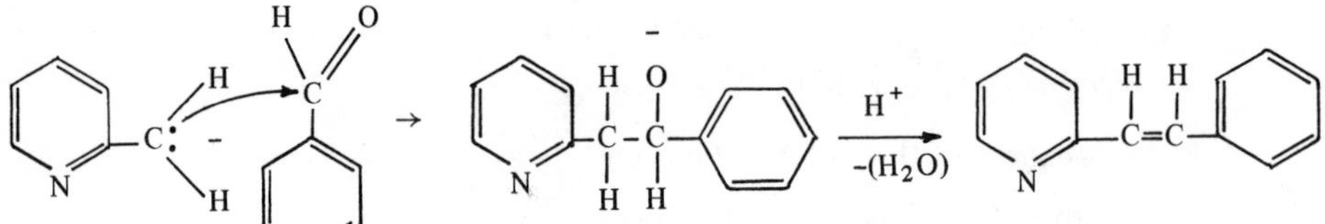

$+ OH^- \rightarrow$ [ ↔ ↔ ↕ ] $H_2O$ +

$\xrightarrow[-(H_2O)]{H^+}$

17.2 3,7,11-Trimethyl-2,6,10-dodecatrien-1-ol

17.3

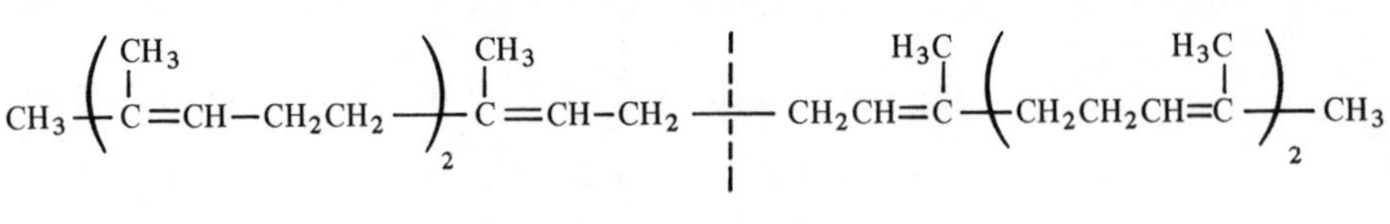

Squalene

17.4

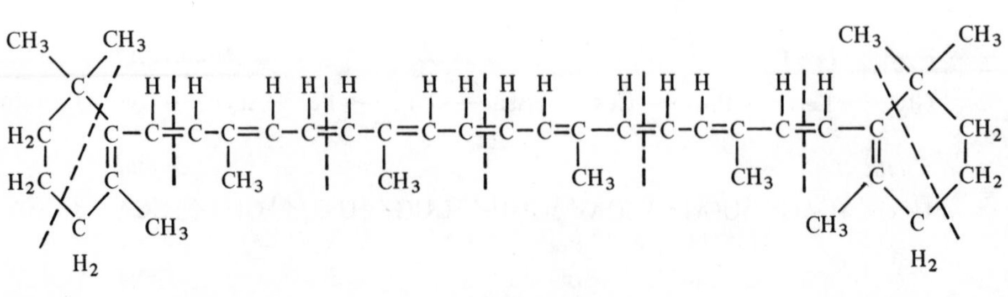

β-Carotene

A head-to-tail arrangement of isoprene units occurs in $\beta$-carotene from both ends of the molecule to meet tail-to-tail in the center of the structure. There are no asymmetric centers.

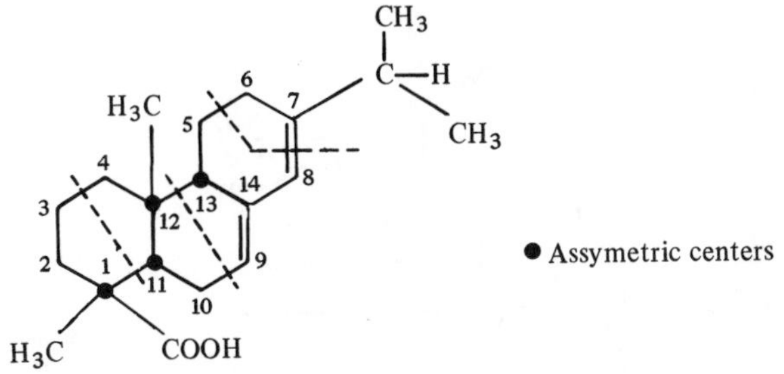

$CH_3$ C—H $CH_3$ $H_3C$ 6 7 5 4 14 8 3 13 12 2 1 9 11 10 $H_3C$ COOH

● Assymetric centers

Abietic acid

The carbon skeleton of abietic acid includes four isoprene units with only three in a head-to-tail arrangement. Carbon atoms 1, 11, 12, and 13 represent asymmetric centers.

17.5 In addition to $\beta$-carotene, there is also an $\alpha$-, $\gamma$-, $\delta$-, $\epsilon$ and a $\zeta$-carotene. These differ as structural isomers—that is, in the position of the double bond in the cyclohexene portion of the structure and also in the structure of each half of the molecule. For example, $\gamma$-carotene has a closed ring at one end, but is open-ended at the other.

17.6 In the planar cyclopentane structure all hydrogen atoms would be completely eclipsed (See Sec. 2.5). This would make for a slightly unstable arrangement. The cyclopentane molecule, therefore, assumes a puckered conformation.

17.7 The isoprene unit, $CH_2{=}C(CH_3){-}CH{=}CH_2$

17.8 (a) Chemical tests:

(1) Abietic acid will dissolve in 5% NaOH and 5% $NaHCO_3$.

(2) Citronellol and abietic acid will react with sodium metal to liberate hydrogen.

(3) Abietic acid, $\alpha$-pinene, and citronellol will decolorize bromine in carbon tetrachloride *without* the liberation of HBr. Camphor will react with liberation of HBr.

(4) Camphor will form a 2,4-dinitrophenylhydrazone or phenylhydrazone.

(b) Spectroscopic test:

(1) Abietic acid will show carboxyl group absorption in the ir and nmr spectra (around $\delta$ 10–13).

(2) Camphor will show the carbonyl group absorption in the ir spectrum (plus a $CH_2$ peak around $\delta$ 2.0 in the nmr spectrum).

(3) The nmr and ir spectra of α-pinene will show the absorption characteristic of the double bond but of no other functional group.

(4) Citronellol will show hydroxyl group absorption in the ir and nmr spectra which may be removed by treatment with $D_2O$.

17.9

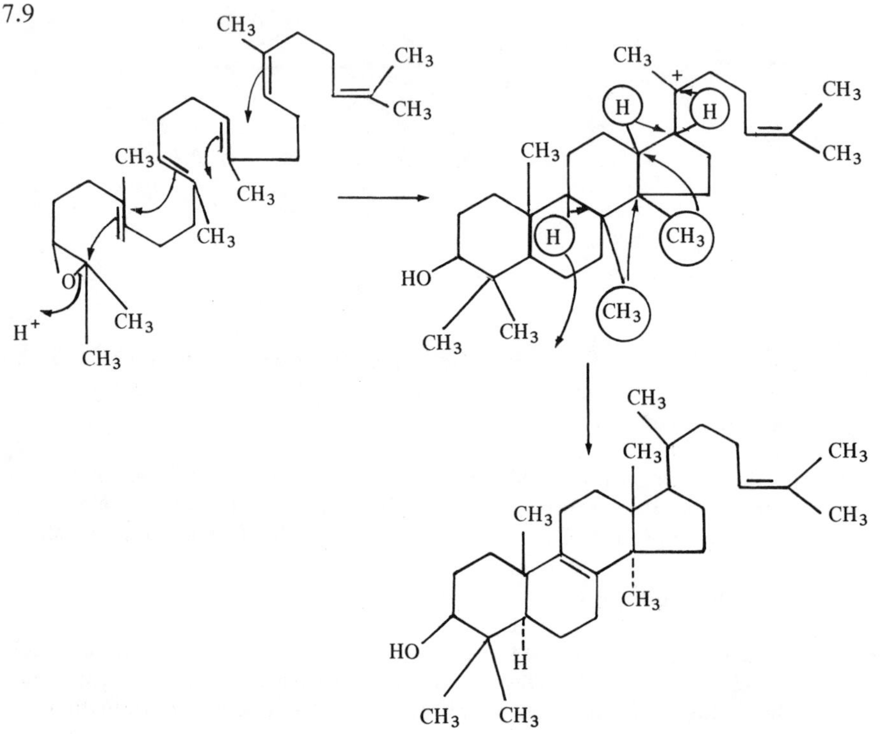

$CH_3$ $CH_3$ $CH_3$ $CH_3$ $CH_3$ $CH_3$ $H^+$ O $CH_3$ $CH_3$

HO $CH_3$ $CH_3$ $CH_3$ + H H $CH_3$ $CH_3$ $CH_3$ H $CH_3$ $CH_3$

HO $CH_3$ $CH_3$ H $CH_3$ $CH_3$ $CH_3$ $CH_3$ $CH_3$ $CH_3$

In the first step, addition of a proton to the oxygen of the ethylene oxide initiates a series of ring closures that proceed from the bottom left to the upper right and leaving a carbonium ion at the top of the molecule. In the second step, a series of rearrangements of protons and methyl cations begins at the upper right and proceeds to the left and downward, culminating in the ejection of a proton.

17.10

$$-CH_2-\overset{\displaystyle CH_3}{\overset{|}{C}}=CH-CH_2-\left(CH_2-\overset{\displaystyle CH_3}{\overset{|}{C}}=CH-CH_2\right)_n-CH_2-\overset{\displaystyle CH_3}{\overset{|}{C}}=CH-CH_2-$$

$\downarrow$ $O_3$ followed by oxidative hydrolysis

$$CH_3-\overset{\displaystyle O}{\overset{\|}{C}}-CH_2-CH_2-CO_2H$$

Levulinic acid
(4-Oxopentanoic acid)

## CHAPTER 18

18.1 Olefins are more volatile than their saturated counterparts and have higher octane values. However, unlike the paraffins, they are subject to attack by oxygen. Oxidative deterioration of gasoline results in gum formation, discoloration, and a loss in octane value.

18.2 (a) $CH_3-CH=CH_2 + H_2SO_4 \rightarrow CH_3-\underset{\displaystyle OSO_3H}{\underset{|}{CH}}-CH_3$

$$CH_3\underset{\displaystyle OSO_3H}{\underset{|}{C}}HCH_3 + H_2O \xrightarrow{\text{heat}} CH_3-\underset{\displaystyle OH}{\underset{|}{CH}}-CH_3$$

(b)

$$CH_2=CH_2 + \tfrac{1}{2}\,O_2 \xrightarrow{\text{catalyst}} H_2C\underset{\diagdown O \diagup}{—}CH_2$$

$$CH_2\underset{\diagdown O \diagup}{—}CH_2 + H_2O \xrightarrow{\text{HCl}} \underset{\displaystyle OH}{\underset{|}{CH_2}}-\underset{\displaystyle OH}{\underset{|}{CH_2}}$$

or

$$CH_2=CH_2 + \underset{HO^-Cl^+}{(Cl_2 + H_2O)} \rightarrow \underset{\displaystyle Cl}{\underset{|}{CH_2}}-\underset{\displaystyle OH}{\underset{|}{CH_2}} \xrightarrow{H_2O,\ Na_2CO_3} \underset{\displaystyle OH}{\underset{|}{CH_2}}-\underset{\displaystyle OH}{\underset{|}{CH_2}}$$

(c)

$$C_6H_4(CH_3)_2 \text{ (o-xylene)} + 2\,K_2Cr_2O_7 + 8\,H_2SO_4 \rightarrow C_6H_4(COOH)_2 + 10\,H_2O$$

$$+\,2\,K_2SO_4 + 2\,Cr_2(SO_4)_3$$

$$C_6H_4(COOH)_2 \xrightarrow{180^\circ} C_6H_4(CO)_2O \text{ (phthalic anhydride)}$$

(d)

$$C_6H_5CH_3 + K_2Cr_2O_7 + 4\,H_2SO_4 \rightarrow C_6H_5COOH + K_2SO_4 + Cr_2(SO_4)_3 + 5\,H_2O$$

(e)

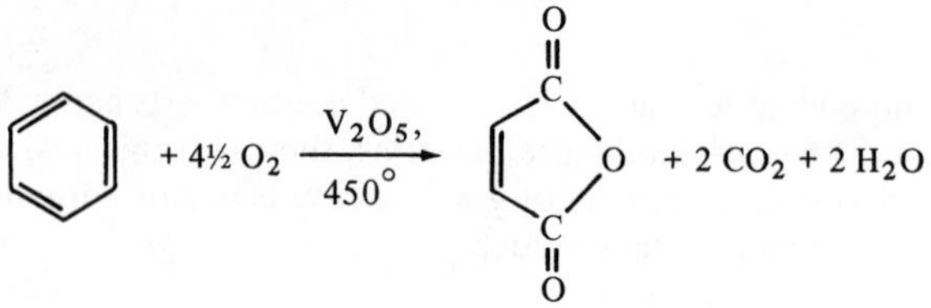

$$C_6H_6 + 4\frac{1}{2}\,O_2 \xrightarrow[450^\circ]{V_2O_5,} \text{(maleic anhydride)} + 2\,CO_2 + 2\,H_2O$$

18.3 100 lb = 45.36 kg = 45360 g

mol. wt. propane = 44

$$\frac{45360}{44} = 1030.9 \text{ moles propane}$$

$$\frac{1 \text{ mole (at S.T.P)}}{22.4 \text{ liters}} = \frac{1030.9 \text{ moles}}{x \text{ liters}}$$

$x$ = 23,092 liters

18.4 (a) 2,6-di-tert-butylphenol: OH at C-1, $C(CH_3)_3$ at C-2 and C-6 ($(CH_3)_3C$–, –$C(CH_3)_3$)

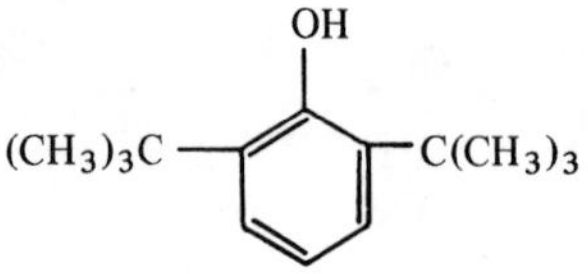

(b) N-phenyl-2-naphthylamine (2-naphthyl–NH–phenyl)

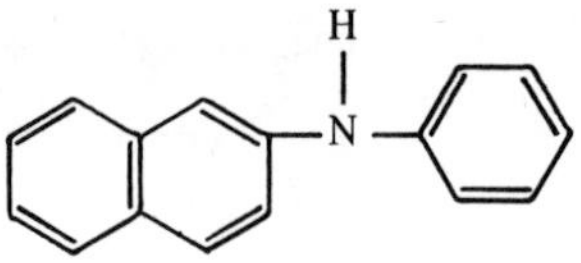

(b) $CH_3$, $CH_3$, $C(CH_3)_3$, OH substituted benzene ring ($CH_3$ at top, $CH_3$ at left, $C(CH_3)_3$ at right, OH at bottom)

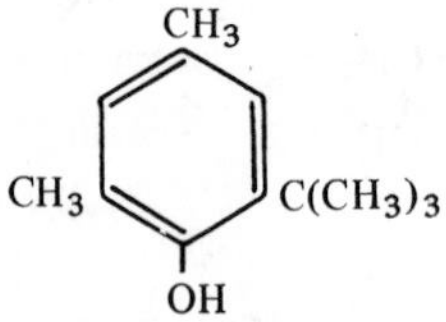

18.5 (a)

$$C + H_2O \xrightarrow{300^\circ} CO + H_2$$

$$n\,CO + (2n + 1)\,H_2 \xrightarrow[250^\circ]{\text{catalyst}} C_nH_{2n+2} + n\,H_2O$$

(b)

$$CH_4 + H_2O \xrightarrow[750^\circ]{\text{catalyst}} CO + 3\,H_2$$

$$CO + H_2O \xrightarrow[750^\circ]{\text{catalyst}} CO_2 + H_2$$

$$n\,C + (n + 1)H_2 \xrightarrow[700 \text{ atm.}]{\text{catalysts, } 400^\circ} C_nH_{2n+2}$$

## CHAPTER 19

19.1 Phenols are acidic and removal of the proton leaves a negatively-charged phenoxide ion. The unshared electrons on the oxygen atom now are able to interact more easily with the benzene nucleus and the nitro group. The resonance of the structure is enhanced with the result that absorption of light occurs at a higher wavelength.

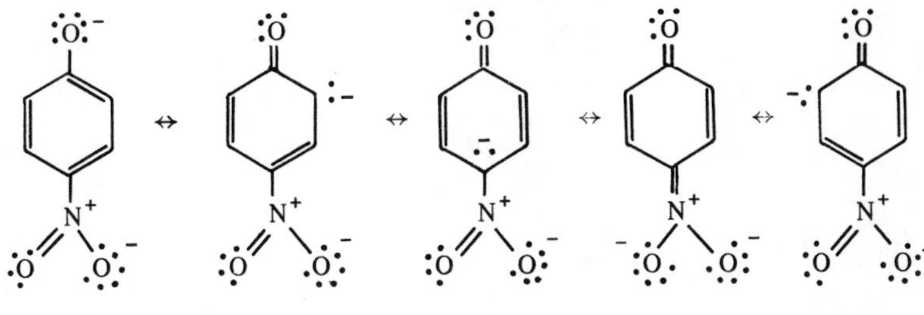

19.2 $5.9\mu = 0.0000059$ m $= 0.00059$ cm $= 59{,}000$ Å. A compound that absorbs light at $5.9\mu$ absorbs radiation in the infrared region, not perceptible to the human eye.

19.3 A thin layer of mineral oil on water does not spread itself in a layer of uniform thickness. As a result, light is refracted to varying degrees when passing through the oil layer. The wavelengths of some light that is reflected from the surface of the oil film and from the surface of the water may be out of phase and cause interference. This results in cancellation of certain wavelengths. The result of such interference is seen as color. The color of birds' feathers and soap bubbles are examples of the same phenomenon.

19.4 A compound absorbing radiation within a wavelength region of 5960-6000 Å would absorb orange. The color of the compound would appear to be greenish-blue.

19.5 The more extensive conjugation of lycopene produces the deeper color. $\beta$-Carotene absorbs at 4,250–4,950 Å, and lycopene absorbs at 4,900–5,000 Å.

19.6 (a)

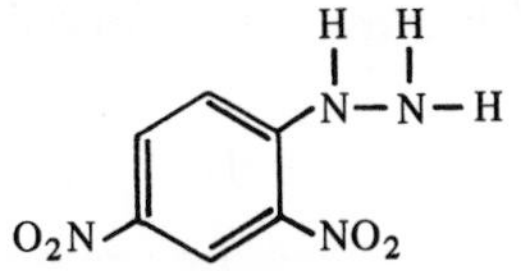

2,4-Dinitrophenylhydrazine (red-orange)

(b) $CH_3-\overset{O}{\overset{\|}{C}}-\overset{O}{\overset{\|}{C}}-CH_3$

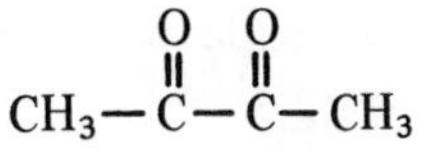

Diacetyl (yellow)

(c)

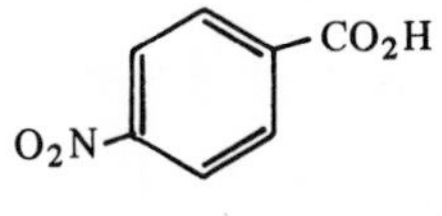

*p*-Nitrobenzoic acid (yellow)

(d)

Quinone (yellow)

All four compounds have a chromophore in their molecular structures and are colored substances but lack "fastness" and would not be usable as dyestuffs.

19.7 (a)

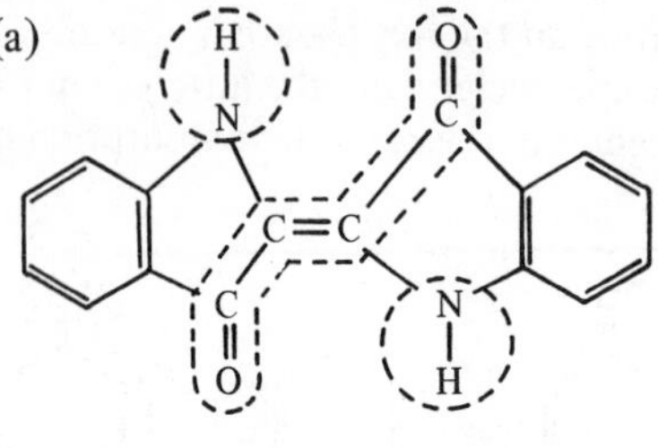

Indigo (vat)

(b)

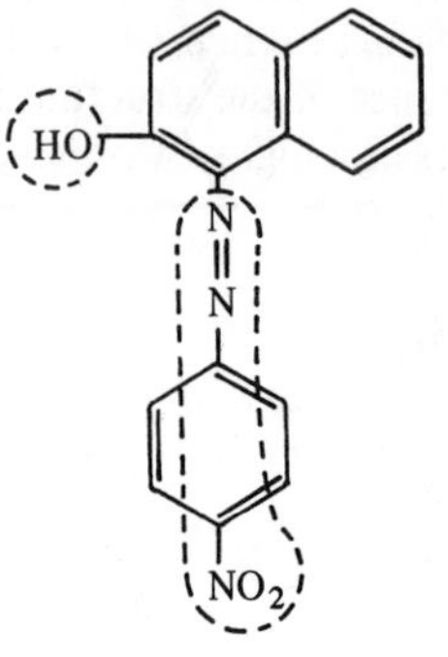

Azo (ingrain)

(c) $H_3C$ $CH_3$ $Cl^-$

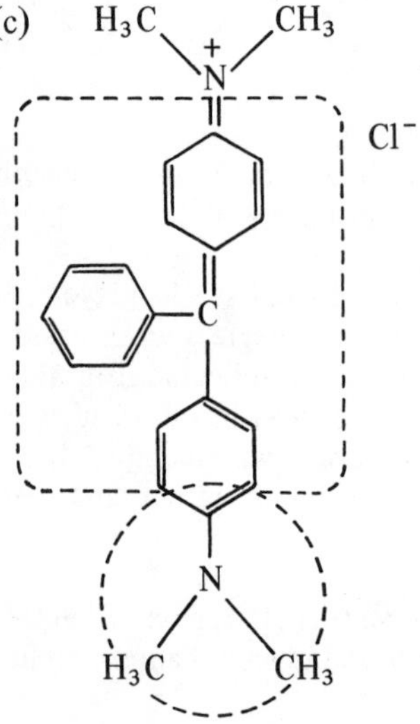

Triphenylmethane (direct or mordant)

19.8 (a) In solutions of a pH less than 8.3 phenolphthalein is colorless because it exists as a substituted phenolic structure. At pH values of 10 or slightly above, phenolphthalein becomes an acid salt and the quinoid chromophore is evident. In strongly basic solutions phenolphthalein is converted to a carbinol and again becomes colorless.

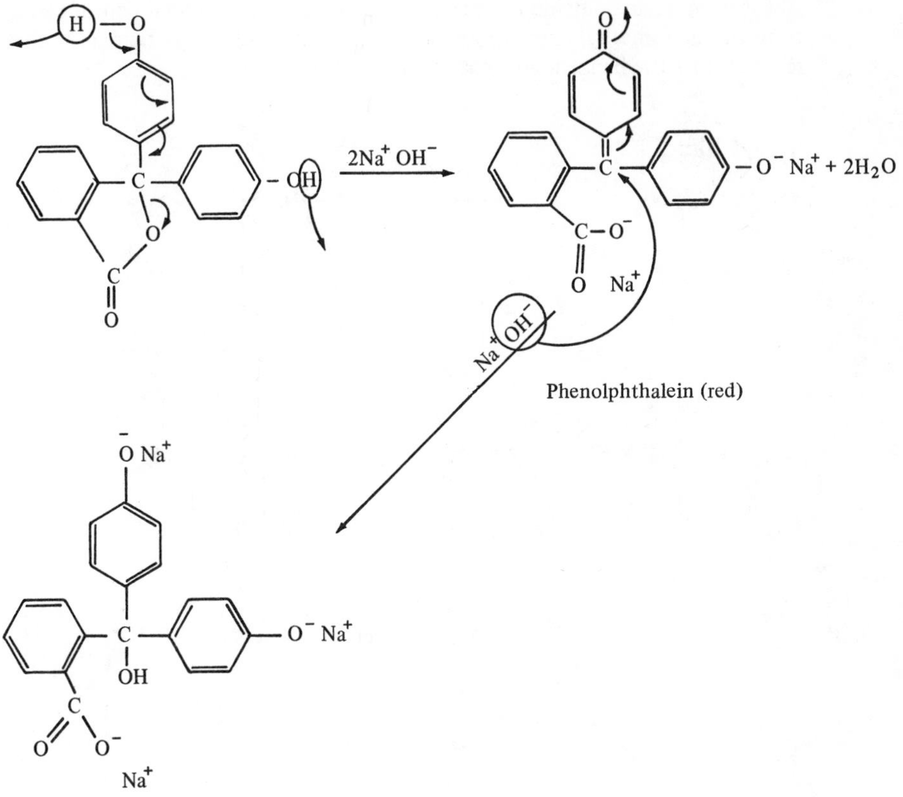

$\xrightarrow{2Na^+ OH^-}$ $-O^- Na^+ + 2H_2O$

Phenolphthalein (red)

$O^- Na^+$

$O^- Na^+$

OH

$Na^+$

(b)

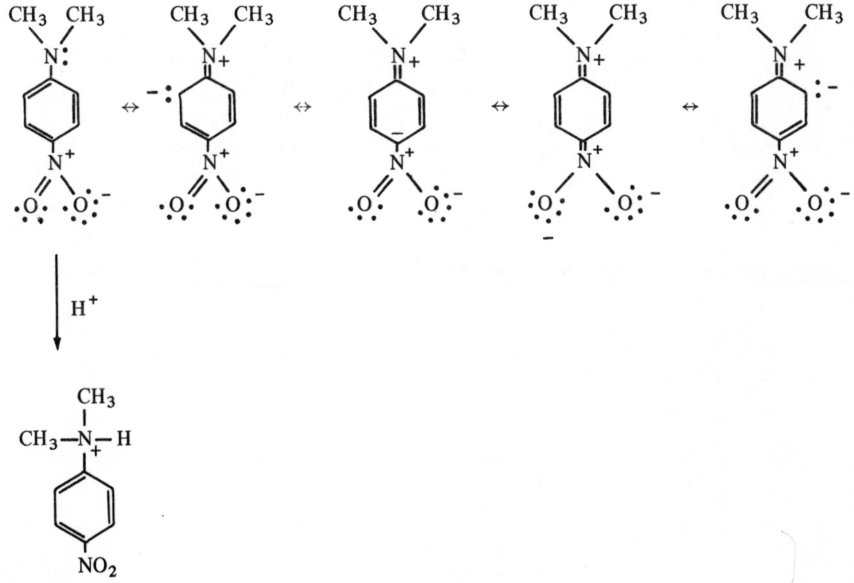

$\xrightarrow{H^+}$

$NO_2$

Absorption occurs outside the visible region when the unshared pair of electrons on the dimethylamino group is no longer available to participate in the resonance of the benzene nucleus and the *p*-nitro group.

19.9

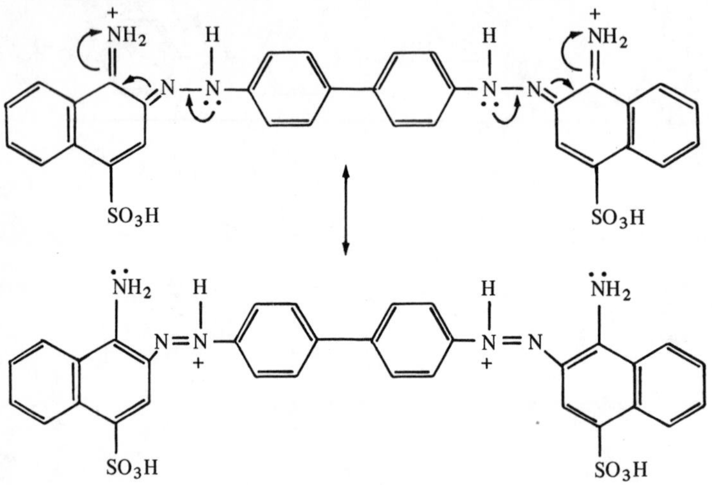

$NH_2$ H H $NH_2$

N—N N—N

$SO_3H$ $SO_3H$

$NH_2$ H H $NH_2$

N=N N=N

$SO_3H$ $SO_3H$

19.10

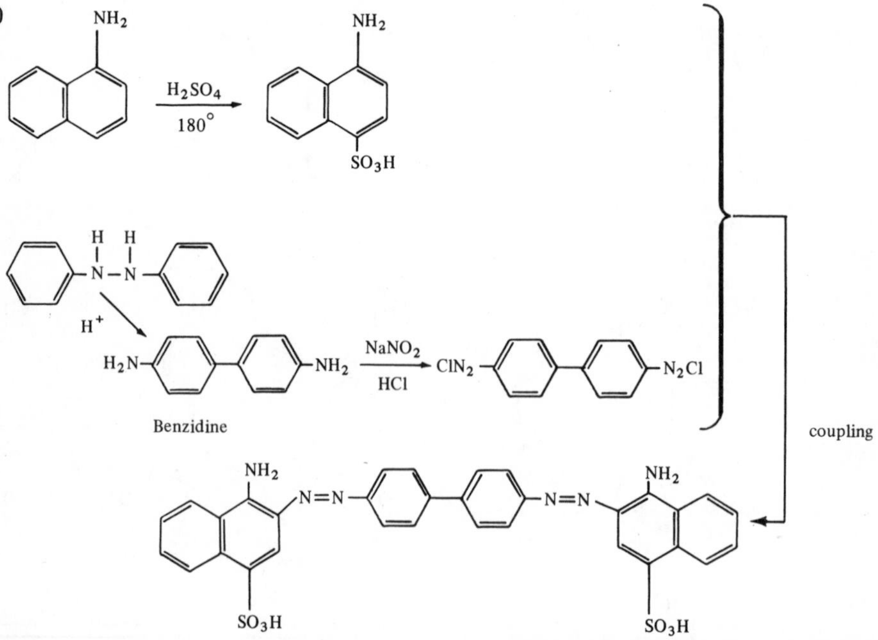

$NH_2$ $\xrightarrow[180^\circ]{H_2SO_4}$ $NH_2$

$SO_3H$

H H

N—N

$H^+$

$H_2N$— —$NH_2$ $\xrightarrow[HCl]{NaNO_2}$ $ClN_2$— —$N_2Cl$

Benzidine

coupling

$NH_2$ N=N N=N $NH_2$

$SO_3H$ $SO_3H$

2 3 4 5 6 7 8 9 10